# PROBLEMS OF PHYLOGENETIC RECONSTRUCTION

Proceedings of an International
Symposium held in Cambridge

THE SYSTEMATICS ASSOCIATION
SPECIAL VOLUME NO. 21

# PROBLEMS OF PHYLOGENETIC RECONSTRUCTION

*Edited by*

## K. A. JOYSEY

### and

## A. E. FRIDAY

*Department of Zoology, University of Cambridge,*

*Cambridge, England*

1982

*Published for the*

SYSTEMATICS ASSOCIATION

by

ACADEMIC PRESS

LONDON   NEW YORK

PARIS   SAN DIEGO   SAN FRANCISCO   SÃO PAULO

SYDNEY   TOKYO   TORONTO

ACADEMIC PRESS INC. (LONDON) LTD.
24—28 Oval Road
London NW1 7DX

U.S. Edition published by
ACADEMIC PRESS INC.
111 Fifth Avenue
New York, New York 10003

British Library Cataloguing in Publication Data

Problems of phylogenetic reconstruction. — (The Systematics Association special volume,
   ISSN 0309-2593: no.21)
1. Evolution—Congresses 2. Genetics—Congresses
I. Joysey, K.A. II. Friday, A.E. III. Series
575.1      QH430

ISBN 0-12-391250-4

PRINTED IN GREAT BRITAIN AT THE ALDEN PRESS, OXFORD

# Contributors

Butler, P.M., *Department of Zoology, Royal Holloway College, Alderhurst, Bakeham Lane, Englefield Green, Surrey TW20 9TY, England*

Cain, A.J., *Department of Zoology, University of Liverpool, P.O. Box 147, Liverpool L69 3BX, England*

Charig, A.J., *Department of Palaeontology, British Museum (Natural History), Cromwell Road, London SW7 5BD, England*

Crane, P.R., *Department of Botany, University of Reading, Whiteknights, Reading RG6 2AS, England*

Crowson, R.A., *Department of Zoology, University of Glasgow, Glasgow G12 8QQ, Scotland*

Forey, P.L., *Department of Palaeontology, British Museum (Natural History), Cromwell Road, London SW7 5BD, England*

Fortey, R.A., *Department of Palaeontology, British Museum (Natural History), Cromwell Road, London SW7 5BD, England*

Halstead, L.B., *Departments of Geology and Zoology, University of Reading, Reading RG6 2AB, England*

Hill, C.R., *Department of Palaeontology, British Museum (Natural History), Cromwell Road, London SW7 5BD, England*

Jefferies, R.P.S., *Department of Palaeontology, British Museum (Natural History), Cromwell Road, London SW7 5BD, England*

Meeuse, A.D.J., *Hugo de Vries Laboratorium, University of Amsterdam, Plantage Middenlaan 2a, 1018 DD Amsterdam, The Netherlands*

Patterson, C., *Department of Palaeontology, British Museum (Natural History), Cromwell Road, London SW7 5BD, England*

Paul, C.R.C., *Department of Geology, University of Liverpool, P.O. Box 147, Liverpool L69 3BX, England*

# Preface

When the Council of the Systematics Association first suggested that we might organize a symposium in the general area of phylogenetic studies we were involved in collaborative work on the evolution of myoglobin, in which we were attempting to achieve a synthesis between biochemical and palaeontological data. Believing that it was appropriate to explore opinions currently held on this side of the Atlantic, and deliberately drawing largely on local talent, we accepted the invitation. While we were still laying the foundations for a Symposium on "Problems of Phylogenetic Reconstruction", the cladistic debate on the relationships of the salmon, the lungfish and the cow burst upon the pages of *Nature*, and we found that our myoglobin studies were being quoted in a context which we believed to be inappropriate. This spurred us to enlarge the Symposium and to include an extra session on "Methods of Phylogenetic Reconstruction".

Eventually we held a three-day meeting in Cambridge in April 1980, attended by about 100 participants. The sessions covered the nature of characters and problems of determining homology, problems of determining direction of change, gains and losses, two sessions on methods of phylogenetic reconstruction and, finally, the relationships between phylogeny and classification.

After all the contributions had been assembled and edited ready for publication, economic circumstances dictated that the proposed volume was too large. It was decided to split the publication into two sections. This volume, carrying the title of the Symposium, is being published in the Systematics Association Special Volume series, and it has been arranged that the remaining ten contributions dealing with "Methods of Phylogenetic Reconstruction" will be grouped together in the Zoological Journal of the Linnean Society of London. We greatly welcome this practical collaboration between the Systematics Association and the Linnean Society and hope that the goodwill generated will be further developed.

*August 1981*                                                      K.A.J.
*Cambridge*                                                        A.E.F.

## Systematics Association Publications

1. BIBLIOGRAPHY OF KEY WORKS FOR THE IDENTIFICATION OF THE BRITISH FAUNA AND FLORA
   *3rd edition* (1967)
   *Edited by* G.J. Kerrich, R.D. Meikle and N. Tebble
2. FUNCTION AND TAXONOMIC IMPORTANCE (1959)
   *Edited by* A.J. Cain
3. THE SPECIES CONCEPT IN PALAEONTOLOGY (1956)
   *Edited by* P.S. Sylvester-Bradley
4. TAXONOMY AND GEOGRAPHY (1962)
   *Edited by* D. Nichols
5. SPECIATION IN THE SEA (1963)
   *Edited by* J.P. Harding and N. Tebble
6. PHENETIC AND PHYLOGENETIC CLASSIFICATION (1964)
   *Edited by* V.H. Heywood and J. McNeill
7. ASPECTS OF TETHYAN BIOGEOGRAPHY (1967)
   *Edited by* C.G. Adams and D.V. Ager
8. THE SOIL ECOSYSTEM (1969)
   *Edited by* J. Sheals
9. ORGANISMS AND CONTINENTS THROUGH TIME (1973)[†]
   *Edited by* N.F. Hughes

LONDON. Published by the Association

## Systematics Association Special Volumes

1. THE NEW SYSTEMATICS (1940)
   *Edited by* Julian Huxley (Reprinted 1971)
2. CHEMOTAXONOMY AND SEROTAXONOMY (1968)[*]
   *Edited by* J.G. Hawkes
3. DATA PROCESSING IN BIOLOGY AND GEOLOGY (1971)[*]
   *Edited by* J.L. Cutbill
4. SCANNING ELECTRON MICROSCOPY (1971)[*]
   *Edited by* V.H. Heywood
5. TAXONOMY AND ECOLOGY (1973)[*]
   *Edited by* V.H. Heywood
6. THE CHANGING FLORA AND FAUNA OF BRITAIN (1974)[*]
   *Edited by* D.L. Hawksworth
7. BIOLOGICAL IDENTIFICATION WITH COMPUTERS (1975)[*]
   *Edited by* R.J. Pankhurst

[*] Published by Academic Press for the Systematics Association
[†] Published by the Palaeontological Association in conjunction with the Systematics Association

8. LICHENOLOGY: PROGRESS AND PROBLEMS (1977)*
   *Edited by* D.H. Brown, D.L. Hawksworth and R.H. Bailey
9. KEY WORKS (1978)*
   *Edited by* G.J. Kerrich, D.L. Hawksworth and R.W. Sims
10. MODERN APPROACHES TO THE TAXONOMY OF RED AND BROWN ALGAE (1978)*
    *Edited by* D.E.G. Irvine and J.H. Price
11. BIOLOGY AND SYSTEMATICS OF COLONIAL ORGANISMS (1979)*
    *Edited by* G. Larwood and B.R. Rosen
12. THE ORIGIN OF MAJOR INVERTEBRATE GROUPS (1979)*
    *Edited by* M.R. House
13. ADVANCES IN BRYOZOOLOGY (1979)*
    *Edited by* G.P. Larwood and M.B. Abbot
14. BRYOPHYTE SYSTEMATICS (1979)*
    *Edited by* G.C.S. Clarke and J.G. Duckett
15. THE TERRESTRIAL ENVIRONMENT AND THE ORIGIN OF LAND VERTEBRATES (1980)*
    *Edited by* A.L. Panchen
16. CHEMOSYSTEMATICS: PRINCIPLES AND PRACTICE (1980)*
    *Edited by* F.A. Bisby, J.G. Vaughan and C.A. Wright
17. THE SHORE ENVIRONMENT: METHODS AND ECOSYSTEMS (2 Volumes) (1980)*
    *Edited by* J.H. Price, D.E.G. Irvine and W.F. Farnham
18. THE AMMONOIDEA (1981)*
    *Edited by* M.R. House and J.R. Senior
19. BIOSYSTEMATICS OF SOCIAL INSECTS (1981)*
    *Edited by* P.E. Howse and J.-L. Clément
20. GENOME EVOLUTION (1982)*
    *Edited by* G.A. Dover and R.V. Flavell
21. PROBLEMS OF PHYLOGENETIC RECONSTRUCTION (1982)*
    *Edited by* K.A. Joysey and A.E. Friday

* Published by Academic Press for the Systematics Association

# Contents

# 1 | On Homology and Convergence

A. J. CAIN

*Department of Zoology, University of Liverpool, Liverpool,
England*

### INTRODUCTION

Perhaps the biggest single problem facing the evolutionist in the
determination of phylogenies from imperfect material (which is all
that one normally has) is to distinguish resemblances due to homologous characters from those due to convergent ones. In the currently fashionable phylogenetic theory of Hennig (1966) this problem
is ignored by means of an "auxiliary principle", that convergence
can be taken to have occurred only when it can be proved to have
occurred, otherwise "phylogenetic systematics would lose all the
ground on which it stands". This so-called principle merely prejudges
the whole issue (Cain 1967). Do we always know when a resemblance
is homologous or convergent? How often are we left with a situation
in which we simply cannot say? It is notable that Hennig's opinion
has been principally accepted by those who are not occupied with the
study of evolution in populations of real animals in the field. To
comparative anatomists, zoogeographers, and taxonomists it may
offer a clearly-marked road to definite phylogenetic conclusions
which give satisfaction; but if the principle itself is wrong, then a
state of suspended judgement, however frustrating, may be more correct scientifically.

Systematics Association Special Volume No. 21, "Problems of Phylogenetic Reconstruction", edited by K. A. Joysey and A. E. Friday, 1982, pp. 1–19, Academic Press,
London and New York.

RECOGNITION OF CONVERGENCE

Although there are considerable difficulties in the concepts of homology and convergence (De Beer 1938; Boyden 1947, 1969), I shall take a simple view, that homology is defined by essential similarity of structure, whether anatomical, physiological or behavioural, the type example being the pentadactyl limb in all its modifications. (If in fact, some of the entities referred to below, such as the gastropod shell, are not homologous throughout their exemplars, this will merely strengthen my thesis.) Analogy can then be taken as resemblance in function without essential similarity of structure. Convergence (including parallelism for the sake of brevity) is even more difficult to define. In phylogenetic theory, it is a resemblance acquired independently in two or more phyletic lines; in practice it must normally be recognized by discordance of characters (see e.g. Le Quesne 1972, 1979). My point, however, is that this gives only a minimum estimate of convergence.

How are the accepted examples of convergence recognized, when there is no good fossil record to help? In every case, we see one or two characters, or character-complexes, appearing in two or more forms of which at least one, on all other characters, is not related to the rest. A classic example from Darwin is the standard fish shape, appearing in so many of the bony fishes, and in the elasmobranchs, but also in the Cetacea. Since in innumerable other characters, the Cetacea agree with the rest of the mammals, and these with the mammal-like reptiles, and these with the primitive reptiles, and these with the primitive amphibia, all of which have no such shape, and only the last tie in with the "true" fishes, it is clear that the fish shape has appeared independently twice (in fact, three times when we add the ichthyosaurs). Generally speaking, if a character $x$ (body shape in this case) shows a particular state $x_1$ (here, fish shape) in two forms which in a vast array of other characters ($y$, $z$, $w$, etc.) differ profoundly, it is taken that the character-state $x_1$ shows convergence (or parallelism).

Where we are dealing with character-states that we feel might be very readily alterable, we may be unable to choose one that we can regard as more likely to be steadily retained than the others. While probably nobody doubts that the massive complex of characters which binds whales to other mammals is a genuine index that they have a common ancestor much more proximal than do bony fishes and whales, one might be much less certain in considering, for example, the colour pattern on the head, as compared with the shape of the wing, in closely related birds.

If we have, say, five forms, with character states of particular characters $x$ and $y$ as follows

| Forms | A | B | C | D | E |
|---|---|---|---|---|---|
| Character $x$ | $x_1$ | $x_1$ | $x_2$ | $x_2$ | $x_2$ |
| Character $y$ | $y_1$ | $y_1$ | $y_2$ | $y_2$ | $y_2$ |

no question would usually arise – most would take the course of evolution without question as

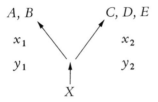

However, since we cannot inspect the common ancestor, this is an inference which may or may not be justified, depending on the likelihood of parallel or convergent variation in one or other of the characters, and we are really prejudging the issue.

If the observed distribution were

| A | B | C | D | E |
|---|---|---|---|---|
| $x_1$ | $x_2$ | $x_1$ | $x_2$ | $x_2$ |
| $y_1$ | $y_1$ | $y_2$ | $y_2$ | $y_2$ |

then there is a choice of courses, either

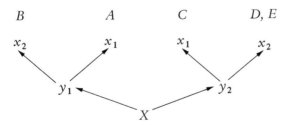

or else

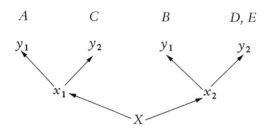

In short, we can say that there has been convergence in *either x or y*, but we cannot say which. The simplest situation in which such convergence could be recognized would be with only four forms – in the above diagram, we can combine forms $D$ and $E$ without loss of this type of information on convergence.

Here, then, we seem to have a suitable rule to guide us in studying the course of evolution in groups without a good fossil record. Whenever we meet such a combination of character-states as

| Forms | $A$ | $B$ | $C$ | $D$ |
|---|---|---|---|---|
| Character $x$ | $x_1$ | $x_1$ | $x_2$ | $x_2$ |
| Character $y$ | $y_1$ | $y_2$ | $y_1$ | $y_2$ |

then we can say that convergence must have occurred in one or other character but we cannot say in which; until a group has been scanned for such character distributions we have not even a minimum estimate of what convergence has gone on in it, even in the forms at present available to us.

Less than four forms cannot give us information of this nature. Take the example

| $A$ | $B$ | $C$ | | | | |
|---|---|---|---|---|---|---|
| $x_1$ | $x_2$ | $x_1$ | i.e. | 1 | 2 | 1 |
| $y_1$ | $y_1$ | $y_2$ | | 1 | 1 | 2 |

While this could be an example of convergence in $x_1$, it could equally well be the retention of this character-state from a common ancestor, with subsequent divergence of $x_2$ in form $B$. In fact, if we simply use the subscripts to write down distributions, and then try out all the possible ones for two characters and three forms, e.g.

| 1 | 2 | 1 |   | 2 | 1 | 1 |   | 1 | 1 | 2 |
|---|---|---|---|---|---|---|---|---|---|---|
| 1 | 1 | 2 |   | 1 | 1 | 2 |   | 1 | 1 | 2 |

and so on, the first being that just discussed above, we find that none of them can give us any certain information on convergence. Without the minimum of four forms and two of each character-state, we cannot distinguish patristic from convergent characters; but not all arrangements of character-states in four forms give us this information. And without the information, we are not entitled to make any assertion either way.

The number of permutations of $n$ objects consisting of groups of which $n_1$ are alike, $n_2$ are alike . . . . is

$$(n_1 n_2 \ldots)^P = \frac{n!}{n_1! n_2! \ldots}$$

This expression will give the number of possible arrangements of those character-states ascertained in a particular example to occur in a group of related forms. If we have four forms, $A$, $B$, $C$, $D$, and a character $x$ is represented in two states, $x_1$, $x_2$, by two examples of each,

i.e.     $A$    $B$    $C$    $D$

          $x_1$    $x_1$    $x_2$    $x_2$

then the number of permutations that could occur (of which that actually observed is one) is

$$\frac{4!}{2! 2!} = 6$$

If we now consider a second character $y$, also with two states twice represented, the total number of permutations is the product of those of each taken separately, i.e. $(x_1 x_2 P)(y_1 y_2 P)$, which in the example considered is $6^2 = 36$. Examples of such permutations are easily written as

| (a) | 1 | 1 | 2 | 2 | (b) | 1 | 2 | 2 | 1 | (c) | 1 | 1 | 2 | 2 |
|-----|---|---|---|---|-----|---|---|---|---|-----|---|---|---|---|
|     | 1 | 1 | 2 | 2, |    | 2 | 1 | 1 | 2, |    | 2 | 1 | 2 | 1, |

and so on, the upper line giving the character $x$, the lower one $y$.

If we now consider what evolutionary interpretation can be placed

Table I. Permutations of character states. $A$ in the seventh column indicates equivalence of distribution of characters $x$ and $y$, $B_x$ equivalence of states in character $x$ (upper line of each pair), $B_y$ equivalence in $y$; the number is the serial number of the permutation listed earlier which is equivalent. $N_1$ marks lines directly and $N_2$ inversely associated. N, no evidence on convergence; C, evidence of convergence.

| | | $A$ | $B$ | $C$ | $D$ | | |
|---|---|---|---|---|---|---|---|
| (1) | $N_1$ | 1 | 1 | 2 | 2 | | |
| | | 1 | 1 | 2 | 2 | | |
| (2) | C | 1 | 1 | 2 | 2 | | |
| | | 1 | 2 | 1 | 2 | | |
| (3) | C | 1 | 1 | 2 | 2 | | |
| | | 1 | 2 | 2 | 1 | | |
| (4) | C | 1 | 1 | 2 | 2 | | |
| | | 2 | 1 | 1 | 2 | | $B_y(3)$ |
| (5) | $N_2$ | 1 | 1 | 2 | 2 | | |
| | | 2 | 2 | 1 | 1 | | $B_y(1)$ |
| (6) | C | 1 | 1 | 2 | 2 | | |
| | | 2 | 1 | 2 | 1 | | $B_y(2)$ |
| (7) | C | 1 | 2 | 1 | 2 | | |
| | | 1 | 1 | 2 | 2 | $A(2)$ | |
| (8) | $N_1$ | 1 | 2 | 1 | 2 | | |
| | | 1 | 2 | 1 | 2 | | |
| (9) | C | 1 | 2 | 1 | 2 | | |
| | | 1 | 2 | 2 | 1 | | |
| (10) | C | 1 | 2 | 1 | 2 | | |
| | | 2 | 1 | 1 | 2 | | $B_y(9)$ |
| (11) | C | 1 | 2 | 1 | 2 | | |
| | | 2 | 2 | 1 | 1 | | $B_y(7)$ |
| (12) | $N_2$ | 1 | 2 | 1 | 2 | | |
| | | 2 | 1 | 2 | 1 | | $B_y(8)$ |
| (13) | C | 1 | 2 | 2 | 1 | | |
| | | 1 | 1 | 2 | 2 | $A(3)$ | |
| (14) | C | 1 | 2 | 2 | 1 | | |
| | | 1 | 2 | 1 | 2 | $A(9)$ | |
| (15) | $N_1$ | 1 | 2 | 2 | 1 | | |
| | | 1 | 2 | 2 | 1 | | |
| (16) | $N_2$ | 1 | 2 | 2 | 1 | | |
| | | 2 | 1 | 1 | 2 | | $B_y(15)$ |
| (17) | C | 1 | 2 | 2 | 1 | | |
| | | 2 | 2 | 1 | 1 | | $B_y(13)$ |
| (18) | C | 1 | 2 | 2 | 1 | | |
| | | 2 | 1 | 2 | 1 | | $B_y(14)$ |

|        |                | $A$ | $B$ | $C$ | $D$ |         |                     |
|--------|----------------|-----|-----|-----|-----|---------|---------------------|
| (19)   | C              | 2   | 1   | 1   | 2   |         |                     |
|        |                | 1   | 1   | 2   | 2   | $A(4)$  | $B_x(13)$           |
| (20)   | C              | 2   | 1   | 1   | 2   |         |                     |
|        |                | 1   | 2   | 1   | 2   | $A(10)$ | $B_x(14)$           |
| (21)   | $N_2$          | 2   | 1   | 1   | 2   |         |                     |
|        |                | 1   | 2   | 2   | 1   | $A(16)$ | $B_x(15)$           |
| (22)   | $N_1$          | 2   | 1   | 1   | 2   |         |                     |
|        |                | 2   | 1   | 1   | 2   |         | $B_x(16)B_y(21)$    |
| (23)   | C              | 2   | 1   | 1   | 2   |         |                     |
|        |                | 2   | 2   | 1   | 2   |         | $B_x(17)B_y(19)$    |
| (24)   | C              | 2   | 1   | 1   | 2   |         |                     |
|        |                | 2   | 1   | 2   | 1   |         | $B_x(18)B_y(20)$    |
| (25)   | $N_2$          | 2   | 2   | 1   | 1   |         |                     |
|        |                | 1   | 1   | 2   | 2   | $A(5)$  | $B_x(1)$            |
| (26)   | C              | 2   | 2   | 1   | 1   |         |                     |
|        |                | 1   | 2   | 1   | 2   | $A(11)$ | $B_x(2)$            |
| (27)   | C              | 2   | 2   | 1   | 1   |         |                     |
|        |                | 1   | 2   | 2   | 1   | $A(17)$ | $B_x(3)$            |
| (28)   | C              | 2   | 2   | 1   | 1   |         |                     |
|        |                | 2   | 1   | 1   | 2   | $A(23)$ | $B_x(4)B_y(27)$     |
| (29)   | $N_1$          | 2   | 2   | 1   | 1   |         |                     |
|        |                | 2   | 2   | 1   | 1   |         | $B_x(5)B_y(25)$     |
| (30)   | C              | 2   | 2   | 1   | 1   |         |                     |
|        |                | 2   | 1   | 2   | 1   |         | $B_x(6)B_y(26)$     |
| (31)   | C              | 2   | 1   | 2   | 1   |         |                     |
|        |                | 1   | 1   | 2   | 2   | $A(6)$  | $B_x(7)$            |
| (32)   | $N_2$          | 2   | 1   | 2   | 1   |         |                     |
|        |                | 1   | 2   | 1   | 2   | $A(12)$ | $B_x(8)$            |
| (33)   | C              | 2   | 1   | 2   | 1   |         |                     |
|        |                | 1   | 2   | 2   | 1   | $A(18)$ | $B_x(9)$            |
| (34)   | C              | 2   | 1   | 2   | 1   |         |                     |
|        |                | 2   | 1   | 1   | 2   | $A(24)$ | $B_x(10)B_y(33)$    |
| (35)   | C              | 2   | 1   | 2   | 1   |         |                     |
|        |                | 2   | 2   | 1   | 1   | $A(30)$ | $B_x(11)B_y(31)$    |
| (36)   | $N_1$          | 2   | 1   | 2   | 1   |         |                     |
|        |                | 2   | 1   | 2   | 1   |         | $B_x(12)B_y(32)$    |

on such distributions of characters relative to one another, we see immediately that the first permutation (a) suggests that $y$ has diverged *pari passu* with $x$; the phylogenetic arrangement suggested by $x$ is that also suggested by $y$, and no hint of convergence appears, i.e. we do not know if it did occur or not. In view of the possibilities of parallel mutation and selection discussed below, every such example of this must be viewed with great caution; it cannot be simply interpreted as showing that convergence has not occurred. The same is true of the second, if we have no reason to regard one of the two states of each character as more primitive. In this case the arrangement based on $x$ is again echoed by $y$ but with $x_1$ associated with $y_2$ this time, an inverse association. In the third permutation, however, it is clear that convergence must have occurred in one or other character, although without other evidence we cannot say which; either the phylogenetic arrangement is truly shown by $x$, in which case there has been convergence in one or other state of $y$, or it is shown by $y$, in which case convergence must have occurred in $x$. If we write out the 36 permutations, we can label each as to whether it gives evidence of convergence $C$, or no evidence, $N$ as in Table I. Now, if we have good reason from other evidence to consider the forms $A, B, C, D$ in a particular order (i.e. we know that some arose later than, or from, others) the order of characters in the forms is biologically meaningful. Also, if we have reason to prefer the evidence of character $x$ phylogenetically, then such a permutation as

$$\begin{array}{cccc} 1 & 2 & 2 & 1 \\ 1 & 1 & 2 & 2 \end{array}$$

is not the same to us as

$$\begin{array}{cccc} 1 & 1 & 2 & 2 \\ 1 & 2 & 2 & 1 \end{array}$$

Moreover if, again from other evidence, we have good reason to believe that a particular character state is more primitive than another, such an arrangement as 1 1 2 2 is not necessarily equivalent to 2 2 1 1. If, therefore, we have grounds for all these discriminations, all 36 permutations will be of interest. As will be seen, 24 give evidence of convergence, 12 do not. If we have no reason to prefer one or other character as phylogenetically more trustworthy than the other, i.e. if the two permutations (2-line) mentioned above are equivalent for our purposes, then we have 21 types of arrangement, 12 giving convergence. If we have no reason to prefer one character-state before the other, i.e. 1 1 2 2 is the same as 2 2 1 1, then the number reduces to 18 arrangements, 12 giving convergence, while if this is the case with both characters we have only 9 and 6. A lack of evidence for order both between and within characters brings us down to 6 arrangements, 3 giving convergence. With a combination of lack of previous evidence for characters, states, and forms — i.e. where we merely have four forms of equally unknown phylogeny and take one of them arbitrarily as our starting-point — we are reduced to two types of arrangement, one giving definite evidence of convergence. This is by far the most usual situation.

In Table I, all entries are permutations containing only two types of combination of characters. For example in No.(1) are two of each of $x_1y_1$ and $x_2y_2$, in No.(5) of $x_1y_2$ and $x_2y_1$. All those giving evidence of convergence are permutations of four combinations of characters, e.g. No.(2) has $x_1y_1$, $x_1y_2$, $x_2y_1$ and $x_2y_2$, as have all the others showing convergence. This rule simply gives an easy method of picking out those permutations useful for the present purpose. In this type of scrutiny, all forms identical for the characters considered at any one time can be represented by a single one of them as with $D$ and $E$ in the example above.

If we take three characters with two states represented twice in four forms, it is easy to see that no further information on convergence will be gained unless the third character has a different arrangement from both the others — if it is the same as one of them or the exact inverse, the case reduces to a two-character arrangement, one of the two now being composite. In such a combination as

$$1 \quad 1 \quad 2 \quad 2$$

$$1 \quad 2 \quad 1 \quad 2$$

$$1 \quad 2 \quad 2 \quad 1$$

all three characters are equally liable to suspicion of convergence.

An example of this state of affairs is given by the snail *Theba pisana* which is common in coastal districts all around the Mediterranean and up the western European shores. This snail certainly seems closely related to the Common Snail *Helix aspersa,* but what that relation is, is very doubtful, *T. pisana* has a typical helicid love-dart, with which it prods its partner in courtship; as the dart has four blades sticking out from a central axis, the species has usually been placed in the same sub-family, Helicinae, as the Common Snail, the four-bladed dart being a sub-family characteristic. Unfortunately other features of the genitalia (e.g. the two mucous glands being unbranched, not bushy) relate it not to the Helicinae but to a different sub-family, the Helicigoninae. And lastly, the ornamentation of the shell (pale ground colour with dark bands split up into dots, dashes, continuous lineolations etc.) relates it to yet another group, the Helicellinae. (Whether the Helicellinae are a distinct sub-family from the Hygromiinae or not does not affect the argument.) While the four-bladed dart has always been given overriding importance in what are really key classifications, no good reasons for preferring it rather than any other character have been given.

## INTERPRETATION OF RESEMBLANCES

It was stated above that the occurrence of the same character-state in two forms is not evidence of its inheritance in common in both of them. Parallel mutation at homologous loci in closely related forms is a well-established phenomenon (see e.g. Vavilov 1922; Alexander 1976). But that the mutations are at homologous loci can only be determined with considerable difficulty (Alexander, 1976) since there are plenty of non-homologous "mimic" genes producing phenotypically indistinguishable effects. De Beer (1938) had already drawn attention to these, quoting albinism in animals. Industrial melanism has arisen independently in over 50 different species of moths in Britain (Kettlewell 1973), and it is almost certain that the same genes are not involved in all the species. De Beer concluded that the homology of phenotypes does not imply the similarity of genotypes; certainly the ordinary inspection of the anatomy of a series of specimens, or even the comparative embryology of two forms does not allow one to establish gene or locus homology and in many cases it may be confidently assumed to be absent. Where forms are very closely allied, and produce very similar structures, considerable doubt may arise. De Beer (1938, p. 65) remarks

It is possible that, in requiring the visible presence of structures as far back as the point of divergence from the common ancestor, the theory of homology may in some instances be over-exacting. There seem to be cases (e.g. horns in titanotheres) in which the independent appearance of closely comparable characters in related but divergent stocks is more satisfactorily explained, as by Huxley . . . on the lines of genetic affinity; the manifestation of the character in the common ancestor having been obscured or delayed for developmental reasons, involving questions of growth and magnitude. Such cases are covered by the expression 'latent homology'.

This does not, however, dispose of the difficulty, because it is impossible to draw any line between "latent homology" and convergence. Cytochrome as a cellular constituent is almost ubiquitous in the animal kingdom, haemoglobin is not. Yet haemoglobin appears in all vertebrates but one, some insects, some worms, and some molluscs (to name only these). If the intracellular role of cytochrome is so important physiologically that natural selection has kept both it and its genetic basis constant throughout the animal kingdom, then haemoglobin may well have arisen from it by homologous mutation, every time, and would be an instance of "latent homology". Yet its appearance so sporadically in such widely separate phyla hardly allows the use of the term homology at all.

Either we must fall back on essential similarity of structure as a definition of homology, denying to haemoglobin a sufficient essentiality of similarity to itself or a sufficient complexity to rule out convergence, or we must insist on evolutionary continuity of a structural complex, which will make titanothere horns in different lines to be examples of convergence. But if the very similar nasal horns of titanotheres, developed in closely related lines, are revealed by the fossil record to be examples of convergence, then how many other similar structures in closely related groups are also convergent? The possibilities of parallel or convergent developments of characters have never been fully explored; in the case of the melanic moths just mentioned (an example virtually ignored by constructors of phylogenies) a comparatively slight change in the environment has affected similarly or nearly identically the phenotypes in many, often not closely related, phyletic lines, and the present cleaning-up of the atmosphere in Britain appears to be now removing the effect. The lines should be represented as all executing the same swerve in parallel and then returning to rectilinearity, or better like a bundle of lines, straight at first, then converging towards one another, and then diverging again. Numerous variations of the environment may have had similar effects, and most phyletic lines should probably be drawn with a very shaky hand.

## ECOLOGICAL LIMITATIONS ON EVOLUTION

If convergent or parallel mutation in related forms (and some not so closely related) is a widespread phenomenon, what is it that determines whether the mutations will be established in a given form or not? Unfortunately, there is still a good deal of controversy over the relative importance of mutation, selection and chance in evolution. A vast amount has been written on the mathematical possibilities in population genetics, very little on what actually occurs in the wild, because of both dogmatic prejudgements and great practical difficulties (Cain 1964, 1977b, 1979). So far as my own studies and a survey of the literature have taken me, it becomes apparent that even the most apparently trivial and variable characters in wild populations when adequately investigated show the influence of considerable natural selection — how much more, then, will characters of obvious physiological or behavioural importance. Animal species are all of them highly specialized for particular modes of life in face of pressures from the rest of the fauna and flora, and the inorganic environment. What keeps most species within bounds of variation must be not a lack of ability to vary but the continual removal of variants by natural selection. In the long term (except where the breeding-system is affected) it is ecology, not genetics, that dictates the course of evolution. The only major generalization that we can make about the course of evolution is that, given enough time, all available niches (i.e. available energy-pools sufficient to maintain a species continuously and made available by following a particular course of action) will be occupied. The beautiful parallels between the extinct South American mammals and those of the rest of the world are almost sufficient in themselves to make this point, still more so when backed up by the extraordinary convergences seen in different bird faunas. Land-snails can now be used as a further example.

The shapes of the shells of land snails, both pulmonates and prosobranchs, are wildly diverse, varying from elongate needle-like through tear-drop-shaped to conoid, subglobular, depressed, discoidal and even slightly ultra-depressed with the apex below the level of the last whorl. It is amazing that no one has referred to them as a classic example of neutral or nonadaptive or random variation. (Perhaps someone has.) At any one place, species of various shapes can be found side by side, living intermixed and not obviously specialized in any way except for a few subterranean burrowers, which are smooth and elongate. In some pulmonate families there is comparatively little variation, in others almost all shapes are found. Shells from different families may show remarkable convergence, others in the same sub-family may be remarkably different. (Those in the same genus are not often very different simply because, however alike they are in their anatomy, two forms differing markedly in shell shape are usually put into separate genera because of that very difference.) It might be said that there is some overall physiological restriction on the range of variation of a snail-shell, therefore much the same repertoire of variation will occur over and over again in different faunas, with no significance but coincidence.

When, however, we plot distributions of mean height $(h)$ and maximum breadth $(d)$ of the adult shell for all the species of a single fauna (Cain 1977a, 1978a,b) we find ourselves in the presence of major regularities that effectively deny any but the slightest role to chance or coincidence. A plot of $h$ and $d$ for the western European pulmonate snail fauna produces two wedge-shaped scatters, one corresponding to high slender to moderately stout shells, the other to more or less equidimensional to discoidal shells, with a marked gap between the two, even between the very small ones only a millimetre or so high. Exactly the same pattern is seen in the taxonomically unrelated faunas of eastern North America, Puerto Rico, and New Caledonia (Cain 1977a) and again in the Philippines and New Guinea but with one sub-family in each forming a definite exception distinct from the others in each fauna (Cain 1978a). Even more remarkable, when prosobranch snails, mainly marine, come out on land in the humid tropics, they too take up a bimodal distribution in $h$ and $d$, although in the sea they are unimodal. Their bimodality is seen in Australasia, south-east Asia, India, the Caribbean, and South America, and is only lacking in Africa and Madagascar, where they are unimodal but not with the same scatter as their close relatives in the sea (Cain 1978b).

These massive regularities become even more intriguing when

looked into. In the European and North American faunas, which are best known, it is found that the two scatters are made up in a sort of mosaic from different families, each occupying a sub-area of the scatter, and combining with others to fill it, with but little overlap. This immediately suggests that there is some sort of interference between them, presumably by means of competition, and leads to suggestions about how this can act (Cain 1977a), which have since been partly verified (Cain and Cowie 1978; Cameron 1978). The same family in different parts of the world may take up different parts of the scatters according to what it is associated with. Thus the Camaenidae occur in the Central Americas and Caribbean, and in the Far East and Australasia. In the Caribbean they coincide with the tall slender Urocoptidae and the tall stout Bulimulidae, and are themselves confined to the lower scatter (subglobular to depressed shells). In the Far East they coincide with the tall slender Clausiliidae and Subulinidae but are without tall stout competitors, and there they themselves produce tall stout forms (the genus *Amphidromus*) remarkably convergent on some American bulimulids. Convergence, in short, in land-snail shells is a highly pervasive and highly organized phenomenon which can be due to nothing else but natural selection, and indicates how closely members of a fauna are limited in respect to one another.

It would follow from what has been said that when a continental area has built up a balanced fauna, further evolution can only produce either parallel trends in large numbers of species as general conditions fluctuate, or minor speciation as areas become separated. If a population of shrikes in Africa, for example, becomes isolated, either by fission of its range by spreading dryness (for example) or by colonization of a newly accessible area — a recent volcanic mountain, or a new vegetated area in mountainous desert as rainfall increases — it is likely to be isolated in company with other birds. If bush-shrikes, sun-birds, orioles, kingfishers, turacos and so on, go with it, the population may be modified somewhat for local conditions, but can only continue to live as shrikes. If on the other hand it colonizes a remote island across the sea with few or no other birds, or persists on a continent after the extinction of most of the avifauna by great climatic change, all sorts of possibilities may be open to it. The greybirds are thrush-like perching birds, insectivorous and frugivorous, which (with few exceptions) persist as such all through the tropics of the Old World from Africa into Australasia. But of four species which I have seen on the island of Guadalcanal, two are normal greybirds (one in main lowland forest, one in riverine forest), one is a huge dark bird acting as a crow in the montane forest where there is no indigenous crow, and one lives like a shrike, and looks like one, in the drier coastal areas.

The limitations to the evolution of a group, then, depend on the available habitats and the associated fauna. But also, what opportunities can be taken by a particular stock depend on its constitution at the time. It is noteworthy that in the example just quoted, the greybirds, which are perching birds, have extended into two other perching-bird niches. If there were no doves on the island it is possible that, given time, they could go on to fill a terrestrial pigeon niche and so on. It is impossible to place any limitations on what can evolve into what, given an ecologically continuous route, and sufficient time. But the history of the earth may well be such that whole phyla cannot have been evolved from very different ancestors, simply because there never has been sufficient isolation for a sufficient length of time. When we are considering closely allied forms, such considerations do not apply. It is precisely closely allied forms with very similar ecological habits and much genetic homology in common, which are liable in different parts of the earth to slip into related habitats, and converge because in them they come under parallel selection pressures. But standard taxonomic practice would regard their convergent characters as equally ancestral with those that were genuinely continuous in time.

### PRIVILEGED CHARACTERS?

All attempts so far to dismiss characters as neutral, or non-adaptive, or merely ancestral, rest only on our ignorance about them. Such characters may well exist but they cannot be identified as such, and, as already mentioned, what positive knowledge we have as the result of field studies suggests that even apparently trivial characters are subject to selection pressure. This would seem to suggest that no class of characters can be relied on more than any other to give a reliable indication of phylogeny — they are all open equally to selection, consequent variation, and convergence. If it could be shown, however, that any class of characters is acted on only or primarily by selection conserving it within groups but keeping a distinction beween groups, this would be more likely than any other to be a reliable guide to ancestors. Three classes of character need to be considered.

In the first place, there are those characters of great physiological importance which can hardly be allowed to vary. The intracellular mechanism of respiration may well be one, but might be so constant as to be useless in classification; and where functional variation is possible it will no doubt be as open to convergence as anything else. Different haemoglobins, for example, have different loading and unloading tensions but these are adapted to the circumstances in which their possessors live and are liable to convergence. Only if we can show that in the ecological history of the earth a given change can have occurred only once can we take the product of it as monophyletic — and if we could show anything like this, then any sort of character would do. We would really be arguing from the ecological history, as when we can be fairly sure that nearly all the South American mammals, once in South America, were isolated for about 70 million years and went through a simple adaptive radiation, complicated only by the arrival of primates and rodents in the Oligocene.

Secondly, and much more promising, is the class of secondary sexual characters. The complex genitalia of insects, oligochaetes, gastropods and other groups have long been used in classification simply because they often offer a wealth of characters. If the various combs, plates, bristles and processes in the genitalia of a male *Drosophila* are wholly unaffected by general ecological pressures and are solely the arbitrary signals by which it stimulates a female of its own species to successful mating, then the prospects are hopeful. Consider a mutant with considerably altered genitalia; it is highly unlikely to get anything to mate with it successfully. Once established, the arbitrary routine must be gone through with only slight variation. In speciation, it is true, some variation must be produced and stabilized which will allow discrimination between the two new forms. A single character in the complex may be altered rapidly, but the rest must stay the same, or the altered individual may go unrecognized. Except when other considerations interfere, as for example when some small birds must be heavily protected by their coloration against predation, and must use song rather than decorated plumage for courtship, a complex of secondary sexual characters may usually consist of one or two highly specific characters and many very conservative ones. The possibility of arbitrary sexual characters being most conservative needs to be examined.

Lastly, in so far as the genetic code itself is arbitrary, it may show a conservatism similar to that of the genitalia. The code is known to be highly redundant (e.g. Herskowitz 1967) which seems to mean that it is in part arbitrary. However, there need be no stabilizing mechanism comparable to courtship, in which case variation in it may even be truly random. Certainly it is not the sort of evidence that can be used for most organisms.

### CONCLUSIONS

(1) It has been taken as normal in taxonomic practice that all similarities between closely related forms are true homologies and could theoretically be seen in the phenotypes of all the intermediate forms back to the common ancestor.

(2) It is, however, closely related forms that are more likely to converge in different parts of the world, and it may be impossible to know which of their characters are convergent and which ancestral.

(3) Overall, broad ecological and historical considerations may suggest to us that on a given continent, evolution is likely to have been a simple adaptive radiation without convergence. This does not help, however, with closely related forms, nor does it exclude such parallelism as is seen in industrial melanism.

(4) The study of discordant characters is extremely valuable in bringing to light examples of independent variation, but the question is, how many examples of convergence are concealed by apparently concordant characters?

(5) It is doubtful whether any class of characters is specially valuable in tracing out phylogenies, but secondary sexual characters may be.

(6) Hennig's so-called heuristic principle has no basis in our present knowledge of actual evolution in the field, and confounds

the numerous cases in which we do not know whether convergence has taken place or not, with those in which by means
of a reasonable fossil record we may believe that it has not.

### ACKNOWLEDGEMENTS

Some general considerations on concordant characters and homology were given
in a paper read to the International Congress of Zoology at Monaco, 1975, the
proceedings of which were not published. The present chapter also contains the
substance of my presidential lecture to the Systematics Association on convergence. I wish to thank those who on these occasions discussed ideas with me,
and Dr C. R. C. Paul and Dr K. A. Joysey for criticism of the present chapter.

### REFERENCES

Alexander, M. L. (1976). The genetics of *Drosophila virilis*. *In* "The Genetics
and Biology of *Drosophila*" (M. Ashburner and E. Novitski eds) vol. 1c,
pp. 1365–6427. Academic Press, London and New York.

Boyden, A. (1947). Homology and analogy. A critical review of the meanings
and implications of these concepts in biology. *Amer. Midl. Nat.* **31**, 648–669.

Boyden, A. (1969). Homology and analogy. *Science, N.Y.* **164**, 455–456.

Cain, A. J. (1964). The perfection of animals. *Viewpoints in biology* **3**, 36–63.

Cain, A. J. (1967). One phylogenetic system. *Nature, Lond.* **216**, 412.

Cain, A. J. (1977a). Variation in the spire index of some coiled gastropod shells,
and its evolutionary significance. *Phil. Trans. roy. Soc. Lond.* **B277**,
377–428.

Cain, A. J. (1977b). The efficacy of natural selection in wild populations. *In*
"The Changing Scenes in Natural Sciences, 1776–1976" (C. Goulden, ed.).
Philadelphia, Academy of Natural Sciences Special Publication **12**, 111–133.

Cain, A. J. (1978a). Variation of terrestrial gastropods in the Philippines in
relation to shell shape and size. *J. Conch., Lond.* **29**, 239–245.

Cain, A. J. (1978b). The deployment of operculate land snails in relation to
shape and size of shell. *Malacologia* **17**, 207–221.

Cain, A. J. (1979). Introduction to general discussion. *Proc. roy. Soc. Lond.*
**B205**, 599–604.

Cain, A. J. and Cowie, R. H. (1978). Activity of different species of land-snails
on surfaces of different inclinations. *J. Conch., Lond.* **29**, 267–272.

Cameron, R. A. D. (1978). Differences in the sites of activity of coexisting
species of land mollusc. *J. Conch., Lond.* **29**, 273–278.

De Beer, G. R. (1938). Embryology and evolution. *In* "Evolution: Essays on
aspects of Evolutionary Biology Presented to Professor E. S. Goodrich"
(G. R. de Beer, ed.), pp. 57–78. Clarendon Press, Oxford.

Hennig, W. (1966). "Phylogenetic Systematics." [Translated by D. D. Davis
and R. Zangerl.] University of Illinois Press, Urbana and London.

Herskowitz, I. H. (1967). "Basic Principles of Molecular Genetics". Little, Brown Co., Boston.

Kettlewell, H. B. D. (1973). "The Evolution of Melanism". Clarendon Press, Oxford.

Le Quesne, W. J. (1972). Further studies based on the uniquely derived character concept. *Syst. Zool.* **21**, 281–288.

Le Quesne, W. J. (1979). Compatibility analysis and the uniquely derived character concept. *Syst. Zool.* **28**, 92–94.

Vavilov, N. I. (1922). The law of homologous series in variation. *J. Genet.* **12**, 47.

# 2. | Morphological Characters and Homology

COLIN PATTERSON

Department of Palaeontology, British Museum (Natural History),
London, England

Abstract: Five ways of defining homology (classical, evolutionary, phenetic, cladistic and utilitarian) are discussed. Defining homology in terms of common ancestry (cladistic, evolutionary) causes problems which are avoided by defining it in terms of monophyly: homology is the relation which characterizes monophyletic groups. The classical, phenetic, cladistic and utilitarian views of homology fall into line with this definition if monophyletic and natural taxa are the same: homologies distinguish natural groups. Corollaries of this definition are that synapomorphy and homology are the same; that every worthwhile proposal of homology is a hypothesis of a monophyletic (natural) group; that homologies form a hierarchy; and that paraphyletic (unnatural) groups are not characterized by homologies. These conclusions are justified. Homologies which do not entail grouping are discussed and differentiated from those which do by a distinction between archetype and morphotype. Tests of homology are of three types: conjunction (the homologues may not coexist in one organism), similarity (topographic, ontogenetic, compositional) and congruence (with other homologies). The third test is the most valuable. Archetypal homologies are immune to these tests. A probabilistic, rather than parsimony-based, resolution of incongruent sets of homologies is suggested. Different categories of homoplasy (non-homology) may be distinguished by their performance in the tests. The only relations useful in systematics are homology, homonomy (mass homology, which fails the conjunction test) and the complement relation (presence vs. absence, which fails the similarity test). Parallelism (fails the congruence test) is, like convergence (fails congruence and similarity), a hindrance in systematics. On weighting, it is argued that the systematist has no role to play: homologies weight themselves. Polarity of homologies (discriminating mono- and paraphyletic groups) is resolved by von Baer's law of ontogeny. In-group and out-group

Systematics Association Special Volume No. 21, "Problems of Phylogenetic Reconstruction", edited by K. A. Joysey and A. E. Friday, 1982, pp. 21—74, Academic Press, London and New York.

comparisons merely order polarity to match that determined by ontogeny. Resolving a hierarchy of monophyletic groups (a classification) by means of ontogenetic information and congruence testing need have no evolutionary implications. If phylogeny is about evolution, and is more than systematics, homology cannot contribute to it. I suggest that the extra element, beyond systematics, is the appeal to extinct paraphyletic groups, which exist only in the minds of those who appeal to them. Recognizing homologies is comparable to discovering new species.

## INTRODUCTION

Homology is "without question the most important principle in comparative biology" (Bock 1974, p. 386). And of comparative biology, Nelson (1970, p. 374) wrote "its distinguishing features are its concern with history and its ability to generate and test historical hypotheses and theories by means of the comparative method". Homology is thus central to any discussion of phylogeny reconstruction. Yet it is possible to write a good book on phylogenetics without mentioning homology, except incidentally or in a historical context (e.g. Crowson 1970; Ross 1974; Løvtrup 1977): why this should be so is one of the questions discussed here. I also discuss, and try to integrate, two recent ideas: that homology is the only law-like hierarchic relationship in biology above the level of the individual organism (Riedl 1979); and that homology and synapomorphy are synonymous. As the word synapomorphy implies, I shall present a cladist viewpoint on homology, one that I hope will complement, if not supersede, recent expositions of the classical (e.g. Boyden 1973), evolutionary (e.g. Bock 1974, 1977) and phenetic (e.g. Jardine 1970; Sneath and Sokal 1973) points of view.

## DEFINITIONS AND USAGE

Definitions do *not* play any very important part in science (Popper 1962, p. 14).

Comments with the same message as the epigraph, but referring to homology in particular, are made by Cracraft (1967, p. 355) and Jardine (1970, p. 331). The aim of this section is not to choose one correct definition (though at the end of the chapter I suggest a definition that might be acceptable to four of the schools discussed), but to use definitions to illustrate, and comment on, various approaches to homology. Wiley (1975, p. 235) suggests that available definitions of homology fall into four classes: classical, evolutionary, phenetic and cladistic. A few examples of each are given, with apologies to authors who feel themselves wrongly classified, or find more or less casual statements quoted as definitions.

## 1. Classical

1. Essential structural correspondence (Boyden 1973, p. 82, paraphrasing Richard Owen's early work).
2. Essentially similar in relative positions and connections, and in adult structure and development (Boyden 1973, p. 140).
3. Homologue, c'est-à-dire foncièrement identique, nonobstant la variété de ses modalités d'expression (Dupuis 1979, p. 35).
4. Homologous structures are structures in different organisms that resemble each other so much that they warrant the same name (Inglis 1966, p. 219, paraphrasing Richard Owen's later work).
5. Anatomical singulars, or homologues (Riedl 1979, p. 52).

The common thread running through these five examples is that of identity, in essentials, fundamentals, or in name. Operationists have asked for definitions of "essential" and "same" (Owen's original definition (1843, p. 374) was "the same organ under every variety of form and function"). Popper says "a definition can only reduce the meaning of the defined term to that of undefined terms" (1962, p. 279), but "essentials" and "the same" beg too many questions. Inglis's paraphrase of Owen, and Riedl's abbreviated "anatomical singulars" (i.e. named anatomical structures of which there is only one, or a bilateral pair, per individual) seem to me to make an important point. When Bock (1969, p. 415) writes "statements such as 'the arm of the gorilla is homologous to the arm of the chimpanzee' . . . are meaningless" and (1977, p. 881) "it makes no sense to say that the femur of the gorilla is homologous to the femur of the chimpanzee", I suggest that such statements are mistaken not only in the sense that Bock intends — that they are incomplete because they lack a phrase specifying the nature (level) of the relation — but also because they are redundant. "Arm" in the first statement is shorthand for a set of named anatomical singulars, or homologues, and "femur" in the second is one named homologue. The conditional phrase that Bock demands is implicit in the use of the anatomical name.

## 2. Evolutionary

1. Structural similarity due to common ancestry (Boyden 1973, p. 82, paraphrasing De Beer, Haas and Simpson).
2. Homologous features (or states of features) in two or more organisms are those that can be traced back to the same feature (or state) in the common ancestor of those organisms (Mayr 1969, p. 85).
3. Structures and other entities are homologous when it is true that they could, in principle, be traced back through a genealogical series to a (stipulated) common ancestral precursor (Ghiselin 1966b, p. 219).
4. Features (or conditions of a feature) in two or more organisms are homologous if they stem phylogenetically from the same feature (or the same condition of the feature) in the immediate common ancestor of these organisms (Bock 1977, p. 881 – "same" here means "self-identical"; Bock 1974, p. 387).

These four definitions are arranged in order of increasing precision. The first, simple form is now avoided because it includes the word "similarity". This has been dropped from the other evolutionary definitions following Bock's (1963) argument that homologues need not be at all similar. To illustrate his point, Bock cited the ear ossicles of mammals and the homologous parts of the visceral arches of sharks; Cracraft (1967) and Mayr (1969, p. 85) use the same example. Taking that example (and no others have been offered), even if we allow that there is no similarity between (say) the mammalian stapes and the shark hyomandibular, Bock seems to have forgotten his insistence that statements of homology are meaningless unless qualified by a conditional phrase saying what is being compared (cf. quotations above). Thus the example has meaning only if put in some such form as "the stapes of mammals is homologous with the hyomandibular of sharks as that element of the gnathostome hyoid arch which articulates with the braincase". When this is done, the conditional phrase specifies the level at which the homologues are similar. And since, in Bock's view (1974, 1977, p. 882) the only test of homology is similarity, homology without similarity is untestable. Other attempts to discriminate between homology and similarity have not been successful (cf. Ghiselin 1969; Inglis 1970), and are open to the same criticism.

Perhaps a more cogent (though unstated) reason for deleting similarity from the evolutionary definition of homology is to escape the charge of circularity in definition – similarity is used to infer the common ancestry, which is used to validate the similarity as homology.

The evolutionary definition of homology seeks clearly to distinguish homologous similarity from convergent similarity without appeals to essentials or fundamentals, which raised problems in classical definitions (Simpson, 1959). But some systematists have sought to include parallelism in homology (e.g. Mayr 1974, p. 116; Dupuis 1979, p. 37). For example, Hecht and Edwards (1977) define homology as "Two or more character states which are derived from a common ancestral character state" (p. 6), and parallelism as "the character is present in the ancestral form but a common derived character state has been independently evolved in each descendent form" (p. 7). They say "By our definitions of parallelism and homology, derived character states which have evolved in parallel are also homologous" (p. 7). It seems to me that by their definitions, much of what is commonly thought of as convergence is also homology, for current interpretations of the relation between the wings of bats and birds, or between the flippers of dolphins and the pectoral fins of sharks, fit their definitions of both homology and parallelism. Something seems to have gone wrong. In my opinion, what has gone wrong is the muddle caused by "character" and "character state". Griffith (1974, p. 108) puts his finger on the mistake: "The only requirement for arranging character statements in sequence is that their meaning must be exclusive, in other words that it be impossible for any object to satisfy more than one of the character statements in the sequence at the same time." Hecht and Edwards (1977) seem to urge that a given feature may occupy more than one position in the sequence, being viewed both as "character" and as "character state", so that dolphin's flipper and the shark's fin are homologous both as character (pectoral appendage) and character state (fin). I agree with Bock (1974, p. 387) "that no distinction exists between characters and character states. The latter are simply characteristics which may be homologous with a more restrictive conditional phrase" (cf. Platnick 1980, p. 543). Bock's definition of homology (cited above) avoids confusion with convergence and parallelism by including the phrase "immediate common ancestor".

Yet there remains what Bock (1969, p. 416) calls a "gray zone" between homology and non-homology, which stems from inclusion of common ancestry in the definition, and the difficulty of specifying common ancestry with any precision. This is discussed under cladistic definitions.

## 3. Phenetic

1. When we say that two characters are operationally homologous, we imply that they are very much alike in general and in particular (Sokal and Sneath 1963, p. 70: Colless 1969, recommends calling this "taxonomic" homology, since "operational" can beg too many questions).
2. The relation between parts which occupy corresponding relative positions in comparable stages of the life-histories of two organisms (Jardine 1967, p. 130).
3. Feature $a_1$ of organism A is said to be *homologous* with feature $b_1$ of organism B if comparison of $a_1$ and $b_1$ with each other, rather than with any third feature, is a necessary condition for minimising the overall difference between A and B (Key 1967, p. 276).
4. The relation between parts of organisms which are regarded as the same (Jardine and Sibson 1971, p. 270).
5. Homology may be loosely described as compositional and structural correspondence (Sneath and Sokal 1973, p. 77).

The concept of operational homology was developed to avoid the circularity implied in evolutionary definitions, where common ancestry is used to define homology, yet the evidence of common ancestry is homology. This circularity was denied by Ghiselin (1966a,b), wielding the theory of definitions to show that the alleged circularity involves a confusion between the definition of words and the criteria that might be used to identify things. This point seems first to have been made by Hennig (1953, p. 11), criticizing Remane. Hennig illustrates the confusion elegantly (1966, p. 94): "As though it mattered for the definition of the concept 'truth' that we cannot recognize truth itself, and everywhere in science are limited to erecting hypotheses concerning truth." That quotation implies a second criticism of operationism, that it demands certainty, or safe, sure knowledge, as is evident from Sokal and Sneath's comment (1963, p. 71): "A phylogenetic concept of homology . . . is not susceptible to direct proof but only to proof-by-inference." Hull (1967) showed that inference and theory (Hennig's "hypotheses concerning truth") are as necessary to phenetic homology as to evolutionary homology, and pheneticists now make weaker claims to operationism (e.g. Sneath and Sokal 1973, p. 18), and have given up trying to define homology precisely, as is shown by Nos 4 and 5 above, and by Jardine's comment (1970, p. 331) on attempts at empirical definition of theoretical terms. Definitions 4 and 5 above return to Owen's original classical concept ("the same organ" or "essential structural correspondence").

## 4. Cladistic

1. Different characters that are to be regarded as transformation stages of the same original character are generally called homologous (Hennig 1966, p. 93).

2. Two (or more) characters are said to be homologous if they are transformation stages of the same original character present in the ancestor of the taxa which display the characters (Wiley 1975, p. 235).

3. Corresponding features in different members of a monophyletic taxon or of a particular phylogenetic lineage are homologous by definition (Bock 1969, p. 415).

4. Homology is the study of the monophyly of structures (Hecht and Edwards 1977, p. 6).

The first two of these definitions resemble evolutionary definitions in appealing implicitly (1) or explicitly (2) to common ancestry. They differ from evolutionary definitions in including the phrase "transformation stages", but this implies only the commonality which can be abstracted as the ancestral condition, as in evolutionary definitions. The third statement (it was not intended as a definition) can be viewed as a classical definition (correspondence) put into a phylogenetic framework by the inclusion of monophyly. This introduces a way of looking at homology in a cladistic context. If, as homologies lead us to believe, all living organisms form a monophyletic group, then corresponding characters in different organisms may be homologous or non-homologous, so that statement 3 covers both concepts. Applying the statement to subgroups of organisms, rather than to life as a whole, we meet the problem of defining monophyly of groups. Bock (1963, 1969) identified a "gray zone" between homology and non-homology which depends on the definition of monophyly adopted. In 1963, following Simpson, Bock allowed the concept of monophyly to be relative, so that the concept of homology became ambiguous. Since then, monophyly has received a good deal of attention (Platnick 1977a, and references cited there), and has been found to be more subtle than was originally supposed. Attempts to define monophyly in terms of ranking of taxa have failed, and attempts to define it in terms of character distribution lead to confusion, because monophyly is to do with groups, not characters (this is what is wrong with statement 4 above). Monophyly is best defined in terms of relationships, as the grouping of all descendants of a common ancestor, or uniquely derived and unreversed group membership (Platnick, 1977a).

But defining monophyly in terms of common ancestry recalls the evolutionary definitions of homology: it seems that both monophyly (hypotheses about groups) and homology (hypotheses about features) are to be defined with respect to the same standard. Yet when applied to homology, that standard itself is ambiguous, as Bock's (1963) discussion shows. Just as common ancestry, without qualification, cannot distinguish monophyletic, paraphyletic and polyphyletic groups (Farris 1975; Platnik 1977a), it cannot distinguish homology from non-homology. This problem can be avoided by defining monophyly in terms of common ancestry, and homology in terms of monophyly: homologous features are those which characterize monophyletic groups. It is easy to modify previous definitions in this way, be they classical ("structural correspondence characteristic of monophyletic groups"), evolutionary ("structural similarity due to monophyly"; "structures and other entities are homologous when it is true that they characterize monophyletic groups"), phenetic ("when we say that two characters are homologous, we imply that they share a commonality which defines a monophyletic group") or cladistic ("two or more characters are said to be homologous if they are transformation stages of a character of a monophyletic group"). Of greater interest than such exercises is a corollary: if homologous features characterize monophyletic groups, non-homologous features must be those which characterize non-monophyletic groups. It has always been acknowledged that polyphyletic groups are characterized by non-homologous features, and the interest centres on paraphyletic groups, those characterized by symplesiomorphies. The definition implies either that paraphyletic groups are not characterized by homologies, or that shared primitive characters are non-homologous, something which is certainly counterintuitive, and which is discussed further in the section on homology as synapomorphy.

## 5. Usage

Classical, evolutionary, phenetic and cladistic definitions do not exhaust the possible meanings of homology. As mentioned in the

Introduction, there are first-rate books on systematics which avoid the term (e.g. Crowson 1970; Ross 1974). In such books, it is evident that "homology" is replaced by "character", so that a definition might be "homologies are taxonomic characters". A related way of defining homology could be called utilitarian: here are two examples.

> 1. Certain structures are homologous, such as the wing of a bird, the foreleg of a horse, and the arm of a man, and these can be compared usefully with each other (Blackwelder 1967, p. 141).
> 2. Features assumed to be homologous may be assumed, simply, to be comparable, and when compared may be expected to result in an hypothesis of ancestral conditions (Nelson 1970, p. 378).

In other words, homology is the relation that systematists and comparative anatomists use in generating hypotheses of relationship.

A final point: as a word, the quality homology has adjectival (homologous), nounal (homologue) and verbal (homologize) forms which may be applied to individual comparisons. Of these, the least used but most interesting is the verb. "I homologize" expresses the hypothetical or conjectural nature of the relationship; that it is an abstract relationship, existing in the mind, not in nature (Nelson 1970, p. 378); and that it is new conjectures of homology that interest us most.

### HOMOLOGY AS SYNAPOMORPHY

Although the equivalence of homology and Hennig's concept of synapomorphy is implicit in Hennig's work (e.g. 1966, p. 95), I did not become fully aware of the equation until early in 1976, through a manuscript by Gareth Nelson. That manuscript is just published (Nelson and Platnick 1981), but appreciation of the equivalence of homology and synapomorphy is fairly widespread (Wiley 1975, 1976; Bonde 1977, p. 779; Bock 1977, p. 888; Szalay 1977, p. 16; Platnick and Cameron 1977; Nelson 1978; Cracraft 1978; Patterson 1978a; Platnick 1979; Gaffney 1979b). Indeed, it is easy to give an elementary account of the technique of cladistics entirely in terms of homology (Patterson 1978b; Anon. 1979). Does this mean that Hennig's idea of synapomorphy is superfluous? At the popular level, perhaps, but to many professional biologists synonymy of homology and synapomorphy will be odious, or at least premature, in view of the different usages summarized in the last section. My aim here is to explore this synonymy: to ask if it is exact, and to follow its consequences.

Bonde (1977, p. 779), considering synapomorphy and homology as " the same relation", wrote

> A term seems to be missing from Hennig's vocabulary, namely one for the sharing of an apomorph feature – whether convergent or homologous. The term "synapomorphy" might in fact be used for this more inclusive concept, while the older term homology could be reserved for Hennig's "synapomorphy".

Farris and Kluge (1979, p. 405) make essentially the same point, but recommend a different usage, in writing that certain authors "confuse synapomorphies with shared derived states". The issue here is that features such as the absence of hind legs in apodans, snakes and whales are shared derived characters, but are neither synapomorphy nor evolutionary or classical homology. Hennig (1966, p. 95) used a similar example (absence of wings in insects) as his reason for distinguishing symplesiomorphy and synapomorphy from homology. He argued that absence of wings is synapomorphous in Anoplura (sucking lice) and Mallophaga (biting lice), symplesiomorphous in apterygote insects, but homologous in both: distinctive terms are necessary to avoid confusion of the two cases (Hennig's point is not, as Bonde (1977, p. 779) supposed, that "it seems artificial to talk about the homology of a non-existent structure").

It is worth following up this example of Hennig's, in order to bring out the equivalence of homology and synapomorphy, and the redundancy of symplesiomorphy, once that concept has been appreciated. As Hennig says, lack of wings is homologous, and synapomorphous, in lice, but symplesiomorphous in apterygote insects. But a symplesiomorphous lack of wings characterizes not only apterygote insects, but other arthropods, other invertebrates, all plants, protists, and so on. The absence of wings in all these organisms is homologous by the conventional evolutionary definition (similarity due to common ancestry), but there is no more interest in that "homology" than in the fact that invertebrates, plants and protists also lack molar teeth, feathers, pentadactyl limbs and an infinity of other homologous structures. In this example, therefore, symplesiomorphy means nothing. Yet there remains a problem in distinguishing the synapomorphous (and homologous) presence of wings in pterygote insects and the synapomorphous (and homologous) absence of wings in lice: this is the question mentioned above (p. 28), whether homologies characterize monophyletic groups only, or monophyletic and paraphyletic groups, for I find that secondary loss is the only circumstance in which a paraphyletic group might legitimately be thought to be characterized by a homology.

Let us imagine that lice (Phthiriaptera) are the only "wingless pterygotes" — that is, that fleas and other wingless forms now included in the Pterygota do not exist, or are winged. If that were so, a systematist might remove the lice from the Pterygota, as a monophyletic wingless group, leaving the remaining Pterygota characterized by the synapomorphous presence of wings. According to our present understanding of insect relationships, the Pterygota would then be a paraphyletic group, characterized by a homology, the presence of wings. Homologies can therefore characterize paraphyletic groups. The argument that follows is intended to counter that proposition.

First, in order for the problem to exist, we have identified absence of wings in lice as due to reduction or suppression (reversal), not to primitive absence, as in apterygotes and plants. Such inferences rely on parsimony (or probabilistic, see p. 38) arguments: in this case, parsimony demands that lice have suppressed the wings, rather than independently acquired the numerous homologies that they share uniquely with psocopterans in particular, and with pterygotes in general (Boudreaux 1979; Seeger 1979). The character of lice is therefore not "no wings" but "wings suppressed or reduced".

Now let us ask how the imaginary systematist justified removing the lice from the Pterygota, i.e. how he characterized the Phthiriaptera. There are two possibilities: the justification is lack of wings alone, or lack of wings in conjunction with other characters (Boudreaux (1979, p. 280) lists 17 synapomorphies, apart from suppression of wings, which differentiate lice from their sister-group, the psocopterans). If the justification is not lack of wings alone, what characterizes the Pterygota is not the presence of wings, but absence of the synapomorphies of lice. Let us imagine what would happen if a winged louse were discovered: that is, an animal presenting some (or all) of the synapomorphies of lice except the lack of wings. Two courses of action are open to the systematist finding such an animal:

to include it in the Phthiriaptera, or to include it in the Pterygota
(the third possibility, placing the new animal neither with the lice
nor with the pterygotes, but in a third taxon, is, in this context,
equivalent to denying its existence). If the new animal is placed with
the lice, we realize that "having wings" was not a homology characterizing Pterygota; they were united only by lacking the synapomorphies of lice. An analogous example is the distinction of
tetrapods from bony fishes (Osteichthyes). According to our current
understanding of vertebrate interrelationships, tetrapods are a
monophyletic group, and osteichthyans (*sensu* Romer 1971, and
other textbooks – excluding tetrapods) are paraphyletic. One of the
characters universal in bony fishes is the presence of lepidotrichia,
bony fin-rays, which are absent in living tetrapods. Lepidotrichia are
homologous and symplesiomorphous in osteichthyans, another
apparent example of a homology characterizing a paraphyletic
group. In 1952, Jarvik described lepidotrichia in the Devonian
tetrapod *Ichthyostega* (the analogy is the discovery of a winged
louse). Whereas previously lepidotrichia appeared to be a homology
defining a paraphyletic group, their discovery in *Ichthyostega*
showed that to be a mistake. It is not the presence of lepidotrichia
that characterizes bony fishes, but the lack of something – hands,
feet and other tetrapod characters.

The other alternative open to the imaginary systematist is to
place the new "winged louse" in the Pterygota (just as Jarvik could
have insisted that *Ichthyostega* is not a tetrapod but a bony fish).
This is to say that the systematist must insist that the only character
of Phthiriaptera is the absence of wings: no louse can have wings, and
no pterygote may lack them. That is the only condition under
which a paraphyletic group may be characterized by a homology –
when the monophyletic group whose detachment produces the
paraphyly is characterized *solely* by absence of the homologue,
by a condition morphologically indistinguishable from the primitive
condition, and distinguished from it only by parsimony analysis. For
if the condition in the monophyletic group is morphologically
distinguishable from the primitive condition, it is recognizable as the
homologue in modified form, and the paraphyletic group is characterized not by the homology, but by having it unmodified – as usual,
by the lack of something. And the justification for the paraphyletic
group must, as usual, rest not on homology but on some other

criterion – tradition, authority, convenience, or assumptions about evolution.

Thus, in theory, it is possible for a paraphyletic group to be characterized by a homology: a systematist may detach a mono-phyletic group *solely* because it lacks that homology. I do not know whether a systematist has ever created such a group (for para-phyletic groups are created by systematists' conventions; unlike monophyletic groups, they do not exist in nature – Patterson 1978a, p. 220; Wiley 1979a, p. 213). But the possibility of such groups is, so far as I can see, the only reason for denying the exact synonymy of homology and synapomorphy or the rationale of defining homology in terms of monophyly. I think that the possibility of idiosyncratic systematics – that a systematist might create a paraphyletic group through detaching a subgroup characterized only by lack of some-thing – is remote, for as Nelson (1978, p. 340) says "the absence of a character is not a character". From here on I shall treat homology and synapomorphy as synonymous.

Eldredge (1979b, p. 181) wrote "synapomorphies are homologies, by definition (the converse is not true, since symplesiomorphies are also homologies)". I have been arguing that the converse *is* true, for symplesiomorphies are also synapomorphies. They are simply hom-ologies that circumscribe a group whose generality is wider than, and therefore irrelevant to, the problem under study. Every hypothesis of homology is a hypothesis of monophyletic grouping and, in any particular context, a symplesiomorphy is a hypothesis of a set, and a synapomorphy is a hypothesis of a subset of that set. Symplesio-morphy and synapomorphy are thus terms for homologies which stand in hierarchic relation to one another.

## HOMOLOGY AND HIERARCHY

Riedl (1979) devoted the longest chapter in his book to homology, and called the chapter "The hierarchical pattern of order". Others have commented, more or less explicitly (e.g. Woodger 1945; Zangerl 1948; Simpson 1961; Hennig 1966), on the hierarchical, or group-in-group pattern of homologies, but Bock seems to have made the point most consistently and clearly (1963, p. 268 – "Homologous features and conditions form a hierarchy that corresponds to the taxonomic hierarchy of groups"; 1969, p. 425; 1974, p. 387; 1977,

p. 881). Bock (e.g. 1969, p. 415) has also insisted that statements of homology are meaningless unless accompanied by a "conditional phrase" stating the conditions of homology. Ghiselin (1966a, p. 129) concurred, writing "It is simply a coincidence that 'the wings of birds are homologous' is not as manifestly absurd as 'John is older than'." As examples of these conditional phrases, Bock (1974, p. 387) states that "the wing of birds and the wing of bats are homologous as the forelimb of tetrapods", and Ghiselin gives "the wings of eagles are homologous *as wings* to those of hawks, and as *derivatives of fore-limbs* to those of bats". So far as I can see, all such statements must specify a hierarchy of groups, either explicitly, by name (birds and bats within tetrapods in Bock's example), or implicitly (hawks plus eagles as winged creatures; eagles plus bats as forelimbed creatures in Ghiselin's).

The idea that every worthwhile hypothesis of homology specifies a hierarchy of groups is all I wish to emphasize here. The force of a hypothesis of homology is that the inclusive group is monophyletic, by virtue of the homology. There are, however, hypotheses of homology which specify, or imply, no hierarchic grouping of organisms. They are discussed in the next section.

TAXA AND TRANSFORMATIONS: MORPHOTYPE AND ARCHETYPE

Eldredge (1979a) distinguishes two approaches to evolutionary theory which he calls "taxic" and "transformational". The first is concerned with the origin of diversity; the second with the process of change. The same distinction can be applied in the study of homologies. The taxic approach (the one I am advocating) is concerned with monophyly of groups. The transformational approach is concerned with change, which need not imply grouping. I suggest that the two approaches can be readily distinguished by considering the "conditional phrase" explicit or implicit in any hypothesis of homology. In the taxic approach, these phrases specify a hierarchy of groups; in the transformational approach they do not.

The transformational approach to homology is not necessarily evolutionary. Consider Owen's (1849) argument that the limbs and limb girdles of vertebrates are transformed parts of the axial skeleton (the scapula is homologous with a rib, the coracoid with a sternal rib, the forelimb with an uncinate process or epipleural bone, the sternum with a haemal spine, etc.). If one tries to imagine conditional phrases for these homologies, all seem to specify the same group, animals with vertebrae and ribs, and girdles and limbs, i.e. gnathostomes. No hierarchy of groups is implied. In the evolutionary context, an example is Gegenbaur's hypothesis (recently revived in modified form by Zangerl and Case 1976) that gnathostome pectoral and pelvic girdles and limbs are each homologous with a branchial arch and branchial rays. This specifies no grouping except animals with gill-arches and gill-rays, and with girdles and limbs: the two are synonymous as gnathostomes. Owen's and Gegenbaur's theories thus seem to be empty transformations, which lead to no new hypotheses of grouping. In Owen's terms, such transformations were "general homologies" — the relation of observed structures to the archetype, the latter conceived as Platonic essence, or as the Creator's blueprint (cf. Simpson 1959). This suggests that we may draw a distinction between archetypes and morphotypes, two concepts which I have previously thought to be the same.

An archetype is an idealization with which features of organisms may be homologized by abstract transformations which entail no hypotheses of hierarchic grouping. A morphotype is a list of the homologies (synapomorphies) of a group.[1] * If this distinction holds good, archetypes are by no means extinct. For example, consider Bjerring's (1967) argument that the polar cartilage of chondrichthyans, actinopterygians and birds is homologous with the subcephalic muscle found in coelacanths and inferred in rhipidistians. There are two possible interpretations of this homology. In taxic terms, the polar cartilage defines a group A (chondrichthyans, actinopterygians, birds), the subcephalic muscle a group B (coelacanths, rhipidistians), and homology of the two structures defines a third, more inclusive group C (gnathostomes). Within C, A and B may be non-overlapping (A and B are subsets of C) or one may be a subset of the other (B includes A or vice versa). In the latter case, C is equivalent to A or B, and one of the two structures homologized must be regarded as a modification of the other (see Fig. 4 and accompanying text). This is not the intention of Bjerring's argument, however. Instead, it is a transformational approach,

---

*Superscript numbers in square brackets refer to numbered notes at the end of the chapter.

seeking general homologies of the polar cartilage and subcephalic muscle in an archetype. Bjerring's (1979) interpretation of the dermal bones of the side of the osteichthyan skull in terms of gill-covers is another instance of the transformational approach. Here the bones are homologized with parts of a hypothetical pro-gnathostome. The homologies proposed lead to no hypotheses of grouping, and the pro-gnathostome is plainly an archetype.

Homology of the ear ossicles of mammals with bones involved in the jaw articulation of other tetrapods is usually cited as the chief triumph of comparative morphology (e.g. Simpson 1959, p. 292 – "most extraordinary result"; De Beer 1971, p. 6 – "sensational"). De Beer (1971) regards this homology as proved "beyond possibility of error". However, it is not beyond doubt, for Bjerring writes of the theory as "ready for cancellation" (1977, p. 155), and Jarvik (1980) argues that the incus and malleus of mammals are homologous not with the quadrate and articular, but with the stylohyal and ceratohyal. I mention this difference of opinion not to take sides, but to point out that Jarvik's view, whether right or wrong, shows up the difference between the taxic and transformational approaches to homology. For in taxic terms, it makes little or no difference which hypothesis is correct. Malleus and incus, as ear ossicles, are two of the synapomorphies of mammals. Articular and quadrate, as bones, are a synapomorphy of osteichthyans, and stylohyal and ceratohyal, as bones, characterize the same group (the only difference between the two hypotheses in taxic terms is that quadrate and articular also occur in two extinct groups, the acanthodians and placoderms, whereas a stylohyal is reported in acanthodians but not in placoderms). That example implies that the transformational approach to homology may be more informative, and a lot more interesting, than the taxic approach, a point I will not deny. Yet as Eldredge (1979a) argues with regard to evolutionary theory, concentrating on transformation at the expense of taxa is not fruitful. Further criticisms of transformational homology are made in the section on testing homologies.

There is a variant in which the transformational approach to homology is combined with a rejection of archetypes (or morphotypes) as speculative, or "going beyond the facts". This is my understanding of the view of arthropod phylogeny currently accepted in this country (e.g. Manton and Anderson 1979; Clarke 1979; Whittington

1979). According to that view, the limbs, jaws, head tagmata, sclerotization etc., are convergent, not homologous, in chelicerates, crustaceans and insects + myriapods. The reason given for this conclusion is, in brief, that the mouth parts and limbs in each group are mutually exclusive in structure and function, so that transformation of one type into another can not be imagined (transformational approach). When that approach is combined with a refusal to imagine archetypes, the result is, naturally enough, refusal to recognize groupings, and the conclusion that arthropods are polyphyletic. In my opinion, and that of others using the taxic approach (e.g. Boudreaux 1979), the real reason behind this conclusion is evaluation or weighting of homologies: apparent homologies between onychophorans (*Peripatus*), myriapods and insects have been rated as true, with the consequence that apparent homologies between chelicerates, crustaceans and insects + myriapods have to be rated as non-homologous. This example introduces the topics of testing and weighting of homologies, and resolving non-homology, which are discussed in the next three sections.

### TESTS OF HOMOLOGY

The notion of testing homologies is a fairly recent one, stemming from the introduction of Popperian philosophy into systematics. There are two opposing viewpoints on tests of homology. Bock (1977, p. 882) writes "shared similarity is the *only valid empirical test of homology*" (Platnick 1978, p. 366, agrees). Wiley (1975), p. 235) writes "the only valid test of homology . . . is to hypothesize that the supposed homology is a synapomorphy" (Cracraft 1979, p. 33, agrees). The two different approaches here are that a homology can be rejected only because of dissimilarity (Bock), or because it specifies a grouping incongruent with those specified by other homologies (Wiley). Bock's argument is that homologies are used to test hypotheses of grouping, not vice versa, but he does allow (1977, p. 889) that homologies may be rejected (rated as convergent) because they conflict with others, so perhaps the conflict of opinion implied by the two opening quotations is more apparent than real. I believe it is connected with the distinction between parallelism and convergence (see below, on non-homology).

Bock emphasizes that evaluation of shared similarity has low resolving power in distinguishing homology and non-homology. I suppose that this is because similarity does not test the hypothesis but validates it as worthy of testing, or evaluates its internal consistency. The similarity evaluated may be of any kind (topographic, ontogenetic, histological, etc.), but "the mode of development itself is the most important criterion of homology" (Nelson 1978, p. 335), a conclusion reached by pre-evolutionary morphologists (Russell 1916). Transformational homologies are hardly subject to the similarity test. For example, Bjerring's (1967) homologization of a muscle with a cartilage can only be judged in terms of topography, for the compositional criterion is denied in the hypothesis, and the ontogeny of the muscle is unknown.

A third test of homology (apart from similarity, and congruence, which is discussed below) may be called the conjunction test. If two structures are supposed to be homologous, that hypothesis can be conclusively refuted by finding both structures in one organism. For example, Bjerring's (1967) homologization of the polar cartilage and subcephalic muscle would be refuted by finding a vertebrate with both structures. But most transformational homologies are not subject to this test. For example, Gegenbaur's and Zangerl's homologization of the pectoral girdle and limb with a gill-arch is not refuted by the fact that a gnathostome has both structures: the test is denied by the hypothesis.

Wiley's test of homology (1975, p. 235) "is to hypothesize that the supposed homology is a synapomorphy". I agree, for synapomorphies are properties of monophyletic groups. And since, as far as I can see, synapomorphies are the *only* properties of monophyletic groups, tests of a hypothesis of homology must be other hypotheses of homology – other synapomorphies. This method of testing, by congruence, is discussed by Wiley (1975, 1976), Løvtrup (1977), Platnick (1977b) and Gaffney (1979b), amongst others. Previously (Patterson 1978a) I argued that this test invoked no more than the principle of parsimony, and was therefore not much of a test. Here I take a different approach.

Suppose that a homology specifying a group X has been identified (or hypothesized, discovered, or what you will). Other homologies may relate to X in five ways (Fig. 1): they may specify part of non-X (Fig. 1A); X and part of non-X (Fig. 1B; i.e. they include X); part of X (Fig. 1C; i.e. included within X); X (Fig. 1D; i.e. the same group as

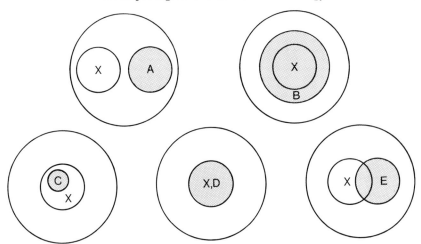

Fig. 1. Diagrams showing the five ways in which one group specified by a homology (A–E, shaded) may relate to another (X). The outer ring in each
diagram encloses the relevant universal set (X, X̄), all organisms.

the original homology); or part of X and part of non-X (Fig. 1E).
The first of these categories (A) is irrelevant to the status of the
original homology. B and C are consistent (congruent) with the
original hypothesis, for they stand in hierarchic relationship to it.
They corroborate X as part of an ordered system, but do not yet
provide any real test of it. The most interesting of the five categories
are D and E: homologies which specify the same group as the original
one, or a group inconsistent with it. Three cases will be considered:
that in which one (or more) D homology and no E is found (Fig. 1D);
that in which no D but one E is found (Fig. 1E); and that in which
one (or more) D and more than one identical E are found (identical
in the sense of specifying the same group – Fig. 2).

In the first case (D, no E), the original grouping is corroborated
and uncontradicted – we accept it. In the second case (no D, one E),
we have the original homology specifying group X, and another
inconsistent with it, grouping part of X and part of non-X. Initially,
we are no wiser, for the original hypothesis is contradicted, but so is
the new one. If there is order (one pattern) in nature, either or both
X and E may be false, but only one of them can be true. Homologies
in category B and C now come into play (Fig. 3). Presumably, there
is a set of homologies of the first type which specify large groups (of
high rank) including both X and E (B$_1$, B$_2$, Fig. 3); these are

*Colin Patterson*

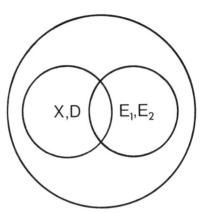

Fig. 2. Diagram showing groups specified by contradictory sets of homologies
(two in each set in this instance).

consistent with both and do not help discriminate between them.
Discrimination between X and E is by homologies which are consistent with one and inconsistent with the other ($B_3$, C, Fig. 3): we
prefer the homology which is intersected by fewer (or no) type B
and C homologies. The reason may be called parsimony, or, following Riedl (1979), order versus disorder.

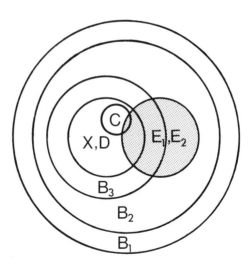

Fig. 3. Diagram showing resolution by parsimony of contradictory sets of
homologies of the type shown in Fig. 2. The incongruent (rejected) set
is shaded; letters refer to the types of group shown in Fig. 1.

In the third case (Fig. 2, one or more D, two or more E), the original hypothesis is both corroborated and contradicted by a set of homologies. Choice of one hypothesis is usually justified by parsimony, "when (say) eight synapomorphies are held to outweigh (say) four" (Patterson 1978a, p. 221). But in terms of order versus disorder, parsimony or corroboration seem words too weak to describe the choice. If there is one pattern in nature, the probability that two or more homologies will specify the same group by chance alone is given by the formula

$$\left(\frac{s!(n-s)!}{n!}\right)^{h-1}$$

where $h$ (homologies) is the number of homologies specifying a group, $s$ (sample) is the number in the group, and $n$ is the total number available for sampling. I suggest that in any particular instance of disagreement, $n$ is the number of groups involved in the disagreement, and about whose individual status there is no disagreement. Thus $n$ is the number of agreed groups within the smallest group containing both X and E ($B_2$, Fig. 3). These numbers are illustrated with examples below, but the minimal case is choosing two groups out of three, where the probability of two matching homologies by chance alone is 1/3, of three is 1/9, and of $n$ is $1/3^n$. Choosing five groups out of ten, the probability of a single repetition is 1/252; for three (or 17) out of 20 it is 1/1140, for 15 (or 5) out of 20, 1/15 504, and so on. Unless $n$, or $s$, or $n - s$ is very small, the chance that two homologies will specify the same group is low, and that three will is negligible. In such a situation, with two contradictory sets of homologies of which only one can be true, the obvious course is to ask whether each of the homologies in the two sets really characterizes the same group.

For example, Manton (1979, p. 391) names three morphological homologies characterizing the Uniramia (onychophorans, tardigrades, myriapods and insects), whereas Boudreaux (1979, p. 45) names 17 morphological homologies characterizing the Arthropoda (crustaceans, chelicerates, myriapods and insects). As in Fig. 2, we have two contradictory sets of homologies. Applying the formula given above, I suggest that a conservative estimate of $n$ here is eight groups (1, Onychophora; 2, Tardigrada; 3, Pentastomida; 4, Trilobita; 5, Crustacea; 6, Pycnogonida; 7, Chelicerata; 8, Tracheata or Atelocerata (myriapods + insects)). Manton's Uniramia groups three of these eight (1, 2, 8), and Boudreaux's Arthropoda groups five of them (4–8). The probability that three homologies should pick the same three or five groups out of eight by chance is $1/56^2$, and that 17 homologies should do so is $1/56^{16}$. The first of these probabilities is much larger than the second, but in practical terms both are very small.[2] We should ask, therefore, whether the homologies in each set really characterize the group named. Manton's three uniramian homologies are uniramous limbs, whole-limb mandibles biting with the tip, and "a unified type of embryonic development". Taking these at face value (i.e. not arguing about the homologies at the level of similarity), uniramous limbs do not characterize uniramians alone, but include other groups such as pycnogonids; and the embryological pattern of uniramians is shared, in part, with annelids, and is unknown in tardigrades (Anderson 1979). However, the Uniramia are not disposed of so easily, for amongst Boudreaux's 17 arthropod synapomorphies, I have been unable to convince myself that there is more than one (jointed exoskeleton) common to all arthropods and only to them (cf. Schram 1980; Kristensen 1980). The problem is that groups like the pycnogonids, pentastomids and tardigrades (the latter often treated as "proarthropods") fragment the pattern implied by Boudreaux's list. However, if tardigrades and pentastomids (two groups that Boudreaux fails to discuss) are included with the arthropods, that group seems to be characterized by more than one homology (Lock and Huie 1977, on the Golgi apparatus; Kristensen 1978, on muscle attachments and hemidesmosomes; cf. Hoyle and Williams 1980, on *Peripatus*). I am not surprised that these homologies are all ultrastructural, for if the problem of arthropod monophyly could be solved by gross morphology, we would presumably have stopped arguing about it by now.

Avoiding further examples, my general point is that of order versus disorder. If several homologies characterize the same group, as hair, mammary glands, ear ossicles, left systemic arch, etc. characterize Mammalia, we have overwhelming evidence of order (Riedl 1979), and currently assume that order, or pattern, to be the product of phylogeny (history), animals possessing some, rather than all, of these homologies having become extinct. If several apparent homologies characterize each of two contradictory groups, only one set can be due to pattern: one must be due to chance — randomness, accident, or in evolutionary terms convergence. Yet the calculations above suggest that the role of accident can generally be neglected. The simplest test, in such cases, is to ask whether each homology in the two sets really characterizes the same group.

A systematist whose group is rejected or threatened by that test may retaliate by appealing to weighting of homologies, or to non-homology (parallelism or convergence), discussed in the next two sections. A final point on testing: we have seen that transformational homologies are generally immune to testing by similarity and conjunction. Since transformational homologies do not specify groups, they are also immune to testing by congruence. So far as they are untestable (make no predictions), transformational homologies are vacuous.

WEIGHTING HOMOLOGIES

Hecht and Edwards (1977; see also Hecht 1976) set out a method of weighting homologous features which, they argue, "allows one to detect true parallelisms and reversals" (p. 17). They give high weight to functional complexes, and higher weight to unique innovations, the more complex the better. In such complex homologies, "parallelism and reversal should be relatively easy to detect" (p. 17) by direct analysis (i.e. by testing for similarity), and a phylogeny based on unrejected homologies of these types should show up parallelisms and reversals in simpler homologies.

Hecht and Edwards's highest weight is given to unique innovations. But with the exception of loss (i.e. absence of characters, Hecht and Edwards's categories 1 and 2, of lowest weight), any homology can be used to circumscribe a group, and so be treated as a unique innovation. Hecht and Edwards's system is therefore designed to prefer some groups to others (cf. Schuh 1978, p. 258). And the reasons for the preference seem to be based on claims to know how evolution works, as is suggested by Hecht and Edwards's remarks on "genetic backgrounds", "coevolution", "adaptive radiations", and so on.

Hecht and Edwards propose that weights be assigned to homologies before the phylogeny is constructed. Bock (1974 p. 390) agrees, since if circular reasoning is to be avoided, "features must be weighed before the taxa are recognized". Yet in my view, to recognize a feature (a homology) is to recognize a group, and I cannot imagine how any systematist could meet Bock's demand. I conclude (with Mayr 1969, p. 217; and Riedl 1979, p. 255) that weighting can only be *a posteriori*. This is obvious enough from Hecht and Edwards's weighting categories 1 and 2, "loss" and "reduction", for these are words describing processes, and so must be inferred from a phylogeny, rather than used as input before a phylogeny is constructed. Thus weighting serves to prefer one grouping or phylogeny over another. It can only be necessary where apparent homologies produce contradictory groupings, as discussed in the previous section. As argued there, I believe that such conflicts are more readily resolved by questions about the groups formed by homologies than by the weight of homologies.

Yet there are homologies which appeal to us of high weight. A few examples from vertebrates include the eye, eye muscles, semicircular canals, amnion, mammalian ear ossicles, feathers, ural centra of teleosts, claspers and placoid scales of chondrichthyans. Have we weighted these homologies, and if so, how? Complexity is not the answer, for placoid scales and ural centra are simple enough. Constancy is one answer: we weight these homologies highly because they are reliable in large numbers of species, and inferred loss is unknown, or rare and easily resolved. But constancy is only to say that the homology remains uncontradicted. Then the real criterion is lack of contradiction, which is, in turn, only to say that these characters are repeatedly corroborated as homologies by others which are congruent with them. I suggest that when we weight homologies highly, it is not because of complexity, or functional importance, but because the groups they form are congruent with those formed by many other characters, and are contradicted by few or none. In short, we do not need to weight homologies: they weight themselves, by associations which are beyond the bounds of chance (Riedl 1979, and preceding section).

Riedl (1979) introduced an alternative way of looking at the weight of homologies with his concept of "burden". He writes "By burden, I mean the responsibility carried by a feature" (p. 80) and "Burden does not necessarily depend on the complexity of the feature itself" (p. 129). He quantifies burden by distinguishing cadre homologues and minimal homologues. Cadre homologues include subordinate homologues, which may be of several hierarchical levels (as the vertebral column includes individual vertebrae, and each vertebra includes subunits such as zygapophyses), whereas minimal homologous can be subdivided only into homonoms, standard parts such as bone cells or Haversian systems. Minimal homologues have burden 1, and the burden of a cadre homologue is the number of minimal homologues and cadre homologues it includes (see also Jefferies 1979).

Riedl argues that burden is correlated with fixation or constancy; with age; with position (e.g. central versus peripheral in the nervous or circulatory system); with taxonomic rank (i.e. with position in the hierarchy); with low variability, or inferred resistance to change; and negatively correlated with convergence (the higher the burden, the less likely is convergence). All these features, except position in an organ system, are characteristic of the examples of high-weight homologies cited above. Like weight, it seems that burden is intrinsic in homology: it is not something that we assign. To repeat, homologies weight, or burden, themselves.

Riedl's positive correlation of burden with hierarchic rank and constancy, and negative correlation with convergence (incongruence), explains why systematics is more difficult at low taxonomic levels than at high, and carries the implication that it is futile to seek for homologies of high weight, and for methods of detecting "true parallelisms and reversals" (Hecht and Edwards 1977, p. 17) in species-level systematics (cf. Mayr 1969, p. 211 — "Weighting becomes the more important the higher the categorical level of the taxon" — I agree, but would write "self-evident" rather than "important").

### NON-HOMOLOGY: HOMOPLASY

The relations which interest us most here are similarities which may be mistaken for homology. Simpson (1961) classes all non-homologous similarity as homoplasy,[3] and distinguishes five sorts of homoplasy: parallelism, convergence, analogy, mimicry and chance similarity. Here, I shall lump the last four as convergence. The term homoiology (Riedl 1979, p. 36: "analogies on a homologous base") also deserves mention, since it discriminates amongst convergences. For example, the relation between the wings of birds, bats and flying fishes is homoiologous, whereas that between those wings and insects wings is not. And the wings of bats and birds share a closer homologous base, the tetrapod forelimb, than they share with the wings of flying fishes, which is the osteichthyan forelimb.

Two topics will be discussed here. First, how convergence, parallelism and other relations are distinguished from each other and from homology; and second, whether parallelism can contribute to phylogeny reconstruction.

In my understanding, the distinction between homology, parallelism and convergence is simple. As discussed in the section on testing homologies, there are three ways of rejecting a potential homology: it fails the test of similarity, it fails the test of congruence with other homologies, or it fails the test of conjunction, when two supposed homologues are found in the same organism. Considering first only the similarity and congruence tests, convergences are those similarities which fail both tests, and parallelisms are those which pass the similarity test but fail the congruence test, as shown in Table I.

Table I.

| Relation | Similarity test | Congruence test |
|---|---|---|
| Homology | Pass | Pass |
| Parallelism | Pass | Fail |
| Complement | Fail | Pass |
| Convergence | Fail | Fail |

In other words, parallelisms are rejected as homologies solely because they fail to characterize monophyletic groups, whereas convergences fail to characterize groups, and are not closely similar, or "not really the same". The complement relation, which passes the congruence test but fails the similarity test, is the presence of a homology versus its absence. If there is no secondary reversion, this relation is congruent with other homologies, which explains why it has the same value to the systematist as homology.

Parallelisms are thus apparent homologies which fail to fit an inferred phylogeny. For example, Hennig (1966, p. 117) discusses eyestalks in Diptera, structures which occur only in acalyperate cyclorrhaphans, and are very similar in all the groups where they occur, but do not characterize a monophyletic group. In the same way, experts on Mesozoic mammals (Crompton, Hopson, Jenkins, Kermack, Parrington) believe that the ear ossicles of monotremes and therians evolved in parallel, because ear ossicles do not characterize a monophyletic group in a phylogeny based on their interpretation of the side wall of the braincase. Since monotreme and therian ear ossicles cannot be shown to differ in structure, ontogeny or topography, this would be an instance of parallelism, not homology.

Non-homologous relations can be categorized in greater detail by utilizing all three tests of homology — congruence, similarity and conjunction, as in Table II.

Table II.

| Relation | Congruence test | Similarity test | Conjunction test |
|---|---|---|---|
| Homology | Pass | Pass | Pass |
| Homonomy | Pass | Pass | Fail |
| Complement | Pass | Fail | Pass |
| Two homologies | Pass | Fail | Fail |
| Parallelism | Fail | Pass | Pass |
| cf. Homoeosis | Fail | Pass | Fail |
| Convergence | Fail | Fail | Pass |
| Endoparasitism ? | Fail | Fail | Fail |

Comments are necessary on some of these relations. The comparisons categorized are proposals of taxic homology — that the structures compared characterize a monophyletic group. Transformational homologies, such as Owen's proposal that gnathosome limb girdles are homologous with parts of the axial skeleton, or Gegenbaur's theory that they are homologous with gill-arches, are not really subject to the congruence or similarity test, but if it is allowed that they pass the similarity test, they emerge as homonomy, whereas if they fail the test they come out as "two homologies". Homonomy (Riedl 1979, p. 38) is mass homology, "anatomical plurals", when there are several or many copies of the homologue in one individual, as with Haversian systems, enucleate erythrocytes or xylem vessels. In metaphytes, most homologies seem to be homonomies. Serial homology, like transformational homologies, appears to be an untestable version of homonomy. Paralogy (Fitch 1977), the term used by molecular biologists for comparisons which require the assumption of a gene duplication to make them congruent with other homologies, is also a form of homonomy. This gene duplication relation recalls the transformations postulated in transformational homologies, a point which will be discussed elsewhere.

The complement relation (presence of a homology versus absence) only passes the conjunction test if the organism is viewed at one time (semaphoront of Hennig 1966). In metazoans and metaphytes, organisms with ontogeny, the conjunction test is not passed if the whole life history is considered, since most homologies present later in life are absent in the zygote (see further under Polarity, p. 54). The "two homologies" relation is that between two unreversed homologies with the inclusion relationship (Fig. 1B, C) such as notochord and feathers, or heart and ear ossicles. The relation between two homologies of this sort, one of which is reversed (lost or suppressed), such as heart and hindlimb, comes out as convergence, since the comparison is incongruent with those homologies resolving the loss or suppression. The relation between two homologies with the exclusion relationship (Fig. 1A), such as plant tracheids and insect tracheae, comes out as convergence. I have been unable to think of a relation which fails the conjunction and congruence tests but passes the similarity test, but the phenomenon known as homoeosis in insects (Ouweneel 1976) may fit this case. Endoparasitism is entered with a query as the relation which fails all three tests. If one mistook an endoparasite for part of its host, comparison between part of the parasite and part of another organism would fail all three tests.

Of the eight relations discriminated by the three tests, the four which pass the congruence test (homology, homonomy, complement and two homologies) are the only ones of value to the systematist. This explains why the congruence test is the most powerful of the three, for it is the only test that discriminates useful comparisons from homoplasy (parallelism, convergence).

However, some systematists take a different view of the role of homoplasy in systematics, arguing that while convergence is a hindrance, parallelism is helpful. In evolutionary systematics, parallelism is often held to be helpful, though less so than homology

(Simpson 1961, p. 106; Mayr 1969, p. 202, 1974, p. 117). Most cladists have denied that parallelism has any relevance to phylogeny reconstruction, and so have shown little or no interest in discriminating convergence from parallelism (e.g. Hennig 1966, p. 121; Løvtrup 1977, p. 35; Bonde 1977, p. 770; Hennig and Schlee 1978, p. 5). But recently there has been an argument on this topic between Brundin (1972, 1976a, b) and Schlee (1975, 1978), two Hennigian chironomid specialists (see also Saether 1979). Schlee argues that parallelism is no different from convergence: both are hindrances to cladistic analysis. Brundin (1976a, p. 140) coins the term "unique inside-parallelism" for "the presence of a unique parallelism inside a group that is supposed to be monophyletic, meaning that a unique character has been developed independently within each of two subgroups making up a major group". In the example of mammalian ear ossicles given on p. 47, the assumed parallel development of ear ossicles in the two monophyletic sister-groups (Prototheria, Theria) making up the Mammalia would be a unique inside-parallelism. According to Brundin (1976b) such a parallelism would "give valuable additional evidence of monophyly" of the Mammalia.

Saether (1979) distinguishes parallelism due to parallel selection from parallelism due to "common inherited genetic factors including parallel mutations", and calls the latter "underlying synapomorphies". The first sort of parallelism he regards as homoplasy, the second as "comparable to true synapomorphy" as evidence of relationship. As an example of underlying synapomorphy, Saether cites the eyestalks of Diptera, mentioned above. I am not sure that Saether's underlying synapomorphy is exactly equivalent to Brundin's unique inside-parallelism, for Saether does not define his concept so precisely as does Brundin, but from Saether's usage it seems that the two concepts are the same.

Using the ear ossicles of mammals as an example, Brundin and Saether's argument implies that the monophyly of mammals would be reinforced by the ear ossicles as a unique inside-parallelism or underlying synapomorphy. But unless the monophyly of mammals had already been inferred by other homologies, how could it be inferred that Prototheria and Theria together made up a major group, rather than one of them being the sister of some other taxon, lacking ear ossicles? More crucial is the fact that if the monophyly of Mammalia was not established, and one were to argue that prototherians are related to birds (as did Lamarck in *Philosophie Zoologique*), ear ossicles would still be a unique inside-parallelism, "developed independently within each of two subgroups [Theropsida, Sauropsida] making up a major group [Amniota]". In the same way, homothermy is a unique inside-parallelism of amniotes according to current belief, and the inference that the monophyly of Amniota is reinforced, or in any way supported, by homothermy in birds and mammals, is shown to be false by the fact that homothermy could be used to reinforce many other groupings, for instance turtles as the sister-group of mammals, and those two as the sister of crocodiles and birds. I therefore agree with Hennig, Schlee and other cladists that parallelism, like convergence, is nothing but a hindrance to phylogenetic analysis.

POLARITY

Most recent discussions of homology and phylogeny reconstruction have given a lot of attention to determining polarity of morphoclines: deciding which of two (or more) homologous conditions is primitive and which derived, which is synapomorphous and which symplesiomorphous.

As noted above in the section on transformation, recognition of two homologous conditions defines two groups (A, B) which may stand in only two relations to each other: the two may be mutually exclusive, so that together they form a more inclusive group C; or one group may be included within the other, so that A is a subset of B, or vice versa, and C is redundant (Fig. 4). The difference between the two relations is that between a morphocline of undetermined polarity and one of determined polarity, and in terms of grouping, it is the difference between two monophyletic groups, and one paraphyletic and one monophyletic group. The second diagram in Fig. 4 also illustrates the simpler case of the complement relation: a homology is recognized, defining a group, and absence of the homology defines a complementary group. In this case, one of the groups must be monophyletic and one paraphyletic, and the problem of polarity is to determine which is which. Figure 5 illustrates undetermined and the two possible patterns of determined polarity in a three-set problem, that is, two homologous conditions and the third defined by absence of the character.[4]

Two methods of determining polarity are widely recommended:

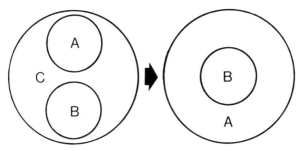

Fig. 4. Diagram illustrating resolution of polarity of two homologous conditions, which may be viewed as conversion of two non-overlapping sets
(groups) into two groups with the subset relation, one paraphyletic and
one monophyletic (B in this instance). In the left-hand diagram, the set C
includes only A and B.

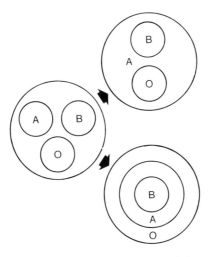

Fig. 5. Diagrams illustrating resolution of polarity of three sets (groups), two
(A, B) defined by homologous conditions and the third (O) by absence of
the structure concerned. There are three possible solutions of the upper
type, where a paraphyletic group includes two monophyletic groups, and
four possible solutions of the lower type, where two paraphyletic groups
include a monophyletic group. (Two additional solutions of the second
type, those in which absence (O) characterizes the intermediate group,
are forbidden because they violate the meaning of homology.)

comparative and ontogenetic. For example, Jefferies (1979) writes of
two comparative methods, in-group and out-group comparison. Of
the first, he says (p. 451) "If a homologous feature is universal in a
monophyletic group, then it will have been present in the latest

common ancestor of the group (e.g. hair in mammals)." But this is jumping the gun, for it assumes that monophyly of the group is already established, and this could only be done by a decision on polarity. Initially, all we can say is "if a homologous feature is universal in a group, then the group will be monophyletic, unless the feature is lost or suppressed within the group". And secondary loss can only be recognized by incongruence with other homologies of determined polarity (see above, p. 31). Of out-group comparison, Jefferies says (p. 452) "If a feature is present among many relatives of a monophyletic group, or best of all in the sister-group, and if its homologue occurs, sometimes or universally, in the members of the monophyletic group, it will have been primitive for the group." But here again, polarity of other homologies must already have been determined ("sister-group", "monophyletic group"). Further, this method does not seem to differ from in-group comparison: a homology is recognized, and unless it is secondarily missing, it will define a monophyletic group. In-group and out-group comparison turn out to be the same thing, a device to make homologies of unknown polarity congruent with others whose polarity has already been determined. And as yet, we have no hint of how polarity can be determined.

The ontogenetic method (Jefferies 1979, p. 453) "is based on Haeckel's law of recapitulation ... the sequence of embryonic inductions in the ontogeny of an organ will usually parallel its phylogenetic development". Suppose our problem is to determine the polarity of two pairs of homologies: the pectoral fin and hyomandibular of a fish, and the wing and stapes of a bird. Ontogeny does not help here, for in the ontogeny of birds and fishes there is no transformation of a fin into a wing, or vice versa, or of a hyomandibular into a stapes. Instead, these ontogenies show divergence from the more similar to the less similar, from the general to the particular — von Baer's law rather than Haeckel's.

But ontogeny does sometimes show transformations from one condition to another, as in Nelson's (1978) example of flatfish eyes, and his (1973) example of open visceral clefts versus closed visceral clefts. The meaning of Haeckel's law of recapitulation (the biogenetic law) in these examples was expressed by Nelson (1978, p. 327) in terms of von Baer's law — generality: open visceral clefts and an eye on each side of the head are more general than closed visceral clefts and both eyes on one side of the head, because animals showing the latter conditions also show the former, in early ontogeny. "General" therefore means more widely distributed. If in Fig. 4 homology A is "an eye on each side of the head" and B is "both eyes on one side of the head", we can adopt the right-hand diagram in Fig. 4 by the observation that condition A is present in all creatures with eyes (only in early ontogeny in flatfishes), whereas condition B is present only in some (flatfishes). Group C in Fig. 4 then becomes redundant (equivalent to A), and group B is contained within A, as a subset. But in the absence of ontogenetic transformation, how else can such decisions be made?

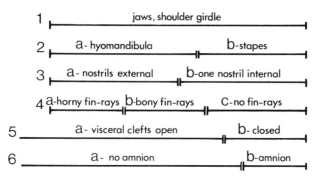

Fig. 6. A series of homologies and their observed distribution in animals. The form of the diagram implies that presence of the two homologies on line 1 characterizes a group indicated by the horizontal line; that presence of a hyomandibular characterizes a group to the left of the break in line 2, and stapes a group to the right of that break, and that these animals also possess the structures in line 1, and so on.

Consider the six sets of homologies in Fig. 6. Taking first the two characters in group 1, we could argue that the presence of these characters is more likely to be derived (less general) than primitive (more general), so that their presence defines a monophyletic group (gnathostomes) and their absence a paraphyletic group (non-gnathostomes, i.e. the rest of life). But this decision depends on prior knowledge that in simple presence versus absence situations the "presence group" is usually monophyletic, and the "absence group" usually paraphyletic, because we have learned that phylogeny usually leads from the simpler to the more complex. And this knowledge is, in turn, derived from systems erected by determining monophyletic groups. In any particular case, or in the absence of any previously established pattern, absence and presence must have equal probability as the less general state.

Or we could argue, with Schlee (1976), that the characters in group 1, jaws and shoulder girdles, are credible as synapomorphies because of their complexity of structure and function: this argument also depends on previously determined polarities, since they alone can be the source of an opinion on credibility. But if adopted, this argument would give us a monophyletic Gnathostomata and a paraphyletic non-Gnathostomata, and the observation that visceral clefts and no amnion (groups 5, 6) also occur in some non-gnathostomes would orientate those two sets of homologies.

Or, we could use the ontogenetic argument, and observe that in ontogeny no jaws precedes jaws, no shoulder girdle precedes shoulder girdle, no hyomandibular or stapes precedes presence of either, no nostrils precedes external or internal nostrils, no fin-rays precedes fin-rays, no visceral clefts precedes visceral clefts, which precedes visceral clefts closed, no amnion precedes amnion – in short, that ontogeny passes from the simple to the complex, from the general to the particular (von Baer's law, not Haeckel's, which only applies to character 5, and to which 6 is an exception). The ontogenetic argument is plainly more widely applicable and therefore more powerful than the "presence *vs.* absence" and "complexity" arguments, for it includes both of them,[5] and in this sample it specifies a congruent out-group for the characters in groups 1–4, and resolves the polarity of groups 5 and 6. Group 4 is also partially resolved by ontogeny, which shows that horny fin-rays (actinotrichia) precede, and coexist with, bony fin-rays (lepidotrichia), so that the two are not homologous (Patterson 1977a). Polarity of groups 2 and 3 can now be resolved by congruence with groups 5 and 6 (stapes and internal nostril sometimes occur with visceral clefts closed and gills lost, but hyomandibular and external nostrils do not). And the one incongruence, absence of fin-rays, is resolved as secondary loss by parsimony or probabilistic analysis. Thus Fig. 6 can be converted into Fig. 7.

It seems, therefore, that ontogeny is the decisive criterion in determining polarity (cf. Nelson 1978). In-group and out-group comparison turn out to be the same thing, a method of ordering homologies

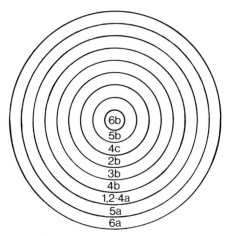

Fig. 7. Diagram showing the pattern of sets formed by resolving the polarity of the homologies in Fig. 6. Numbers and letters refer to homologies named in Fig. 6.

of unknown polarity so that they are congruent with polarities determined by ontogeny. Further, what ontogeny resolves is not necessarily primitive and derived conditions, but a pattern of more inclusive and less inclusive sets (Fig. 7), or the general and the particular: it depends not on Haeckel's law but on von Baer's. If "primitive" and "derived" can be dropped, and if our guide is von Baer, not Haeckel (cf. Gould 1977), then belief in, or knowledge of, evolution is clearly unnecessary for the analysis of homologies. This is obvious enough from the work of pre-evolutionists, who recognized and named many, if not most, of the homologies and monophyletic groups on which we now rely.

### THE ROLE OF HOMOLOGY IN PHYLOGENETIC RECONSTRUCTION

By identifying homologies of greater and lesser generality, primarily through correlation with ontogeny, and by testing for congruence, it is possible to resolve one pattern of uncontradicted, or least contradicted, monophyletic groups, as shown by the example in Figs 6 and 7. Such a pattern (Fig. 7) can then be converted into a cladogram (Fig. 8). But a cladogram is hardly a phylogeny in the usual sense. Instead, it is a summary of the pattern of homologies, a synapomorphy scheme, or, at best, a hierarchic classification. The classification specified by Fig. 8 is:

Chordata ......................................... homology 5a

 (Agnatha)
 Gnathostomata ............................ homologies 1, 2a, 3a, 4a

  (Chondrichthyes)
  Osteichthyes ............................ homology 4b

   (Actinopterygii, Actinistia)
   Choanata .............................. homology 3b

    (Dipnoi)
    Tetrapoda ........................... homology 2b

     (Amphibia)
     Amniota ........................... homology 6b

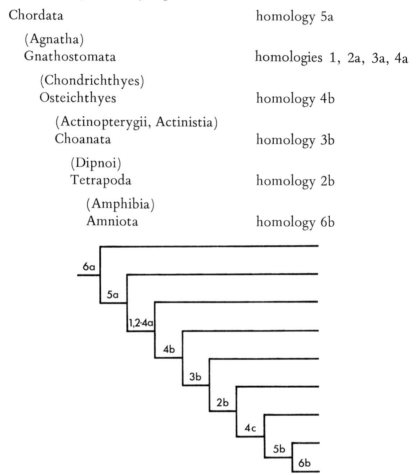

Fig. 8. The pattern of sets of Fig. 7 in cladogram form. Numbers and letters refer to homologies named in Fig. 6.

The names in brackets are the sister-taxa of those defined by homologies in Fig. 8. That such taxa exist is implied by the pattern of distribution of homologies in Fig. 6, but that sample contains no homologies bearing on their monophyly. Three of the characters in Fig. 8 find no place in the classification, numbers 4c (absence of fin-rays), 5b (visceral clefts closed) and 6a (no amnion). The third of these characterizes the whole of life except amniotes, whereas the first and second specify smaller groups (4c – tetrapods minus ichthyostegids; 5b – tetrapods minus perennibranchiate amphibians), but those groups have not been named. That merely confirms the argument on p. 31 above, that monophyletic groups are never characterized solely by the absence of a homology, nor paraphyletic groups by the presence of one. Thus neither presence (4b) nor absence (4c) of bony fin-rays defines a group.[6] Closure of visceral clefts (5b) seems to be the presence rather than absence of something (cf. Lundberg 1973), but it is currently assumed that perennibranchiate amphibians are neotenous, and Nelson (1978, p. 339) shows that neoteny is absence of something – adult characters. Closure of visceral clefts would thus characterize a monophyletic group (Tetrapoda), if some tetrapods were not inferred to have secondarily lost the homology.

The most that homology can contribute to phylogenetic reconstruction seems to be a cladogram or a classification. We expect more than that from phylogeny, a concept defined by Gould (1977, p. 484) as "The evolutionary history of a lineage", and by Mayr (1969, p. 409) as "The study of the history of the lines of evolution in a group of organisms; the origin and evolution of higher taxa", a definition which Mayr contrasts with that of classification – "delimitation, ordering and ranking of taxa". Thus phylogeny has to be about evolution, as we would expect from a word coined by Haeckel, whereas homology seems only to be about groups and hierarchies – systematics. And, as shown in the preceding section, belief in, or knowledge of, evolution is superfluous to homology analysis. This conclusion is reinforced by the classification (above) derived from the example used in the preceding section, for every group in that classification was recognized and named (not always with the name used here) by pre-evolutionary systematists.

Recently (e.g. Eldredge 1979b), it has been suggested that phylogenetic reconstruction can be thought of as comprising three activities: the construction of cladograms, trees, and scenarios. Homology is relevant only to the first, making cladograms. For a tree is a cladogram modified by eliminating some of the branches, so that taxa lie at or between nodes, and are so designated as ancestors. Sister-group relationship, the relationship between terminal taxa in a cladogram, is specifiable by homology: the homologies characterizing the node linking the taxa. Ancestor–descendent relationship, the relationship between a terminal taxon and one at or between the nodes of a tree, is not specifiable by homology, for the qualifications of an ancestor are that it be identical with, or lack something present in, its descendant(s). If Dollo's law is relaxed, and it is allowed that an ancestor may have something lacking in a descendant, then the ancestor–descendent relationship becomes reversible, and something other than morphology must be used to justify it. The role of homology in phylogeny reconstruction is therefore limited to cladograms; it can play no part in trees, except in a tree which is identical with a cladogram, having no taxa that are not terminal. But, as shown above, cladograms are not necessarily about evolution, and if phylogenies have to say something about evolution, then it is evident that homologies can have no role in them. The input that converts a cladogram (the pattern of homologies) into a tree (a pattern of descent) must come not from homology, or from systematics, but from some other source. There is a paradox here, for "our knowledge of phylogeny stems from our knowledge of taxonomy" (Platnick 1980, p. 545). In other words, what we know about "the evolutionary history of a lineage" (Gould 1977) must come from the systematics of that lineage, and so ultimately from homologies. The paradox is what we add to homology to justify phylogeny. Is it stratigraphy, biogeography, or simply the belief that organic phenomena must have a natural, historical explanation?

RECOGNIZING HOMOLOGY

One could argue, with Popper, that hypotheses of homology are conjectures whose source is immaterial to their status. Or one could argue with Riedl (1979, p. 32), that children and primitive peoples are pretty good at homologizing, but that experience helps. Or one could accept Jefferies' (1979, p. 453) point that our brains function by recognizing things as repetitions. There is much to be said for each of these viewpoints, but I will suggest another.

Inglis (1966) emphasized the point that recognizing homology depends on a one-to-one comparison between organisms. Riedl (1979, p. 41) writes "If we were to call two mammals homologous we would be overstretching the concept of homology." I disagree, for recognizing two things, whether two mammals, or an amoeba and an oak tree, as organisms is the necessary condition for recognizing homologies between them. Two mammals, or the oak tree and the amoeba, *are* homologous as organisms, with the implication that they belong in a monophyletic group "life", and share the homologies of that group — feeding, growth, reproduction, DNA, etc. Having recognized two things as organisms, we can then look for more restrictive homologies between them. Beyond that point, I suggest that recognizing a homology is directly comparable to discovering a new species. The systematist who believes he has discovered a new species characterizes it, and christens it with a proper name. In my view, the morphologist who believes he has recognized a new homology thereby characterizes a monophyletic group, which may or may not be new. If the group is new, he can best communicate his discovery by naming the group, either informally, by means of a "qualifying phrase" in the statement of homology, or formally, by christening it with a proper name. I regard Mammalia, Vertebrata, Pterygota, etc., the names of monophyletic groups, as proper names of individuals, and Reptilia, Invertebrata, Apterygota, etc., the names of non-monophyletic groups, as improper names of universals (Patterson 1978a). In other words, monophyletic groups, like species, are real: they exist outside the human mind, and may be discovered. Homology, which exists only in the human mind, is the relation through which we discover them. Non-monophyletic groups, unlike species, have no existence in nature. They have to be invented, are not recognizable by homology, and exist only in the human mind.

Van Valen (1978, p. 289) argues the opposite view, that paraphyletic groups are justified because they exist in the absence of taxonomists, and may be discovered, because they are "adaptively unified". As an example, he uses Insectivora, which "are almost always small for mammals, always terrestrial or at most barely semiaquatic or gliding, almost always with a diet predominantly of individually caught invertebrates, almost always (where known) with less flexible learning than most other mammals". May the reader have more success than I did in deciding what organisms fit this description (opposums, caenolestids, notoryctids, lizards, toads, pitcher plants?); how the group was discovered; and how other paraphyletic groups, named (Holostei, Invertebrata, Apsychota) or unnamed (mammals minus platypus and man — Platnick 1980, p. 544), are "adaptively unified".

Gaffney (1979b, p. 96) criticizes my point of view, homology as discovery, writing "If one systematist "discovers" a synapomorphy that contradicts another systematist's "discovery", how can we rationally choose one over the other? ... a reversion to authoritarianism is a step backward." The rationality of the choice is explained in the section on testing homologies, but it, and the question of authority, may be further explored by pursuing the analogy with discovering a new species. One systematist discovers a new species. Another may assert that the "new" species is one already named, or contains parts of two already named, or is insufficiently characterized, so that its name is a *nomen nudum*. Such mistakes and disagreements are commonplace drudgery in systematics, and yet systematists do agree on faunal and floral lists. If systematists working in the same area repeatedly find "the same" species, and learn that they are also recognized by folk-taxonomy, those species exist outside the authority of any individual systematist. And if observers, whether systematists or not, repeatedly discover (by naming) groups like birds, bats, frogs, spiders, etc., those groups exist beyond the authority of the systematist. The reason is that all these groups are characterized by homologies, whether species-level homologies ("which have all remained nameless" — Riedl 1979, p. 119) or more general homologies. Of course, folk taxonomies also name non-monophyletic groups (worms, shellfish), as do systematists (Pisces, Invertebrata); returning to Gaffney's remark, "how can we rationally choose one over the other?" By means of hypotheses of homology, and their critical evaluation.

### CONCLUSIONS

> Homology is a mental concept obtained by comparison, which under all circumstances retains its validity, whether the homology finds its explanation in common descent or in the laws that rule organic development (Oskar Hertwig, quoted and translated by Russell 1916, p. 355).

I began this chapter by discussing five ways of defining, or thinking about, homology — classical, evolutionary, phenetic, cladistic and utilitarian (homologies as taxonomic characters, or useful comparisons). I settled on a cladistic definition, homology as the relation characterizing monophyletic groups, and have pursued that line through the rest of the chapter. My cladistic interpretation may be unacceptable to evolutionists, pheneticists, and others, as constrained by the "Hennigian straitjacket" (Fryer 1978). So here I want to consider how the other approaches to homology differ from my own cladistic version, and to suggest a reconciliation.

Classical homology, of Owen, Agassiz and other pre- or anti-evolutionary morphologists, served as a key to the plan of the Creator, or as Platonic essence. The only part of my analysis that such workers would deny is, I believe, my use of evolutionary terms like phylogeny, common ancestry and monophyly. As shown in the sections on polarity and phylogenetic reconstruction, cladistic homology analysis does not demand belief in, or knowledge of, evolution, and does not contribute to phylogeny as something distinct from systematics, so Owen and Agassiz would have nothing to complain of. Their remaining objection, to the use of monophyly and paraphyly, terms which have an evolutionary connotation because they are defined in terms of common ancestry, can easily be met by dropping those terms, substituting "natural" for monophyletic, "unnatural" for non-monophyletic, defining those terms by reference to nodes on a cladogram, rather than common ancestors on a tree (Platnick 1977a), and defining homology as the relation characterizing natural groups. With those amendments, I suggest that my analysis would be fully acceptable to pre- or anti-evolutionists. Modern classicists, like Boyden (1973), object to "ancestor worship" (p. 126) in homology, and their objections can be met by the above modifications.

As for pheneticists, they aim, like pre-evolutionists, for natural taxa, and contend that phylogeny, as something additional to, or beyond, systematics, is "in the vast majority of cases . . . unknown and possibly unknowable" (Sneath and Sokal 1973, p. 53). On the latter point, I have come to the same conclusion, and on natural groups, I believe that Farris (1977, 1979, 1980) has shown that monophyletic taxa are best able to meet the demands that pheneticists make of natural taxa: monophyletic taxa are more informative on similarity, and more useful (efficient, parsimonious) than non-monophyletic taxa. So I anticipate no quarrels with pheneticists.

Utilitarians, like Blackwelder (1967), Crowson (1970) and Ross (1974), who equate homologies with taxonomic characters, or with

useful comparisons, can hardly disagree with my analysis, for I have
argued that homologies are the characters of monophyletic, or natural
taxa, and that those taxa are the only useful ones.

As for cladists, Platnick (1980) writes of "classical" and "modern"
cladists. My viewpoint is, I believe, that of modern cladists, for it
closely resembles those of Nelson (1978) and Platnick (1980), as was
to be expected, for my ideas have been much influenced by fifteen
years of discussion with Nelson, and by his 1976 manuscript, now
published as part of Nelson and Platnick (1981). So I expect no disagreement with modern cladists. As for classical cladists, Platnick
(1980, p. 545) suggests that modern cladistics is not classical cladistics
transformed, but ideas that have been "part and parcel of cladistics all
along, even if they were perhaps not very carefully or clearly enun-
ciated".

I have left the evolutionists until last, for here I do expect disagreement. As I see it, there are only three substantial differences between
my approach and that of evolutionists: they wish to recognize
anagenesis (degree or rate of adaptive change); they wish to use
paraphyletic groups; and they claim that phylogeny, in the sense of
something additional to systematics ("the origin and evolution of
higher taxa") is knowable, and, in many cases, known. I believe that
these three factors, anagenesis, paraphyly and phylogeny, are closely
interrelated (for example, paraphyletic groups are a simple way of
expressing the anagenetic advance of the monophyletic groups
removed from them). But if we isolate anagenesis, it causes no real
problem. The number of homologies distinguishing a monophyletic
group is an index of anagenetic advance, and more precise techniques
for evaluating anagenesis are available (e.g. Saether 1979; Anderson
1980). It is the second and third differences, paraphyletic groups and
phylogeny, that raise real difficulty. Because evolutionists regard
paraphyletic groups as monophyletic, I cannot expect them to accept
my definition of homology. I have shown (pp. 31, 57) that paraphyletic groups cannot be characterized by homologies. Again, I
cannot force evolutionists to agree with me, but I have to conclude
that paraphyletic groups are uncharacterizable inventions, limited at
best by convention, and at worst by an individual systematist's whim.
Why should evolutionists (e.g. Mayr 1974; Van Valen 1978;
Gingerich 1979; Bretsky 1979) want to retain them? As devices to
express anagenesis in derived groups is one answer; as devices to

express phylogeny is another, for I suggest that claims that phylogeny is more than mere systematics, and that in many cases it is known, rest entirely on paraphyletic groups.

Knowledge of phylogeny, as distinct from systematics, cannot come from monophyletic groups, as shown in the section on phylogenetic reconstruction. Nor, I am sure, can it come from polyphyletic groups. Two sources remain, paraphyletic groups and stratigraphy—the fossil record. I have already published my view on the role of fossils in phylogeny (Patterson 1977b) and in classification (Patterson and Rosen 1977). Fossils, and extinct monophyletic groups, can be placed in cladograms, by means of homologies, and can be classified in extant monophyletic groups. Those which cannot be so treated are uninformative. In evolutionary terms, fossils serve to order anagenesis — to show the sequence in which the homologies of extant monophyletic groups were developed, or the relative age of these characters. But the end result is still systematics, no more than a classification. When evolutionists talk of "the origin and evolution of higher taxa" they have more than that in mind. What is meant, on the morphological level, is, I suggest, statements like:

> As is usual in well-documented transitions, *Ichthyostega* shows a mosaic of primitive, intermediate, and advanced features linking it phenetically to earlier rhipidistian crossopterygians, on one hand, and to later labyrinthodont amphibians, on the other . . . . The important point is that, looking backwards in time, primitive reptiles converge with anthracosaurs in the lower Carboniferous . . . . The evidence indicates unequivocally that *Archaeopteryx* evolved from a small coelurosaurian dinosaur and that modern birds are surviving dinosaurian descendants (Gingerich 1979, pp. 63–66).

This is the stuff of phylogeny. What all those statements have in common is the names of extinct paraphyletic (ancestral) groups — rhipidistians, anthracosaurs, coelurosaurs. The mysterious additional element, the extra information that transforms systematics into phylogeny, is extinct paraphyletic groups. Recent paraphyletic groups are merely a hindrance. They are uncharacterizable by homology, but at least one can understand what is meant by fish, reptile or invertebrate: one can learn the conventions, what has to be lacking for membership and when in doubt, one can dissect. But with extinct paraphyletic groups the situation is much worse. Not only can no statement of homology be made about the group; often no statement of any kind can be made about it, except perhaps one of minimum stratigraphic or geographic range, for often there will be fossils assigned to the group which are so incomplete that they cannot be checked to see if they lack what is necessary for membership of the paraphyletic group, or have what is necessary for inclusion in the monophyletic group of which the paraphyletic group is part. In short, extinct paraphyletic groups are phenetic associations of scraps, whose limits are arcane, agreed or disputed among a few palaeontologists.

The reader who doubts that statement — that the limits of extinct paraphyletic groups are not common knowledge — may put it to the test by trying to determine, from the literature or any other source, what character of a newly discovered fossil would place it in the Rhipidistia, Anthracosauria, Coelurosauria, Palaeoniscoidei, Pholidophoridae, Eosuchia, or any other such group, but not in an included subgroup. To save wasted effort, recall that to place a fossil in one of these groups, but not in an included subgroup, would mean knowing the homology that linked those groups. And since paraphyletic groups have no homologies of their own, the task is impossible. If it is possible, the group is not paraphyletic.

Is it not strange that the justification of phylogeny, as something beyond systematics, resides in extinct paraphyletic groups? For those groups are the inventions of evolutionists, those who appeal to them as demonstrating the path of descent. So far as I know, such groups did not exist in pre-Darwinian taxonomy, for palaeontologists were then preoccupied with the real problem of allocating fossils to Recent groups (Patterson 1977b, p. 596). Nor do I find any extinct paraphyletic groups in Haeckel's (1866) trees. Such groups are therefore a later invention, imagined by evolutionists, those most committed to the confirmation of Darwin's views. The power of this mystery, extinct paraphyletic groups as the source of phylogeny, is shown by the fact that we still have no cladogram, or series of nested homologies, for tetrapods, the group in which phylogeny is supposed to be best known. Tetrapod relationships will be sorted out, I suppose, not by fossils but by comparative morphology — recognition of homologies, hence of groups, and by discrimination of monophyletic (natural) and paraphyletic (unnatural) groups through correlation with embryology. In Nelson's (1978, p. 336) words, "the concept of evolution is an extrapolation, or interpretation, of the orderliness of ontogeny". Or in the words of a great nineteenth century palaeontologist (Agassiz 1874, p. 92) "It is worth our while to ask if there is any such process as evolution in nature. Unquestionably, yes. But all that is actually known of this process we owe to the great embryologists." Agassiz then names von Baer.

SUMMARY

1. Classical definitions of homology suffer from the problem of defining concepts like "same" or "essential". In evolutionary and cladistic definitions "same" is defined as "self-identical" (Bock 1974, p. 387) in a common ancestor, but there is still a problem in defining common ancestry, as is shown by the fact that Simpson (1961) and Hecht and Edwards (1977) define homology in terms of common ancestry, in similar words, yet Simpson's definition excludes parallelism, and Hecht and Edwards's includes it. These difficulties can be avoided by defining homology as similarity characterizing monophyletic groups. Monophyly can then be defined in terms of a tree (Hennig 1966) or a cladogram (Platnick 1977a).

2. If monophyly is defined with reference to a cladogram, or set theory, or graph theory, then "monophyletic" and "natural" are synonymous terms for groups, since monophyletic groups best meet all the demands of natural groups (Farris 1980). Homology can then be defined as the relation characterizing natural groups, a definition that should be acceptable to classicists, pheneticists, cladists and utilitarians (those who equate homologies and taxonomic characters).

3. If homology is the property of monophyletic groups, several consequences follow. First, homology and synapomorphy must be the same thing. Secondly, homologies must form a hierarchy. And thirdly, non-monophyletic groups (para- and polyphyletic) cannot be characterized by homology. These conclusions are justified. The only circumstance under which a paraphyletic group can be characterized by homology is if the paraphyly is caused by the removal of a monophyletic group characterized *solely* by the absence of that homology.

4. But not all homologies characterize hierarchic groupings (the latter are here called taxic homologies). Homologies like Owen's interpretation of the limbs and girdles in terms of the axial skeleton, and Gegenbaur's interpretation of them as gill-arches, are distinguished as transformational homologies. Taxic homologies refer to morphotypes (ground plans of monophyletic groups); transformational homologies refer to archetypes.

5. Taxic homologies may be tested in three ways: by conjunction (if the supposed homologues occur together in the same organism, the homology is refuted); by similarity (in ontogeny, topography, composition); and by congruence with other homologies. Of the three tests, the last is the strongest.

6. The congruence test can be viewed in terms of probability. When two sets of homologies conflict (are incongruent), both can be shown to be improbable, and only one can be true. The simplest resolution is by querying the grouping specified by each homology in each set.

7. The systematist whose pet group is threatened by the congruence test is unjustified in appealing to weighting of homologies, since weighting by the systematist is an *a posteriori* exercise, designed to prefer one group to another. Homologies weight themselves, by congruences which are beyond the bounds of chance. Homologies of high weight, or burden, characterize groups of high taxonomic rank.

8. Homoplasy (non-homology) is categorized by means of performance in the three tests of homology: congruence, similarity and conjunction. Homologies pass all three tests. Relations which pass two tests but fail the third are homonomy (fails conjunction), complement (presence versus absence, fails similarity) and parallism (fails congruence). Only those relations which pass the congruence test are of use to the systematist.

9. Polarity of homologies is usually thought of as deciding which of two conditions is primitive and which derived. Here the problem is treated in a simpler way, as discriminating paraphyletic and monophyletic groups. Comparative methods (in-group and out-group comparison) are devices to render homologies of unknown

polarity congruent with others of known polarity. Polarity can only be ordered by von Baer's law, the orderliness of ontogeny, which leads to recognition of general and special conditions. Evolutionary interpretations are not a necessary consequence of ontogenetic order, and knowledge of evolution is unnecessary for analysis of homologies.

10. The role of homology in phylogenetic reconstruction is limited to the production of cladograms, or classifications. These activities need have no evolutionary connotations. If phylogeny has to be about evolution, homology has nothing to contribute to it. The extra element that transforms classification into phylogeny can only be extinct paraphyletic (ancestral) groups, phenetic associations of fossils, which are themselves dependent on belief in phylogeny.

11. Recognizing homologies demands that we recognize the things being compared as organisms. Beyond that, the recognition of homologies (and hence of monophyletic groups) is directly comparable to discovering new species.

ACKNOWLEDGEMENTS

For comment and discussion during the preparation of this chapter, I am grateful to colleagues, particularly Dick Jefferies and Dick Vane-Wright. For criticism of the manuscript I thank Gareth Nelson, Donn Rosen, Ed Wiley, and especially Roger Miles.

NOTES

[1] Other terms with the same connotation as morphotype and archetype are ground plan, structural plan and Bauplan. When introducing the word morphotype into English, Zangerl (1948) sought to distinguish it from structural plan. If I understand him, he viewed a particular structural plan as a concept of greater generality than the corresponding morphotype, so that a structural plan characterizes a group whose subgroups are each characterized by a morphotype. In Zangerl's usage, therefore, the two concepts stand in the same relationship as synapomorphy and symplesiomorphy: each morphotype is the structural plan of the included subgroups. Nelson (1978) equates structural plan and morphotype, whereas Gutmann (1977) uses Bauplan for what, in my view, is an archetype. Riedl (1979) equates ground plan and Bauplan.

[2] A less conservative estimate of $n$ in this instance comes from Whittington (1979, fig. 2), whose "diagram of the pattern of evolution in arthropods" includes 70 independent lineages. In that case, the probability that two homologies would characterize the Uniramia (3/70) by chance is 1/54 740, and that three should is $1/3 \times 10^9$ — long odds indeed.

[3] This agrees with current evolutionary usage, but not with Lankester's (1870) intention when introducing the term homoplasy, proposed as a subdivision of homology, a term he deprecated for its idealistic connotations. Homologous (in the pre-evolutionary sense) relations were to be classed either as homogeny (similarity due to common ancestry) or homoplasy (similarity not due to common ancestry, including serial homology, for example).

[4] I suggest that three-state homologies do not exist, for the reason that the relation of homology is absolute, not partial (Inglis 1966; van der Klaauw 1966, p. 56, disagrees, mistakenly in my view). The only apparent exceptions to two-state homologies are quantitative characters, either meristic (e.g. number of branchiostegal rays) or proportional (e.g. second pre-ural neural spine full-length, half-length or a low crest — the examples are abstracted from the 70 character-state trees in Zehren 1979). But these are readily treated as two-state, or two-state with a third state representing absence. The exception to this is circular character state trees (Zehren 1979, table 82), which are not resolvable into monophyletic subgroups by any method.

[5] It also explains Wiley's (1979b, p. 315) conundrum, how a cranial capacity of 1200 cc can be characteristic of *Homo sapiens,* when no newborn individual has the character. The character of *H. sapiens* is not to have that cranial capacity, but both to lack it (general condition) and to have it (special condition), whereas all other organisms exhibit only the general condition. This is why the complement relation (presence *vs.* absence) applies only to semaphoronts, not to life-histories (p. 48).

[6] Gaffney (1979a) has recently named the group "tetrapods minus ichthyostegids" as Neotetrapoda, and characterized it by five characters, not all of which are loss of homologies.

REFERENCES

Agassiz, J. L. R. (1874). Evolution and permanence of type. *Atlantic Monthly* 33: 92–101.
Andersen, N. M. (1980). Phylogenetic inference as applied to the study of evolutionary diversification of semiaquatic bugs (Hemiptera: Gerromorpha). *Syst. Zool.* 28, 554–578.
Anderson, D. T. (1979). Embryos, fate maps, and the phylogeny of arthropods. *In* "Arthropod Phylogeny" (A. P. Gupta, ed.), pp. 59–105. Van Nostrand Rheinhold, New York.

Anonymous (1979). "Dinosaurs and Their Living Relatives". British Museum (Natural History) and Cambridge University Press, London.

Bjerring, H. C. (1967). Does a homology exist between the basicranial muscle and the polar cartilage? *Colloques int. Cent. natn. Res. scient.* **163**, 223–267.

Bjerring, H. C. (1977). A contribution to structural analysis of the head of craniate animals. *Zool. Scripta* **6**, 127–183.

Bjerring, H. C. (1979). Quondam gill-covers. *Zool. Scripta* **8**, 235–240.

Blackwelder, R. E. (1967). "Taxonomy. A Text and Reference Book". John Wiley, New York.

Bock, W. J. (1963). Evolution and phylogeny in morphologically uniform groups. *Am. Nat.* **97**, 265–285.

Bock, W. J. (1969). Comparative morphology in systematics. *In* "Systematic Biology", pp. 411–448. National Academy of Sciences, Publ. 1692, Washington.

Bock, W. J. (1974). Philosophical foundations of classical evolutionary classification. *Syst. Zool.* **22**, 375–392.

Bock, W. J. (1977). Foundations and methods of evolutionary classification. *In* "Major Patterns in Vertebrate Evolution" (M. K. Hecht, P. C. Goody and B. M. Hecht, eds), pp. 851–895. Plenum Press, New York.

Bonde, N. (1977). Cladistic classification as applied to vertebrates. *In* "Major Patterns in Vertebrate Evolution" (M. K. Hecht, P. C. Goody and B. M. Hecht, eds), pp. 741–804. Plenum Press, New York.

Boudreaux, H. B. (1979). "Arthropod Phylogeny with Special Reference to Insects". John Wiley, New York.

Boyden, A. (1973). "Perspectives in Zoology". Pergamon Press, Oxford.

Bretsky, S. (1979). Recognition of ancestor-descendant relationships in invertebrate paleontology. *In* "Phylogenetic Analysis and Paleontology" (J. Cracraft and N. Eldredge, eds), pp. 113–163. Columbia University Press, New York.

Brundin, L. (1972). Evolution, causal biology, and classification. *Zool. Scripta* **1**, 107–120.

Brundin, L. (1976a). A Neocomian chironomid and Podonominae-Aphroteniinae (Diptera) in the light of phylogenetics and biogeography. *Zool. Scripta* **5**, 139–160.

Brundin, L. (1976b). Parallelism and it phylogenetic significance. *Zool. Scripta* **5**, 186.

Clarke, K. U. (1979). Visceral anatomy and arthropod phylogeny. *In* "Arthropod Phylogeny" (A. P. Gupta, ed.), pp. 467–549. Van Nostrand Rheinhold, New York.

Colless, D. (1969). Phylogenetic inference: a reply to Dr Ghiselin. *Syst. Zool.* **18**, 462–466.

Cracraft, J. (1967). Comments on homology and analogy. *Syst. Zool.* **16**, 355–359.

Cracraft, J. (1978). Science, philosophy, and systematics. *Syst. Zool.* **27**, 213–216.

Cracraft, J. (1979). Phylogenetic analysis, evolutionary models and paleontology.

*In* "Phylogenetic Analysis and Paleontology" (J. Cracraft and N. Eldredge, eds), pp. 7–39. Columbia University Press, New York.

Crowson, R. A. (1970). "Classification and Biology". Heinemann, London.

De Beer, G. R. (1971). Homology, an unsolved problem. *Oxford Biology Readers*, **11**, 1–16. University Press, Oxford.

Dupuis, C. (1979). Permanence et actualité de la systematique. La 'Systématique Phylogénétique' de W. Hennig. *Cah. Nat.* **34**, 1–69.

Eldredge, N. (1979a). Alternative approaches to evolutionary theory. *Bull. Carnegie Mus. nat. Hist.* **13**, 7–19.

Eldredge, N. (1979b). Cladism and common sense. *In* "Phylogenetic Analysis and Paleontology" (J. Cracraft and N. Eldredge, eds), pp. 165–198. Columbia University Press, New York.

Farris, J. S. (1975). Formal definitions of paraphyly and polyphyly. *Syst. Zool.* **23**, 548–554.

Farris, J. S. (1977). On the phenetic approach to vertebrate classification. *In* "Major Patterns in Vertebrate Evolution" (M. K. Hecht, P. C. Goody and B. M. Hecht, eds), pp. 823–850. Plenum Press, New York.

Farris, J. S. (1979). On the naturalness of phylogenetic classification. *Syst. Zool.* **28**, 200–214.

Farris, J. S. (1980). The information content of the phylogenetic system. *Syst. Zool.* **28**, 483–519.

Farris, J. S. and Kluge, A. G. (1979). A botanical clique. *Syst. Zool.* **28**, 400–411.

Fitch, W. M. (1977). The phyletic interpretation of macromolecular sequence information: simple methods. *In* "Major Patterns in Vertebrate Evolution" (M. K. Hecht, P. C. Goody and B. M. Hecht, eds), pp. 169–204. Plenum Press, New York.

Fryer, G. (1978). Arthropod morphology and evolution. *Nature Lond.* **273**, 172–173.

Gaffney, E. S. (1979a). Tetrapod monophyly: a phylogenetic analysis. *Bull. Carnegie Mus. nat. Hist.* **13**, 92–105.

Gaffney, E. S. (1979b). An introduction to the logic of phylogeny reconstruction. *In* "Phylogenetic Analysis and Paleontology" (J. Cracraft and N. Eldredge, eds), pp. 79–111. Columbia University Press, New York.

Ghiselin, M. T. (1966a). An application of the theory of definitions to taxonomic principles. *Syst. Zool.* **15**, 127–130.

Ghiselin, M. T. (1966b). On psychologism in the logic of taxonomic controversies. *Syst. Zool.* **15**, 207–215.

Ghiselin, M. T. (1969). The distinction between similarity and homology. *Syst. Zool.* **18**, 148–149.

Gingerich, P. D. (1979). The stratophenetic approach to phylogeny reconstruction in vertebrate paleontology. *In* "Phylogenetic Analysis and Paleontology" (J. Cracraft and N. Eldredge, eds), pp. 41–77. Columbia University Press, New York.

Gould, S. J. (1977). "Ontogeny and Phylogeny". Belknap Press, Cambridge, Mass.

Griffith, G. C. D. (1974). On the foundations of biological systematics. *Acta biotheor.* **23**, 85–131.

Gutmann, W. F. (1977). Phylogenetic reconstruction: theory, methodology and application to chordate evolution. *In* "Major Patterns in Vertebrate Evolution" (M. K. Hecht, P. C. Goody and B. M. Hecht, eds), pp. 645–669. Plenum Press, New York.

Haeckel, E. (1866). "Generelle Morphologie der Organismen". G. Reimer, Berlin.

Hecht, M. K. (1976). Phylogenetic inference and methodology as applied to the vertebrate record. *Evolut. Biol.* **9**, 335–363.

Hecht, M. K. and Edwards, J. L. (1977). The methodology of phylogenetic inference above the species level. *In* "Major Patterns in Vertebrate Evolution" (M. K. Hecht, P. C. Goody and B. M. Hecht, eds), pp. 3–51. Plenum Press, New York.

Hennig, W. (1953). Kritische Bemerkungen zum phylogenetischen System der Insekten. *Beitr. Ent.* **3** (Sonderheft), 1–85.

Hennig, W. (1966). "Phylogenetic Systematics". University of Illinois Press, Urbana.

Hennig, W. and Schlee, D. (1978). Abriss der phylogenetischen Systematik. *Stuttg. Beitr. Naturk.* (A) **319**, 1–11.

Hoyle, G. and Williams, M. (1980). The musculature of *Peripatus* and its innervation. *Phil. Trans. R. Soc.* **B288**, 481–510.

Hull, D. L. (1967). Certainty and circularity in evolutionary taxonomy. *Evolution, Lancaster, Pa.* **21**, 174–189.

Inglis, W. G. (1966). The observational basis of homology. *Syst. Zool.* **15**, 219–228.

Inglis, W. G. (1970). Similarity and homology. *Syst. Zool.* **19**, 93.

Jardine, N. (1967). The concept of homology in biology. *Br. J. Phil. Sci.* **18**, 125–139.

Jardine, N. (1970). The observational and theoretical components of homology: a study based on the morphology of the dermal skull-roofs of rhipidistian fishes. *Biol. J. Linn. Soc.* **1**, 327–361.

Jardine, N. and Sibson, D. (1971). "Mathematical Taxonomy". Wiley, London.

Jarvik, E. (1952). On the fish-like tail in the ichthyostegid stegocephalians, with descriptions of a new stegocephalian and a new crossopterygian from the Upper Devonian of East Greenland. *Meddr Grønland* **114**, 12, 1–90.

Jarvik, E. (1980). "Basic Structure and Evolution of Vertebrates". Academic Press, London and New York.

Jefferies, R. P. S. (1979). The origin of chordates – a methodological essay. *In* "The Origin of Major Invertebrate Groups" (M. R. House, ed.), pp. 443–477. Academic Press, London and New York.

Key, K. H. L. (1967). Operational homology. *Syst. Zool.* **16**, 275–276.

Klaauw, C. J. van der (1966). Introduction to the philosophic backgrounds and prospects of the supraspecific comparative anatomy of conservative characters in the adult stages of conservative elements of Vertebrata with an enumeration of many examples. *Verh. K. ned. Akad. Wet.* (2) **57**, 1, 1–196.

Kristensen, N. P. (1980). (Book review) *Syst. Zool.* **28**, 638–643.

Kristensen, R. M. (1978). On the structure of *Batillipes noerrevangi* Kristensen 1978. 2. The muscle-attachments and the true cross-striated muscles. *Zool. Anz.* **200**, 173–184.

Lankester, E. R. (1870). On the use of the term homology in modern zoology. *Ann. Mag. Nat. Hist.* (4) **6**, 35–43.

Lock, M. and Huie, P. (1977). Bismuth staining of Golgi complex is a characteristic arthropod feature lacking in *Peripatus*. *Nature, Lond.* **270**, 341–343.

Løvtrup, S. (1977). "The Phylogeny of Vertebrata". Wiley, London.

Lunderberg, J. C. (1973). More on primitiveness, higher level phylogenies and ontogenetic transformations. *Syst. Zool.* **22**, 327–329.

Manton, S. M. (1979). Functional morphology and the evolution of the hexapod classes. *In* "Arthropod Phylogeny" (A. P. Gupta, ed.), pp. 387–465. Van Nostrand Rheinhold, New York.

Manton, S. M. and Anderson, D. T. (1979). Polyphyly and the evolution of arthropods. *In* "The Origin of Major Invertebrate Groups" (M. R. House, ed.), pp. 269–321. Academic Press, London and New York.

Mayr, E. (1969). "Principles of Systematic Zoology". McGraw-Hill, New York.

Mayr, E. (1974). Cladistic analysis or cladistic classification? *Z. Zool. Syst. EvolForsch.* **12**, 94–128.

Nelson, G. J. (1970) Outline of a theory of comparative biology. *Syst. Zool.* **19**, 373–384.

Nelson, G. J. (1973). The higher-level phylogeny of vertebrates. *Syst. Zool.* **22**, 87–91.

Nelson, G. J. (1978). Ontogeny, phylogeny, paleontology and the biogenetic law. *Syst. Zool.* **27**, 324–345.

Nelson, G. J. and Platnick, N. I. (1981). "Systematics and Biogeography". Columbia University Press, New York.

Owen, R. (1843). "Lectures on Comparative Anatomy". Longman, Brown, Green, and Longmans, London.

Owen, R. (1849). "On the Nature of Limbs". J. van Voorst, London.

Ouweneel, W. J. (1976). Developmental genetics of homoeosis. *Adv. Genet.* **18**, 179–248.

Patterson, C. (1977a). Cartilage bones, dermal bones and membrane bones, or the exoskeleton versus the endoskeleton. *In* "Problems in Vertebrate Evolution" (S. M. Andrews, R. S. Miles and A. D. Walker, eds), pp. 71–121. Academic Press, London and New York.

Patterson, C. (1977b). The contribution of paleontology to teleostean phylogeny. *In* "Major Patterns in Vertebrate Evolution" (M. K. Hecht, P. C. Goody and B. M. Hecht, eds), pp. 579–643. Plenum Press, New York.

Patterson, C. (1978a). Verifiability in systematics. *Syst. Zool.* **27**, 218–222.

Patterson, C. (1978b). "Evolution". British Museum (Natural History), London.

Patterson, C. and Rosen, D. E. (1977). Review of ichthyodectiform and other Mesozoic teleost fishes and the theory and practice of classifying fossils. *Bull. Am. Mus. nat. Hist.* **158**, 81–172.

Platnick, N. I. (1977a). Paraphyletic and polyphyletic groups. *Syst. Zool.* **26**, 195–200.

Platnick, N. I. (1977b). Cladograms, phylogenetic trees and hypothesis testing. *Syst. Zool.* 26, 438—442.

Platnick, N. I. (1978). Classifications, historical narratives and hypotheses. *Syst. Zool.* 27, 365—369.

Platnick, N. I. (1979). Gaps and prediction in classification. *Syst. Zool.* 27, 472—474.

Platnick, N. I. (1980). Philosophy and the transformation of cladistics. *Syst. Zool.* 28, 537—546.

Platnick, N. I. and Cameron, H. D. (1977). Cladistic methods in textual, linguistic and phylogenetic analysis. *Syst. Zool.* 26, 380—385.

Popper, K. R. (1962). "The Open Society and Its Enemies. Vol. 2. Hegel and Marx" (4th edn). Routledge & Kegan Paul, London.

Riedl, R. (1979). "Order in Living Organisms". J. Wiley, Chichester.

Romer, A. S. (1971). "The Vertebrate Body" (4th edn). W. B. Saunders, Philadelphia.

Ross. H. H. (1974). "Biological Systematics". Addison-Wesley, Reading, Mass.

Russell, E. S. (1916). "Form and Function". John Murray, London.

Saether, O. A. (1979). Underlying synapomorphies and anagenetic analysis. *Zool. Scripta* 8, 305—312.

Schlee, D. (1975). Das Problem der Podonominae-Monophylie; Fossiliendiagnose und Chironomidae-Phylogenetik (Diptera). *Entomologica germ.* 1, 316—351.

Schlee, D. (1976). Structures and functions, their general significance for phylogenetic reconstruction in Recent and fossil taxa. *Zool. Scripta* 5, 181—184.

Schlee, D. (1978). Anmerkungern zur phylogenetischen Systematik: Stellungnahme zu einigen Missverständnissen. *Stuttg. Beitr. Naturk.* (A) 320, 1—14.

Schram, F. R. (1980). (Book review) *Syst. Zool.* 28, 635—638.

Schuh, R. T. (1978). (Book review) *Syst. Zool.* 27, 255—260.

Seeger, W. (1979). Spezialmerkmale an Eihüllen und Embryonen von Psocoptera im Vergleich zu anderen Paraneoptera (Insecta); Psocoptera als monophyletische Gruppe. *Stuttg. Beitr. Naturk.* (A) 329, 1—57.

Simpson, G. G. (1959). Anatomy and morphology: classification and evolution: 1859 and 1959. *Proc. Am. phil. Soc.* 103, 286—306.

Simpson, G. G. (1961). "Principles of Animal Taxonomy". Columbia University Press, New York.

Sneath, P. H. A. and Sokal, R. R. (1973) "Numerical Taxonomy". W. H. Freeman, San Francisco.

Sokal, R. R. and Sneath, P. H. A. (1963). "Principles of Numerical Taxonomy". W. H. Freeman, San Francisco.

Szalay, F. S. (1977). Phylogenetic relationships and a classification of the eutherian Mammalia. *In* "Major Patterns in Vertebrate Evolution" (M. K. Hecht, P. C. Goody and B. M. Hecht, eds), pp. 315—374. Plenum Press, New York.

Van Valen, L. (1978). Why not to be a cladist. *Evolut. Theory* 3, 285—299.

Whittington, H. B. (1979). Early arthropods, their appendages and relationships. *In* "The Origin of Major Invertebrate Groups" (M. R. House, ed.), pp. 253—268. Academic Press, London and New York.

Wiley, E. O. (1975). Karl R. Popper, systematics, and classification: a reply to Walter Bock and other evolutionary taxonomists. *Syst. Zool.* 24, 233–243.

Wiley, E. O. (1976). The phylogeny and biogeography of fossil and Recent gars (Actinopterygii: Lepisosteidae). *Misc. Publs Mus. nat. Hist. Univ. Kans.* 64, 1–111.

Wiley, E. O. (1979a). Ancestors, species, and cladograms – remarks on the symposium. *In* "Phylogenetic Analysis and Paleontology" (J. Cracraft and N. Eldredge, eds), pp. 211–225. Columbia University Press, New York.

Wiley, E. O. (1979b). An annotated Linnaean hierarchy, with comments on natural taxa and competing systems. *Syst. Zool.* 28, 308–337.

Woodger, J. H. (1945). On biological transformations. *In* "Essays on Growth and Form Presented to D'Arcy W. Thompson" (W. E. LeGros Clark and P. B. Medawar, eds), pp. 95–120. University Press, Oxford.

Zangerl, R. (1948). The methods of comparative anatomy and its contribution to the study of evolution. *Evolution, Lancaster, Pa.* 2, 351–374.

Zangerl, R. and Case, G. R. (1976). *Cobelodus aculeatus* (Cope), an anacanthous shark from Pennsylvanian black shales of North America. *Palaeontographica* (A) 154, 107–157.

Zehren, S. J. (1979). The comparative osteology and phylogeny of the Beryciformes (Pisces: Teleostei). *Evolutionary Monographs, Chicago*, 1, 1–389.

## NOTE ADDED IN PROOF

Since the manuscript of this chapter was submitted in May 1980, I have realized that E. O. Wilson's (1965, *Syst. Zool.* 14, 214–220) "Consistency test for phylogenies" is a more rigorous formulation of my congruence testing of homologies, described on p. 38. I failed to appreciate this earlier because Wilson's paper ostensibly concerns phylogeny, whereas I was investigating homology. However, I suggest that Wilson's method is really a test of homology, since inferences about phylogeny follow only if cladograms, or sets and subsets, must be read as trees (i.e. evolution is axiomatic). N. Eldredge and J. Cracraft (1980, "Phylogenetic Patterns and the Evolutionary Process", Columbia University Press, New York) have published an account of homology similar to my own, defining homology as a property of sets or groups. A collection of papers on homology is found in R. B. Masterton, W. Hodos and H. Jerison (eds)(1976), "Evolution, Brain, and Behaviour: Persistent Problems", Lawrence Erlbaum Associates, Hillsdale, N. J. Also of interest is O. Rieppel's (1980) paper, "Homology, a deductive concept" (*Z. Zool. Syst. EvolForsch.* 18, 315–319).

# 3 | The Adequacy of the Fossil Record

C. R. C. PAUL

*Department of Geology, University of Liverpool*
*Liverpool, England*

Abstract: Analysis of gaps, the occurrence of monotypic taxa, uniformitarian arguments and survivorship analysis suggest that the fossil records of echinoderms and vertebrates are at least 25% and 40% incomplete at familial and generic levels, respectively. Growth of our knowledge has resulted in the numbers of cystoid families, genera and species increasing by 10, 50 and 100%, respectively, since 1900.

Assuming random preservation and uniform sedimentation, the frequency distribution of intervals between conspecific specimens in a measured section follows the curve $y = ke^{-\lambda x}$, where $y$ is frequency, $x$ stratigraphic interval, $k$ total intervals and $\lambda$ is a constant. The initial assumptions may be tested using the probability that the largest interval observed would occur in a sample of $k$ intervals.

Fossils can only be preserved in the wrong stratigraphic order if their original ranges overlapped, hence most fossils must be in the correct historical sequence. The probability of two randomly found specimens being preserved in the wrong order cannot exceed 50% and is always less than the probability that they are in the correct order. Known ranges of fossils put minimum ages on latest common ancestors and these ages should not be ignored in phylogenetic reconstruction.

The fossil record is much less incomplete than is generally accepted. Its incompleteness is largely irrelevant to the sequence of preserved fossils and hence to phylogenetic reconstruction, provided only that we confine our phylogenies to known organisms. We ignore the fossil record at our peril.

Systematics Association Special Volume No. 21, "Problems of Phylogenetic Reconstruction", edited by K. A. Joysey and A. E. Friday, 1982, pp. 75—117, Academic Press, London and New York.

This contribution is largely concerned with the usefulness of the fossil record in phylogenetic reconstruction. It would therefore seem wise to stress at the outset one overriding limitation of the fossil record. However reliable, or otherwise, the relative stratigraphic order of fossils may be, this alone cannot yield a phylogeny. Phylogenies can only be based on morphological comparisons, using whichever methods one may prefer. Indeed the very first decision as to which organisms shall be included within the phylogeny is based on morphological similarity (at least as expressed in classification). We normally decide to construct a phylogeny for a taxon (e.g. a family or an order) rather than for a collection of fossils from a particular stratigraphic horizon or a sample of living organisms from a particular locality.

The fossil record preserves two types of information: absolute information, for example the former existence of extinct species, and relative information, of which the sequence of first appearances of fossil species is the most important for phylogenetic reconstruction. Other relative information includes, for example, abundance, diversity, longevity and geographic distribution. While it is accepted that ignoring the absolute information from the fossil record will yield incomplete phylogenies, we usually reject the relative information, particularly in phylogenetic reconstruction, on the grounds that the fossil record is incomplete. I wish to show that notwithstanding the incompleteness of the fossil record, the relative stratigraphic order of fossil species is overwhelmingly correct and can be used as an independent test of phylogenies reconstructed on the basis of morphological comparisons. Indeed the best reconstructions are a compromise between both lines of evidence.

The fossil record suffers from an almost universal, and in my view largely unjustified, prejudice. To be sure it is incompletely known (new species of fossils are being described all the time) and even when all the fossil species there are have been discovered, the fossil record will still be incomplete. Soft bodied organisms such as wormfish, jellyfish and many algae are most unlikely to be preserved. As a result it is usually argued that the fossil record is inevitably unreliable, so that most contributions to evolutionary theory and many phylogenetic reconstructions are based entirely on the neontological record. Yet the latter is also incompletely known (new species of living organisms are still being described at an immense rate) and inevitably incomplete, since it does not include extinct species. In reconstructing the phylogeny of any family or even genus, it is inherently likely that some extinct species will need to be considered. Somehow this incompleteness of the neontological record has never been seen as a barrier to its use in evolutionary theory or phylogenetic reconstruction. So although both records are inevitably incomplete, this is seen as a tremendous drawback in the case of fossils, but no handicap at all in the case of living organisms. Presumably this is because living organisms preserve so much more information (about soft tissue, behaviour, development) than fossils do. However, to ignore the fossil record in phylogenetic reconstruction, even where it is acknowledged to be very incomplete, seems to me like attempting to reconstruct the history of the British Isles on the basis of current media publications. Undoubtedly there will be an immense amount of detailed information to be read, heard and seen in 1980, compared with which our historical records and archaeological discoveries are meagre in the extreme. Nevertheless, this meagre historical record gives a much better outline of *the sequence of events* than the most rigorous analysis of the most extensive contemporary records can ever do. In my view the same is true of the use of the fossil record in phylogenetic reconstruction.

Since the incompleteness of the fossil record has been stressed in the past, I shall consider this before discussing the reliability of the preserved sequence of events. This chapter is therefore divided into three main sections; first, tests of the incompleteness of the fossil record, then tests of our knowledge of the record and finally the adequacy of the resulting known record in phylogenetic reconstruction. The last section is the most important in the context of this symposium.

TESTS OF THE INCOMPLETENESS OF THE FOSSIL RECORD

## 1. General Tests

(a) *Analysis of gaps.* To my mind the real problem about the incompleteness of the fossil record lies in the obvious fact that many forms of life may have existed in the past which have not been preserved as fossils and hence about which we know nothing. Does the fossil record represent the tip of the iceberg of life or does it represent a fish's view of the iceberg with only the tip hidden? To test this idea we need a situation where we know a form of life existed but has not apparently been preserved or discovered. Gaps in the record provide just such a test. For the purposes of developing this argument, let us ignore problems of identification and correlation. We will return to them later. A gap occurs when a taxon is known from below and above, but not actually within, a given stratigraphic interval. Each such gap represents a situation where at least one species must have existed to carry on the line, but has not been preserved or is yet to be discovered. Unfortunately this is so obvious that very few range charts show any gaps at all. We all make the logical deduction and plot total known range irrespective of whether it is based on two, or two million, occurrences. It has proved difficult to gather adequate data to test this idea widely (at least using macrofossils), but some information is presented here. Figure 1 represents the known ranges, including gaps, plotted at series level for all cystoid families which are known from at least two stratigraphic series. Obviously families known from a single series may also have gaps in their records, but we cannot detect them. Equally, there may be additional gaps above or below the known ranges of the families plotted which we cannot detect. For these (and other) reasons, estimates of the incompleteness of the fossil record using gaps are bound to be minimum estimates. The total range of all families in Fig. 1 is 108 stratigraphic series, the gaps total 27 series, hence the fossil record of cystoids at family/series level is at least 25% incomplete. The fossil record of cyclocystoids is 25% incomplete at family level and 36% incomplete at generic level using slightly more refined stratigraphic intervals (Smith and Paul 1982, p. 609). That of extant families of amphibians plotted at stage level throughout the Mesozoic and Tertiary is approximately 30% incomplete (Carroll 1977).

This test may be extended. For example, by reading across Fig. 1 it is possible to see which families have the most, and which the least, complete records. Six families have no detectable gaps in their records. I think they are less likely to have gaps beyond their known ranges than other families with less complete records because the

|  | ORDOVICIAN | | | | | | SILURIAN | | | | DEVONIAN | | | | | |  | |
|---|---|---|---|---|---|---|---|---|---|---|---|---|---|---|---|---|---|---|
|  | Tr | Ar | Lv | Ld | Ca | As | Ll | We | Lu | Do | Ge | Si | Em | Ei | Gi | Fr | G | T |
| 1 |  |  |  |  |  |  |  |  |  |  |  |  |  |  |  |  | 1 | 4 |
| 2 |  |  |  |  |  |  |  |  |  |  |  |  |  |  |  |  | 0 | 6 |
| 3 |  |  |  |  |  |  |  |  |  |  |  |  |  |  |  |  | 6 | 13 |
| 4 |  |  |  |  |  |  |  |  |  |  |  |  |  |  |  |  | 0 | 2 |
| 5 |  |  |  |  |  |  |  |  |  |  |  |  |  |  |  |  | 0 | 2 |
| 6 |  |  |  |  |  |  |  |  |  |  |  |  |  |  |  |  | 3 | 11 |
| 7 |  |  |  |  |  |  |  |  |  |  |  |  |  |  |  |  | 1 | 5 |
| 8 |  |  |  |  |  |  |  |  |  |  |  |  |  |  |  |  | 0 | 5 |
| 9 |  |  |  |  |  |  |  |  |  |  |  |  |  |  |  |  | 2 | 7 |
| 10 |  |  |  |  |  |  |  |  |  |  |  |  |  |  |  |  | 0 | 5 |
| 11 |  |  |  |  |  |  |  |  |  |  |  |  |  |  |  |  | 0 | 5 |
| 12 |  |  |  |  |  |  |  |  |  |  |  |  |  |  |  |  | 1 | 5 |
| 13 |  |  |  |  |  |  |  |  |  |  |  |  |  |  |  |  | 1 | 3 |
| 14 |  |  |  |  |  |  |  |  |  |  |  |  |  |  |  |  | 2 | 4 |
| 15 |  |  |  |  |  |  |  |  |  |  |  |  |  |  |  |  | 3 | 6 |
| 16 |  |  |  |  |  |  |  |  |  |  |  |  |  |  |  |  | 5 | 14 |
| 17 |  |  |  |  |  |  |  |  |  |  |  |  |  |  |  |  | 1 | 3 |
| 18 |  |  |  |  |  |  |  |  |  |  |  |  |  |  |  |  | 1 | 8 |
| G | 0 | 1 | 1 | 3 | 0 | 2 | 7 | 2 | 2 | 3 | 1 | 2 | 2 | 1 | 0 | 0 | Totals | |
| T | 5 | 10 | 12 | 13 | 15 | 14 | 8 | 8 | 4 | 4 | 4 | 3 | 3 | 3 | 1 | 1 | 27 | 108 |

Fig. 1. Ranges of cystoid families known from more than one series, to show gaps (G) as well as total range (T). Reading across the columns gives the proportion of gaps for each family; reading down gives the proportion of gaps for each stratigraphical interval. In both cases G/T yields the proportion of gaps.

greater the ratio of gap to record the less likely it is that the first and last species (or specimens) are known and hence the greater the likelihood of gaps occurring beyond the known range.

The Gomphocystitidae (Family 15 in Fig. 1) with 50% gap has the worst record equalled by only one other family. This is particularly interesting because, although I have not examined any genus myself, I suspect that there is nothing other than spiral ambulacra to unite the Ordovician genus *Pyrocystites* with the Silurian genera *Gomphocystites* and *Celticystis*. This raises the problems of identification and classification which are inherent in the use of gaps at relatively high taxonomic levels. Nevertheless, in this context it is important to note that analysis of gaps may actually direct attention to possible errors of identification or systematics. At the same time, attention should be drawn to the rhombiferan family Pleurocystitidae (Family 3 in Fig. 1) and the diploporite family Sphaeronitidae (Family 16) which are well documented from the Ordovician and Devonian, but so far totally unknown from the Silurian. These are the two most dramatic gaps in the cystoid fossil record. I have examined Ordovician and Devonian representatives of both families personally and I am completely satisfied that these gaps are not due to misidentification, misclassification or convergence. The pleurocystitids are the most distinctive cystoids with a unique morphology and in both families a single genus carries through from the Upper Ordovician to the Devonian.

By reading down Fig. 1, it is possible to see how the completeness of the cystoid fossil record varied with time. In general the Ordovician record is better than average (about 10% gap) while the Silurian (58% gap) and Devonian (40% gap) are worse than average. In particular, the Lower Silurian (Llandovery) with seven gaps from eight known families has by far the worst record of any series and might be mistaken for a life crisis. There is certainly an element of life crisis between the uppermost Ordovician, with 14 known families, 12 of which are actually represented in the Ashgill Series, and younger rocks from which only eight of the Upper Ordovician families are known. However, this is not a life crisis in the sense of the ammonite fossil record at the Triassic/Jurassic boundary when all but one or possibly two families became extinct. The apparent crisis in cystoid evolution results from the exceptionally poor record in the Lower Silurian. It is more due to preservation failure (or lack of discovery) than to near extinction of the cystoids. Thus again, analysis of gaps directs attention to particularly good or bad episodes in the fossil record of a group. For example, if these gaps were due to miscorrelation, this fact would surely soon be recognized once the anomaly had been realized. Incidentally, Nakazawa and Runnegar (1973) have already used this argument in considering the fossil record of bivalves across the Permo–Triassic boundary. Why the fossil record of cystoids should be so exceptionally bad in the Lower Silurian is a moot point, but it may be related to the evolutionary recovery after the stress of the Hirnantian (latest Ordovician) glaciation. Suffice it to say here that the analysis of gaps directs attention to anomalies, both taxonomic and stratigraphic.

(b) *Monotypy and related tests.* Another generalized test is monotypy. If the fossil record were extremely incomplete, most fossils would be so different from each other that we would assign them to relatively high taxa. For example, suppose that on average only one species in a hundred were preserved and that the average family contained 20 species. Small families would stand very little chance of being represented in the fossil record, while only a very few of the largest families would be represented by more than one fossil species. In this case most families would be represented by a unique fossil species and even then only about one in five families would be represented at all. Thus the occurrence of monotypic families or genera could be used as an estimate of the completeness of the fossil record. The main difficulty with the method is that one might expect a certain proportion of families or genera to be genuinely monotypic. In this I suggest we take the extreme view that no taxon should be monotypic, for two reasons. First, it simplifies procedures, but secondly and more importantly, because we can only recognize the founder of a new lineage by hindsight. If *Archaeopteryx* were the only known bird, would we not classify it as an aberrant reptile with feathers rather than assign it to a separate class of vertebrates? Can the same not be said of the first species of new genera or the first genera of new families? Most taxonomists would agree that when we discover a single species (fossil or recent) with a very distinctive morphology, other unknown related species must occur or have occurred in the past. Different taxonomists may vary in their interpretation of what constitutes "very distinctive morphology" and to what taxonomic level the species should be assigned, but at least there would be general agreement that something was missing. Accepting this extreme argument that no truly monotypic families occur, then eight of the 24 known cystoid families are monotypic and hence have incomplete records.

This argument about monotypy may be extended to stratigraphic horizons (or geographical localities). Again, if the fossil record were extremely incomplete, most fossil taxa would be known from a single stratigraphic interval or a single locality. Only the longest ranging or most widely distributed taxa would stand much chance of being found at different levels or localities and even these would have much more gap than record. As with monotypy, a proportion of taxa might be expected to have been genuinely short-lived or genuinely restricted in their distribution and we have no means of estimating how many. Taking the extreme view on stratigraphic range, six of the 24 families of cystoids are known from a single series and probably have incomplete records. So many species are genuinely restricted in their distribution at the present day that I do not think it is justified to take the extreme view in this case. However, the mere occurrence of some families known from several continents (see Paul 1976, p. 553 *et seq.* for cystoid distribution) implies that the record cannot be so poor as all that. By no means all fossils have restricted distributions and in many cases their known distribution is unlikely to be due to post mortem transport.

(c) *Uniformitarian arguments.* Comparisons of the relative abundance of living groups of organisms with their known fossil abundances is a crude test of the completeness of the fossil record. Clearly this is a limited argument, but in certain cases it may yield informative results. Suppose, for example, that the present day fauna were completely known. Then we could extend a uniformitarian comparison back in time just as Kier (1977) has done with regular and irregular echinoids. Kier's basic argument was that whereas just over half the recent species of sea urchins are regulars, only about a quarter of the fossil species are in the period since the irregulars first evolved, i.e. since the late Lower Jurassic. He concluded that the fossil record of regular echinoids was much poorer than one would expect by comparing the relative diversities of living regular and irregular sea urchins. Once again the argument cannot be rigorous. There is no way of telling that the present ratio of irregular to regular species obtained at any time in the past, let alone was typical of long periods in the past.

(d) *Survivorship analysis.* Lyell's subdivision of the Tertiary on the proportions of extant mollusc species found at different levels is one example of a group of methods used in survivorship analysis. Basically survivorship analysis involves plotting the frequencies of taxa which survive for different periods of time as in Fig. 2. This idea can also be used as a non-quantitative test of the completeness of the fossil record. If the fossil record were extremely incomplete, almost all taxa would be known from a single horizon and hence have very short survivorship. Analysis would show initially oversteepened curves with very short "tails", particularly if done for all groups irrespective of when they first appeared in the record. Once again this can only be a generalized method since it is not possible to predict accurately the expected outline of any survivorship curve for fossils.

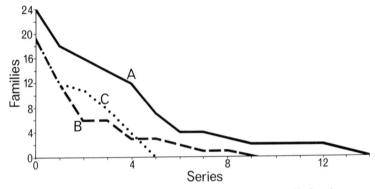

Fig. 2. Cystoid family survivorship. Curve A represents all families, curve B represents survivorship of Caradoc families, curve C represents the Caradoc census.

Survivorship analysis can be done in various ways. The cohort method involves tracing the survivorship of all groups which appear at a particular time (e.g. Fig. 3), such as tracing the histories of all people born in, say, 1900. The census method involves tracing the ancestry of all groups alive at a particular time, like plotting the ages of all people alive in 1980, while Lyell's method traced the percentages of living groups that still existed progressively further back in time. In all three methods a very steep initial portion of the curve would result if the fossil record were extremely incomplete. Survivorship analysis for cystoid families is shown in Fig. 2 and for Palaeozoic crinoid genera in Fig. 3. All curves have the expected concave-up outlines and are by no means grossly oversteepened initially.

To summarize, I hope I have shown that there are numerous common-sense ways in which the completeness of the fossil record

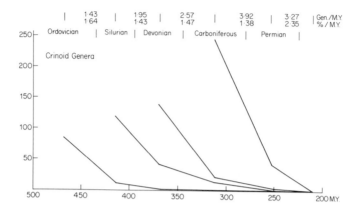

Fig. 3. Palaeozoic crinoid genera survivorship. Cohorts for each geological period are plotted from the midpoints of the periods. Figures at the top give mean extinction rates expressed in genera and percentage per million years for each period.

may be tested, if only in a qualitative manner. The three methods applied to cystoid families which yield crude quantitative estimates of incompleteness all suggest that the record is at least 25–33% incomplete at family/series level, although the similarity of the figures may be coincidental.

## 2. Rigorous Tests

*(a) Shaw's tests for accurately determining stratigraphic range.* Shaw (1964) is the only person I have come across who has considered the completeness of the fossil record rigorously. He did so in the context of correlation and the determination of the stratigraphic ranges of fossils. Since use of the fossil record in both correlation and phylogenetic reconstruction depends on the sequence in which fossils are preserved, we may consider rigorous tests in the same general context. Shaw argued that the total stratigraphic range of a fossil species was the best measure of time for correlation purposes and that this could be determined accurately by careful collecting. In particular he demonstrated how to test the probability that a selected species had been overlooked in a sample and hence how to estimate the reliability of observed ranges. Shaw's method is also

appropriate in determining the size of an "adequate" sample and the probability that two fossil species are preserved in the wrong order (see p. 110). Since the method is so useful, it is worth reiterating it here. Shaw used an example in which several collections of fossils were made from a measured section. Over part of the section samples yielded one particular species, but above a certain level that species was apparently absent. Shaw questioned what the chances were that this species was absent in the first collection made above its known range simply because it had been overlooked. For phylogenetic reconstruction it is more important to determine the first occurrence of a fossil species, but the arguments are exactly the same. Now suppose this species was in fact present but very rare, say only 1% of the fossils preserved at that level. If we collect a single specimen there are only two possible outcomes, either it is a specimen of the desired species or it is something else. Thus the probabilities of either outcome are in the form:

$$p + q = 1 \qquad (1)$$

In this particular case $p = 0.01$ and $q = 0.99$. Now if we repeat the trial $n$ times (by collecting $n$ specimens), the total probability $(Q)$ of not finding a specimen of the desired species declines as $n$ increases according to the relationship:

$$Q = q^n = (1 - p)^n \qquad (2)$$

Even starting with very high initial values of $q$ as in this case, $q^n$ declines substantially as $n$ increases. For example, after 299 trials $Q$ is approximately 0.05, after 459 trials $Q$ is 0.01. That is to say, if we have a sample of 299 randomly collected fossils we can be 95% certain, and with 459 fossils 99% certain, that one will be of the desired species if it is present as 1% of the fossil population. In this context we know the size of our sample $(n)$ and can solve the relationship for $Q$ and $q$. Clearly in other contexts we could use appropriate proportions $(q)$ and probabilities $(Q)$ to determine sample size. (I shall return to this in the next section.) Shaw (1964, table 18–1, p. 109) produced a table of values for $Q$, $q$ and $n$. Independently I have produced a similar table concentrating on the higher values of $q$ (Table I). Newell (1959, fig. 2, p. 495) has also illustrated this point in another way.

Table I. Values of $n$

|   |       | 0.1 | 0.05 | $q$ 0.01 | 0.005 | 0.001 | 0.0005 |
|---|-------|-----|------|------|-------|-------|--------|
|   | 0.9   | 2   | 2    | 3    | 3     | 4     | 4      |
|   | 0.8   | 2   | 2    | 3    | 4     | 5     | 5      |
|   | 0.7   | 2   | 3    | 4    | 5     | 6     | 7      |
|   | 0.6   | 3   | 4    | 6    | 6     | 8     | 9      |
|   | 0.5   | 4   | 5    | 7    | 8     | 10    | 11     |
|   | 0.4   | 5   | 6    | 10   | 11    | 14    | 15     |
| Q | 0.3   | 7   | 9    | 13   | 15    | 20    | 22     |
|   | 0.2   | 11  | 14   | 21   | 24    | 31    | 35     |
|   | 0.1   | 22  | 29   | 44   | 51    | 66    | 73     |
|   | 0.05  | 45  | 59   | 90   | 104   | 135   | 149    |
|   | 0.01  | 230 | 299  | 459  | 528   | 688   | 757    |
|   | 0.005 | 460 | 598  | 919  | 1058  | 1379  | 1517   |
|   | 0.001 | 2302| 2995 | 4603 | 5296  | 6905  | 7598   |

In this table $q$ represents the proportion that any one species comprises of the total preserved fauna ($1.0 = 100\%$); Q represents the probability of *not* finding one specimen of that species in a sample of $n$ specimens. Thus only one in a hundred samples of 459 specimens will not contain a specimen of any species present as 1% of the fossil population.

(b) *Rigorous analysis of gaps.* Shaw's method was developed to determine total ranges of fossils as accurately as possible. I believe that analysis of gaps may be used in a similar exercise, but with smaller samples, and this time using the intervals between specimens of a single species in a measured section. First let us consider the theory and then its application.

Assume that the preservation of individuals of a species is a random process, then for each species there will be an average interval of time between the preservation of two successive specimens, which I propose to call the specific preservation time ($T$). $T$ may vary from minutes or even seconds in the case of an abundant skeletized marine protozoan living in an environment of active sedimentation, to hundreds of millions of years for rare, soft-bodied organisms living where erosion is active. Indeed it is theoretically possible for $T$ to exceed the "lifetime" of a species, in which case that species has a preservation potential of zero. Where sedimentation rates are uniform, $T$ will convert into a specific preservation interval ($I$) which can be determined in a measured section of uniform width. If we assume null hypotheses that preservation was indeed random and

and sedimentation rate uniform, then the frequency distribution of actual intervals between specimens of a single species in a measured section should follow an exponentially declining curve (Fig. 4) of the form:

$$y = ke^{-\lambda x} \tag{3}$$

where $y$ = frequency, $x$ = stratigraphic interval, $k$ and $\lambda$ are constants and $e$ is the base of natural logarithms. When $x = 0$, $e^{-\lambda x} = 1$, so that $k$ is the $y$ intercept of the curve. This curve is identical with the one used to describe radioactive decay and may be treated in a similar fashion. (However, it should be noted that radioactive decay is itself a different process since all the nuclides exist at the beginning of the decay, whereas any one fossil is unlikely to have been alive when the preceding one in the section became preserved.)

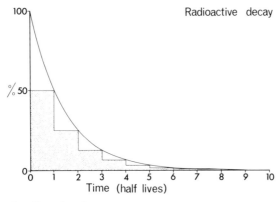

Fig. 4. Idealized radioactive decay curve.

Let $k$ = the total number of intervals ($N - 1$, where $N$ is the number of specimens). Then by ordering the intervals according to increasing size, the median interval ($i$) is an estimate of the half-life of radioactive decay (which I propose to call the half-interval, $h$), since half the intervals will be larger and half smaller than $i$. If the frequency distribution is really of the form in equation 3, then half the intervals will be less than $i$, a quarter will lie between $i$ and $2i$, an eighth between $2i$ and $3i$, and so on. We may test our original hypotheses by comparing the actual frequency distribution with the theoretical curve. If $k$ is normalized to 1 (or 100%) and stratigraphic

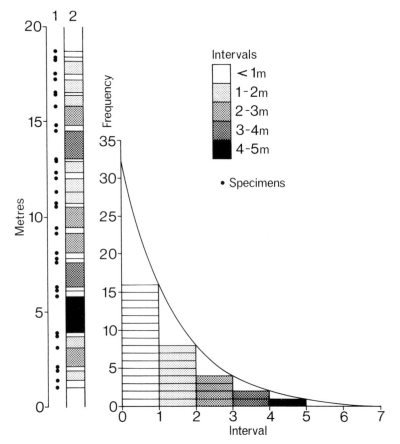

Fig. 5. Derivation of frequency curves from intervals in a measured section. The actual occurrences of specimens are plotted in column 1. The intervals between specimens are shaded according to size (column 2), in classes that are multiples of the median interval (i). The frequency histogram is derived from these intervals.

interval is expressed in terms of $i$, then all actual curves may be compared with the same theoretical curve. Details of this test are given below and in Fig. 5.

Two results may come from such a test. First, if the observed and theoretical curves are similar we have no reason to reject our original hypotheses. While it is theoretically possible for non-random preservation and non-uniform sedimentation rates to combine in such a way as to give the expected frequency curve, the chances against this

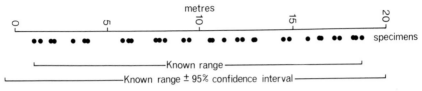

Fig. 6. Derivation of confidence intervals from the data in Fig. 5. $I_{95} \simeq 4\,\text{m}$ above and below the known range.

are very large indeed. If we accept the initial hypotheses, we can place confidence limits on the known stratigraphic ranges of fossils. A total of 93.75% of the area under the curve in Fig. 4 lies between $x = 0$ and $x = 4\,i$, while 99.22% lies between $x = 0$ and $x = 7\,i$. Thus 95% and 99% confidence limits ($I_{95}$ and $I_{99}$) lie at approximately $4\,i$ and $7\,i$ above and below the known range of any species (Fig. 6). It goes without saying that the more thoroughly a section is collected the narrower these confidence intervals will become.

The alternative result of the test is that the actual and theoretical distributions differ by so much that we are forced to reject our original hypotheses. In this case it will usually be possible to determine which one requires rejection. Sedimentological episodes, such as storm or lag deposits, should be detectable and should affect all species in the section. On the other hand, if one species shows the predicted frequency distribution and we accept that sedimentation was more or less uniform for that species, it cannot have been non-uniform for any other species in the same section. These arguments are amplified below in considering some actual examples.

Before we do, it is necessary to consider the best test for comparisons between actual and theoretical distributions. Consider the effects of three clearly non-uniform, but by no means unlikely, sedimentation rates. Suppose a storm deposits a large thickness of sediment so quickly that no fossils are buried within it. This horizon will show up as an anomalously large interval. We would expect frequency distributions for fossils passing through this interval to resemble those of *Haplosphaeronis multifida* (Fig. 7a) or *Grandagnostus falanensis* (Fig. 9b) where one isolated large interval occurs far beyond the expected end of the curve. Secondly, suppose winnowing removed sediment to form a lag deposit crowded with fossils. In this case many intervals would be very small (i.e. those between specimens within the lag deposit) and the median interval ($i$)

*C. R. C. Paul*

would be correspondingly lowered. The resulting distribution would have many values lying beyond the largest expected interval. Finally, consider regular alternations of, say, shales and sandstones where some fossils are confined to the shales. In this case the frequency distribution of intervals would be bimodal, with many small intervals between specimens within each shale unit and many larger intervals represented by the intervening sandstone beds. Again $i$ will tend to be small due to the lower mode, while the largest interval will seem anomalously large and merely reflect the thickest sandstone bed. In all these cases the largest recorded interval will seem too large, either because it is genuinely large or because $i$ is an underestimate of the half-interval, $h$. Furthermore, the same effects will occur if a species migrated out of the area of deposition for part of its range or was genuinely abundant some of the time but rarer for the rest of its range, or if species are preserved in discrete clusters. Thus the largest recorded interval may be used to test the actual distributions and we can predict exactly the minimum sample size required for a given maximum interval from the theoretical curve.

Consider again idealized radioactivity decay. In this case we want to know how many nuclides we must start with for one to survive beyond a given number of half-lives. With 2 ($2^1$) nuclides one will survive beyond one half-life; with 4 ($2^2$) one should survive beyond two half-lives; with 8 ($2^3$) one should survive beyond three half-lives and so on. Thus we must start with at least $2^n$ nuclides for one to survive beyond $n$ half-lives. Returning to fossil distributions, we can predict the minimum number of intervals theoretically necessary to include one as large as the largest observed interval ($I_{max}$). If we regard this number as a statistical population and the actual number of intervals ($k$) as a sample from that population, then the probability ($p$) that the largest observed interval ($I_{max}$) would be included in the sample is given by:

$$\frac{k}{2^n} \geqslant p \geqslant \frac{k}{(2^{n+1}) - 1} \tag{4}$$

where $ni < I_{max} < (n+1)i$. For example, in the analysis of distribution of *Haplosphaeronis multifida* (Table II, Fig. 7a), the largest observed interval is exactly $10i$. The predicted population which would include such an interval lies between $2^{10}$ (1024) and $2^{11} - 1$ (2047), while the actual sample size ($k$) is 22. So

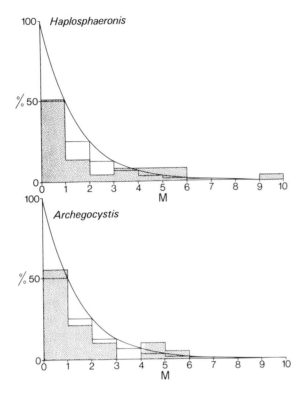

Fig. 7. Frequency distributions for *Haplosphaeronis* and *Archegocystis* in the Sholeshook Limestone (U. Ordovician), near Haverfordwest, S. Wales. Data from Paul 1973, p. 7, fig. 5.

$22/1024 \geqslant p \geqslant 22/2047$ or $0.0213 \geqslant p \geqslant 0.0107$. This is too small to accept our original hypotheses. However, *Archegocystis,* which occurs in the same section, has an $I_{max}$ between $5\,i$ and $6\,i$ (Table II, Fig. 7b) and therefore the predicted population lies between $2^5$ (32) and $2^6 - 1$ (63). The actual sample size is $k = 20$, so $20/32 \geqslant p \geqslant 20/63$, or $0.63 \geqslant p \geqslant 0.315$. If we accept our hypotheses for *Archegocystis,* sedimentation was more or less uniform in the section and the unacceptably large interval for *Haplosphaeronis* must have been due to non-random preservation. In other words, *Haplosphaeronis* probably did not exist in the area during this interval. In these two examples sample sizes are small, indeed near the lower limit for the method, and therefore the possibility of the large interval in the case of *Haplosphaeronis* being due to collection failure

Table II. Ranked intervals in metres for *Haplosphaeronis* and *Archegocystis* in the Sholeshook Limestone at Sholeshook, S. Wales

| Intervals (m) | | | | | | | | | | | |
| --- | --- | --- | --- | --- | --- | --- | --- | --- | --- | --- | --- |
| Rank no. 1 | 2 | 3 | 4 | 5 | 6 | 7 | 8 | 9 | 10 | 11 | 12 |
| *Haplosphaeronis* 0 | 0 | 0.13 | 0.13 | 0.13 | 0.17 | 0.25 | 0.25 | 0.33 | 0.33 | 0.38 | 0.42 |
| *Archegocystis* 0 | 0.19 | 0.19 | 0.33 | 0.38 | 0.38 | 0.56 | 0.56 | 0.65 | 0.69 | 0.69 | 0.82 |
| Rank no. 13 | 14 | 15 | 16 | 17 | 18 | 19 | 20 | 21 | 22 | 23 | 24 |
| *Haplosphaeronis* 0.63 | 0.75 | 1.00 | 1.50 | 1.58 | 1.67 | 1.92 | 2.08 | 2.17 | 4.00 | — | — |
| *Archegocystis* 0.82 | 1.04 | 1.13 | 1.53 | 2.06 | 3.07 | 3.38 | 3.94 | — | — | — | — |

*Haplosphaeronis*: $k = 22$, $i = 0.40$ m; *Archegocystis*: $k = 20$, $i = 0.69$ m. Raw data in Paul (1973, p. 7, fig. 5). See text for further explanation.

cannot be easily dismissed. However, further collection should confirm its reality or show that the distribution is in fact more nearly random.

Thus in summary, if a section is measured, the occurrences of individuals of any one species recorded, and the intervals between specimens ranked in order of size to determine the median interval ($i$) and assigned to size classes which are multiples of $i$, then the resulting frequency distribution curve should approximate to the radioactive decay curve. Real and theoretical curves may be compared by testing the probability that the largest observed interval would be included in the sample, using the relationship in equation 4. Where this probability is not unacceptably small, confidence limits may be added to the known ranges of fossils because $I_{95} \simeq 4i$ and $I_{99} \simeq 7i$.

Now let us consider some further actual examples. Again it has been difficult to obtain adequate data because in most cases total stratigraphic range is presented rather than the original data. I can present three reasonable examples from macrofossils and can add a fourth specifically measured for this purpose.

(i) Paul (1973, fig. 5, p. 7) illustrated the type section of the Sholeshook Limestone (Upper Ordovician) at Sholeshook, South Wales and gave actual occurrences of cystoids in the section. Two genera, *Haplosphaeronis* and *Archegocystis* occur in sufficient numbers to yield just enough data for this test. The intervals are ranked in Table II and their frequencies analysed in Table III and Fig 7a–b.

(ii) Stokes (1977, fig. 2, p. 818) gives actual occurrences of species of *Micraster* in the Upper Chalk of Kent. One species, *M. decipiens* (= *M. cortestudinarium* auctt.) is present in adequate numbers. Intervals are ranked in Table IV and analysed in Table V and Fig. 8.

(iii) Rushton (1978, fig. 1, pp. 246–7) gives data for the actual occurrences of fossils in the Merevale No. 3 borehole through the Middle-Upper Cambrian boundary. In this case three taxa are sufficiently common to repay analysis. They are *Linguella*, *Grandagnostus falanensis* and *Agnostus pisiformis*. Intervals are ranked in Table VI and analysed in Table VII and Fig. 9.

Table III. Frequency distribution of intervals for *Haplosphaeronis* and *Archego-cystis* in the Sholeshook Limestone, Sholeshook, S. Wales

| | Intervals ($i$) | | | | | | | | | |
|---|---|---|---|---|---|---|---|---|---|---|
| | $i$ | $2i$ | $3i$ | $4i$ | $5i$ | $6i$ | $7i$ | $8i$ | $9i$ | $10i$ |
| *Haplosphaeronis* | 11 | 3 | 1 | 2 | 2 | 2 | – | – | – | 1 |
| *Archegocystis* | 11 | 4 | 2 | 0 | 2 | 1 | – | – | – | – |

*Haplosphaeronis*: $0.0213 \geqslant p \geqslant 0.0107$; *Archegocystis*: $0.6333 \geqslant p \geqslant 0.3167$. $p$ is the probability that the observed largest interval would be included in a sample of $k$ individuals under the random preservation and uniform sedimentation rate hypotheses.

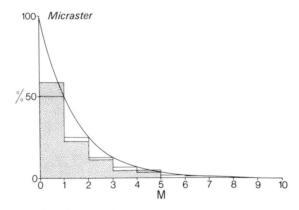

Fig. 8. Frequency distributions for *Micraster decipiens* (= *M. cortestudinarium* auctt.) in the Upper Chalk of south-east England. Data from Stokes 1977, p. 818, fig. 2.

(iv) Finally, I have been able to measure a section in the Elton beds (basal Upper Silurian) north of Much Wenlock, in which the brachiopod *Atrypa reticularis* occurs in sufficient numbers to yield adequate data. In this case the section was measured with this test specifically in mind, whereas the Sholeshook section (Paul 1973) was measured long before I had any such ideas. I made certain that collection was equally thorough throughout the section, which was 20 cm wide and continued through 2.4 m of the Elton Beds. Intervals are ranked in Table VIII and analysed in Table IX and Fig. 10.

Table IV. Ranked intervals for *Micraster decipiens* in the Upper Chalk at Dover, Kent

Intervals (m)

| Rank no. | 1 | 2 | 3 | 4 | 5 | 6 | 7 | 8 | 9 | 10 | 11 | 12 |
|---|---|---|---|---|---|---|---|---|---|---|---|---|
| *Micraster* | 0 | 0 | 0 | 0 | 0 | 0 | 0 | 0 | 0 | 0.30 | 0.30 | 0.30 |
| Rank no. | 13 | 14 | 15 | 16 | 17 | 18 | 19 | 20 | 21 | 22 | 23 | 24 |
| *Micraster* | 0.60 | 0.60 | 0.60 | 0.60 | 0.75 | 0.90 | 0.90 | 0.90 | 1.05 | 1.05 | 1.65 | 1.80 |
| Rank no. | 25 | 26 | 27 | | | | | | | | | |
| *Micraster* | 1.80 | 2.10 | 3.00 | | | | | | | | | |

*Micraster*: $k = 27$, $i = 0.60$ m. Raw data in Stokes (1977, p. 818, fig 2).

Table V. Frequency distribution of intervals for *Micraster decipiens* in the Upper Chalk, Dover, Kent

|              | Intervals $(i)$ | | | | |
| --- | --- | --- | --- | --- | --- |
|              | $i$ | $2i$ | $3i$ | $4i$ | $5i$ |
| *Micraster* | 16 | 6 | 3 | 1 | 1 |

*Micraster*: $0.8438 \geqslant p \geqslant 0.4219$.

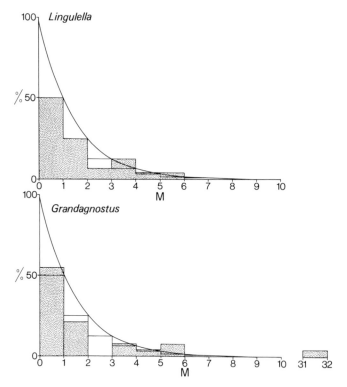

Fig. 9. Frequency distributions for *Lingulella* and *Grandagnostus falanensis* from the Middle/Upper Cambrian of the Merevale No. 3 borehole. Data from Rushton 1978, pp. 246–7, fig. 1.

Inspection shows that three frequency distributions, those for *Archegocystis* $(0.6333 \geqslant p \geqslant 0.3167)$, *Micraster decipiens* $(0.8438 \geqslant p \geqslant 0.4219)$ and *Lingulella* $(0.50 \geqslant p \geqslant 0.25)$, are reasonably close to the theoretical distributions, while others are not. The records for the Merevale borehole are particularly instructive.

Table VI. Ranked intervals for *Lingulella*, *Grandagnostus falanensis* and *Agnostus pisiformis* in the Merevale No. 3 borehole, Middle-Upper Cambrian

Intervals (m)

| Rank no. | 1 | 2 | 3 | 4 | 5 | 6 | 7 | 8 | 9 | 10 | 11 | 12 |
|---|---|---|---|---|---|---|---|---|---|---|---|---|
| *Lingulella* | 0.23 | 0.26 | 0.53 | 0.95 | 1.05 | 1.11 | 1.32 | 1.32 | 1.32 | 1.32 | 1.58 | 1.58 |
| *Grandagnostus* | 0.23 | 0.23 | 0.23 | 0.23 | 0.23 | 0.23 | 0.23 | 0.23 | 0.23 | 0.45 | 0.45 | 0.56 |
| *Agnostus* | 0.30 | 0.30 | 0.30 | 0.30 | 0.30 | 0.30 | 0.30 | 0.30 | 0.30 | 0.30 | 0.30 | 0.30 |

| Rank no. | 13 | 14 | 15 | 16 | 17 | 18 | 19 | 20 | 21 | 22 | 23 | 24 |
|---|---|---|---|---|---|---|---|---|---|---|---|---|
| *Lingulella* | 1.74 | 1.74 | 2.24 | 2.37 | 2.89 | 2.89 | 3.11 | 3.42 | 3.53 | 3.68 | 4.11 | 4.47 |
| *Grandagnostus* | 0.67 | 0.67 | 0.78 | 0.78 | 0.83 | 0.90 | 0.90 | 0.90 | 1.23 | 1.34 | 2.42 | 2.95 |
| *Agnostus* | 0.30 | 0.30 | 0.30 | 0.30 | 0.30 | 0.30 | 0.30 | 0.30 | 0.30 | 0.30 | 0.30 | 0.30 |

| Rank no. | 25 | 26 | 27 | 28 | 29 | 30 | 31 | 32 | 33 | 34 | 35 | 36 |
|---|---|---|---|---|---|---|---|---|---|---|---|---|
| *Lingulella* | 6.84 | 7.63 | 8.16 | 8.68 | 9.47 | 9.47 | 12.5 | 14.5 | – | – | – | – |
| *Grandagnostus* | 3.88 | 4.15 | 4.15 | 24.2 | – | – | – | – | – | – | – | – |
| *Agnostus* | 0.30 | 0.30 | 0.30 | 0.30 | 0.30 | 0.30 | 0.30 | 0.30 | 0.30 | 0.30 | 0.30 | 0.30 |

| Rank no. | 37 | 38 | 39 | 40 | 41 | 42 | 43 | 44 | 45 | 46 | 47 | 48 |
|---|---|---|---|---|---|---|---|---|---|---|---|---|
| *Agnostus* | 0.30 | 0.30 | 0.30 | 0.30 | 0.30 | 0.30 | 0.30 | 0.30 | 0.30 | 0.50 | 0.70 | 0.70 |

| Rank no. | 49 | 50 | 51 | 52 | 53 | 54 | 55 | 56 | 57 | 58 | 59 | 60 |
|---|---|---|---|---|---|---|---|---|---|---|---|---|
| *Agnostus* | 0.70 | 1.00 | 1.00 | 1.00 | 1.00 | 1.20 | 1.30 | 1.30 | 1.30 | 1.50 | 1.50 | 2.00 |

| Rank no. | 61 | 62 | 63 | 64 | 65 | 66 | 67 | 68 | 69 | 70 | 71 | 72 |
|---|---|---|---|---|---|---|---|---|---|---|---|---|
| *Agnostus* | 2.00 | 2.00 | 2.40 | 2.50 | 2.50 | 2.50 | 2.50 | 3.00 | 3.00 | 3.00 | 4.00 | 4.00 |

| Rank no. | 73 | 74 | 75 | 76 | 77 | 78 | 79 | 80 | 81 | 82 | 83 | 84 |
|---|---|---|---|---|---|---|---|---|---|---|---|---|
| *Agnostus* | 4.00 | 4.50 | 4.50 | 5.50 | 5.50 | 6.00 | 7.50 | 7.50 | 8.50 | 9.50 | 11.5 | 18.5 |

*Lingulella*: k = 32, i = 2.63; *Grandagnostus*: k = 29, i = 0.78 m; *Agnostus*: k = 84, i = 0.30 m[*] (*minimum measureable value). Raw data in Rushton (1978, figs 1–2, pp. 246–7).

Table VII. Frequency distribution of intervals for *Lingulella*, *Grandagnostus* and *Agnostus* in the Merevale No. 3 Borehole

| | | | | | Intervals ($i$) | | | | | | |
|---|---|---|---|---|---|---|---|---|---|---|---|
| | $i$ | $2i$ | $3i$ | $4i$ | $5i$ | $6i$ | $7i$ | $8i$ | $9i$ | $10i$ | $11i$ |
| *Lingulella* | 16 | 8 | 2 | 4 | 1 | 1 | – | – | – | – | – |
| *Grandagnostus* | 16 | 6 | 0 | 2 | 1 | 2 | 0 | 0 | 0 | 0 | 0 |
| *Agnostus* | 45 | 1 | 3 | 5 | 5 | 0 | 1 | 4 | 3 | 0 | 3 |
| Intervals | $12i$ | $13i$ | $14i$ | $15i$ | $16i$ | $17i$ | $18i$ | $19i$ | $20i$ | $21i$ | $22i$ |
| *Grandagnostus* | 0 | 0 | 0 | 0 | 0 | 0 | 0 | 0 | 0 | 0 | 0 |
| *Agnostus* | 3 | 0 | 2 | 0 | 0 | 0 | 2 | 0 | 1 | 0 | 0 |
| Intervals | $23i$ | $24i$ | $25i$ | $26i$ | $27i$ | $28i$ | $29i$ | $30i$ | $31i$ | $32i$ | $33i$ |
| *Grandagnostus* | 0 | 0 | 0 | 0 | 0 | 0 | 0 | 0 | 0 | 1 | – |
| *Agnostus* | 0 | 2 | 0 | 0 | 0 | 1 | 0 | 0 | 1 | 0 | 0 |
| Intervals......$38i$. | .$61i$ | | | | | | | | | | |
| *Agnostus* | 1 | 1 | | | | | | | | | |

*Lingulella*: $0.50 \geqslant p \geqslant 0.25$; *Grandagnostus*: $29/2^{31} \geqslant p$; *Agnostus*: $84/2^{61} \geqslant p$.

Table VIII. Ranked intervals for *Atrypa reticularis* in the Elton Bed near Much Wenlock, Salop

| | | | | | Intervals (cm) | | | | | | | |
|---|---|---|---|---|---|---|---|---|---|---|---|---|
| Rank no. | 1 | 2 | 3 | 4 | 5 | 6 | 7 | 8 | 9 | 10 | 11 | 12 |
| *Atrypa* | 0 | 0 | 0 | 0 | 0 | 0 | 0 | 0 | 0 | 0 | 0 | 0 |
| Rank no. | 13 | 14 | 15 | 16 | 17 | 18 | 19 | 20 | 21 | 22 | 23 | 24 |
| *Atrypa* | 0 | 1 | 2 | 3 | 6 | 8 | 10 | 11 | 12 | 14 | 15 | 17 |
| Rank no. | 25 | 26 | 27 | 28 | 29 | 30 | 31 | 32 | 33 | 34 | 35 | 36 |
| *Atrypa* | 22 | 28 | 31 | 36 | – | – | – | – | – | – | – | – |

*Atrypa*: $k = 28$, $i = 0.25$ cm.

Table IX. Frequency distribution of intervals for *Atrypa reticularis* in the Elton Beds, Upper Silurian, Much Wenlock, Salop

| | | | | | Intervals ($i$) | | | | | | | |
|---|---|---|---|---|---|---|---|---|---|---|---|---|
| | $i$ | $2i$ | $3i$ | $4i$ | $5i$ | $6i$ | $7i$ | $8i$ | $9i$ | $10i$ | $11i$ | $12i$ |
| *Atrypa* | 14 | 1 | 1 | 1 | 3 | 1 | 2 | 0 | 2 | 0 | 0 | 1 |
| | $13i$ | $14i$ | $15i$ | $16i$ | | | | | | | | |
| *Atrypa* | 1 | 0 | 1 | – | | | | | | | | |

*Atrypa*: $18/2^{15} \geqslant p$.

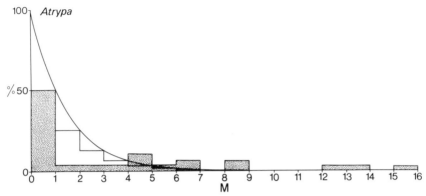

Fig. 10. Frequency distribution for *Atrypa reticularis* from the Elton Beds (U. Silurian) near Much Wenlock, Salop.

All three species are small and came from the core of the borehole which presumably had the same cross-sectional area throughout. Hence it is a reasonable assumption that collection was equally thorough through the core. The frequency distribution of intervals for *Lingulella* is similar to the expected pattern and there is no reason to suppose that sedimentation rates varied by so much that the null hypothesis must be rejected. The other two species, both agnostid trilobites, range through the same interval as *Lingulella* and so their frequency distributions, which differ in the extreme from the expected pattern, cannot be entirely due to sedimentological features. *G. falanensis* has a single enormous gap within which the other trilobite occurs quite commonly. It seems extremely unlikely, therefore, that diagenesis has destroyed one trilobite but not the other, or that Rushton overlooked one species but found many examples of the other. The most reasonable explanation is that *G. falanensis* is absent from the section because it was not present in the area during the interval of time represented by this one large gap. Indeed, if we ignore that single interval, the relevant data change as follows: $k = 28$, $i = 0.73$ m, $I_{max}$ lies between $5i$ and $6i$, and $p$ between 0.875 and 0.437. Thus the frequency distribution of intervals for *G. falanensis* is very close to the expected pattern except for that one anomalously large interval. In the case of *A. pisiformis*, however, quite a different picture emerges. More than half the intervals are at the lowest resolvable level (approximately 0.3 m) and inevitably the median interval is likely to be an underestimate of the half-interval, *h*. As a result there are many intervals, not just one, far beyond the largest theoretically expected. The most likely explanation is that *A. pisiformis* was genuinely very common at certain horizons and hence that preservation was non-random. This is known to be the case with the *Atrypa* specimens from the Elton Beds, which occur in local dense clusters. Each time the measured section passed through such a cluster on a bedding surface, several values of zero were recorded. Once again we find that analysis of gaps directs attention to anomalous horizons and sometimes allows us to interpret the anomalies.

Incidentally, where it is known or suspected that *i* underestimates *h* and some estimate of confidence intervals is still desired, this can be calculated from the theoretical relationship between *h* and the mean interval (which is the specific preservation interval, *I*, mentioned at the beginning of the section.) This relationship depends on sample size, as shown in Fig. 11. As $k \rightarrow \infty$, $I/h \rightarrow 1.5$.

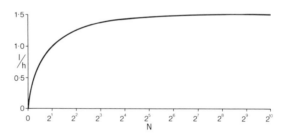

Fig. 11. Graph to show the relationship between the ratio of mean interval (*I*) and half-interval (*h*) as sample size (*N*) increases.

## TESTS OF THE INCOMPLETENESS OF OUR KNOWLEDGE OF THE FOSSIL RECORD

### 1. Adequacy of Collecting

At least part of the incompleteness of our knowledge of the fossil record depends on how good our collecting has been to date. I and others (e.g. Durham 1967) are inclined to say that it has been pretty poor in the past, but again tests can be devised. The most obvious test is replication, a point on which Durham (1967, p. 561) has argued strongly, although in slightly different terms. In most sciences a single experiment is often inconclusive by itself, so we repeat the experiment several times. If all the replicates yield the same result, we become progressively more confident that the result is correct. If we collect twice from the same site and obtain exactly the same suite of fossils both times, it is reasonable to assume that our first collection was adequate. In the present context, if a phylogenetically important species occurs but is very rare, we should take very large samples. However, if it is common, much smaller samples will be adequate. We usually have no way of knowing whether a phylogenetically important species will occur at all, let alone if it will be common or rare. Nevertheless, using the data in Table I, we can determine sample size on the probability of detecting a species of a given rarity. Hence I suggest that if collecting is undertaken with no predetermined pattern in mind, appropriate probabilities and proportions should be decided beforehand and used to determine sample sizes. For example, if we wish to be 95% certain of finding a species present as 1% or more of the collection, 299 specimens is the minimum sample size required.

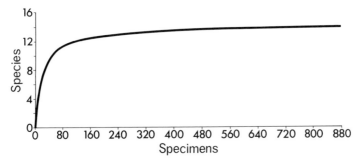

Fig. 12. Species/specimens curve for a sample of Late-glacial molluscs.

On similar lines another test of our sampling can be used. As the number of specimens collected increases, the number of species represented in the collection also increases but progressively more slowly (Fig. 12). If we stop collecting while still on the steeply rising part of the curve our sampling will be inadequate. On the other hand once we reach the flat part of the curve, greatly increased collecting will add little or nothing to our knowledge of the fauna present. Both abundance (no. of specimens) and diversity (no. of species) are important characteristics of samples. Irrespective of true diversity, large samples tend to contain more species than smaller ones. Taking a standard sample size (which in palaeontology is frequently impossible) obscures important information about abundance, which in palaeoecological analysis, for example, is as vital as diversity. I believe that both abundance and diversity can be recorded in comparable samples if we collect until we reach an equivalent point on the curve – for example, if we collect until doubling the number of specimens does not add a single new species. The method simply involves recording the number of specimens collected each time a new species is added to the faunal list for a site. Each time a new species is discovered, a new target sample size is set. Suppose the 30th specimen found was the first example of the 10th species on the faunal list, then the new target sample size would be 60. If we collect 60 specimens without finding an 11th species we stop collecting. Alternatively, if the 50th specimen represented an 11th species, we have a new target sample size of 100. Applying this method literally, if the first two specimens are of the same species we stop collecting. To overcome this difficulty, I suggest an absolute minimum of 20 or 30 specimens always be taken. Recording the time taken to reach the required level of sampling gives an independent measure of abundance provided a similar method of collecting has been undertaken in all samples. This idea is not new and was argued independently by Cain (1938) with respect to species/area curves in plant ecology. However, as far as I can determine, Cain's method can only be used *after* the species/ area curve has been plotted. This requires collecting from a large initial sample area to determine the shape of the particular curve involved. My own suggestion is independent of the precise form of the curve.

## 2. Growth of Our Knowledge

A species/specimens curve, such as that shown in Fig. 12, is also in effect a species/time curve, because the longer one collects the more specimens one finds. The idea may be extended to assess the growth of our knowledge of the fossil record. I have already applied it to

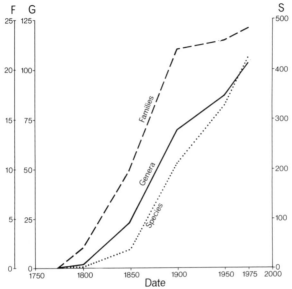

Fig. 13. Curves to show the growth in knowledge of cystoid families, genera and species. Note that whereas since 1900 the number of families has scarcely increased, the number of genera has gone up by 50% and the number of species has doubled.

the cystoid fossil record (Paul 1980, p. 206 *et seq.*). To avoid problems of taxonomic treatment, all known species of cystoids were assigned to their current classification and then analysed to see what proportions of families, genera and species were known at various times in the past (Fig. 13). Assignment of species to their current classification avoids all the problems of differing taxonomic treatment in the past. Merely counting the number of genera in Bather (1900) for example, would not be sufficient since a few may since have been shown to be synonyms, while others will have been subdivided several times. Figure 13 shows that whereas the numbers of genera and species have increased significantly in recent years, the number of families has hardly increased at all since 1900. Twenty-two of the currently accepted 24 families were known in 1900. The two that have been described since are both based on a single species and the material for one of them, the Thomacystidae (Paul 1969) was actually collected last century, but lay undescribed in the British Museum (Natural History) for about 75 years. This curve implies, but cannot prove, that there are really very few more families of cystoids to be discovered, but the numbers of genera and especially species can confidently be expected to increase in the foreseeable future. Furthermore, the curve implies that in 1900 Bather already had a good basis on which to outline the phylogeny of the main cystoid families.

THE ADEQUACY OF THE KNOWN FOSSIL RECORD

*1. Sample Sizes and Sampling Intervals*

The adequacy of the fossil record is, to my mind, the most crucial point as far as its use for comparative purposes is concerned. The fossil record is only a sample of all past life and the known record merely a further subsample. Notwithstanding the obvious resulting incompleteness, I believe that the known fossil record is adequate for all uses and particularly for phylogenetic reconstruction. To illustrate the point let me reiterate an artificial example I have used elsewhere (Paul 1980, pp. 204–5). Suppose a factory produced coloured balls and dated samples were taken periodically for quality control. Let us also assume that the colours were produced continuously and that once production ceased no colour was repeated again. Using the dated samples we are trying to reconstruct which colours were produced and when. What is the smallest sample that would enable us to build a complete picture of production? In other words what constitutes "an adequate sample?" The answer depends on how many colours were being produced and in what proportions. *The total number of balls produced is irrelevant.* If only one colour was produced, one ball would be adequate. If 10 colours were produced in equal proportions, then less than 10 balls could not possibly detect all colours, 20 almost certainly would and 100 would probably give a reasonable indication of the proportions of each colour. If one ball of a unique colour were produced, a sample as large as half the total production would still only give a 50% chance of detecting it. The size of an adequate sample is determined by the proportion of the rarest colour and the probability of including one ball of that colour in a sample of a given size. Again Table I may be used to determine the size of an adequate sample.

So we can decide what constitutes an adequate sample (of balls or fossils) by selecting predetermined probabilities and proportions. However, this only settles how large any one sample should be. We also set ourselves the task of finding out *when* each colour was produced. This can obviously be done using the dated samples and, even if many samples are missing, we can still reconstruct periods of production because we assume that production of colours was continuous. Providing the duration of production of colours was long compared with the sampling interval, the former can be determined with reasonable accuracy. Once again, the sampling interval is not determined by the volume of production of any colour, but by the *period* of production. Returning to the fossil record, providing we take adequate samples of fossils at stratigraphic intervals which are small compared to the total range of species, we should be able to determine the stratigraphic range of species reasonably accurately. Clearly the closer the sampling interval, the more accurately ranges can be determined.

## 2. The Sequence of Events

In the particular context of this symposium, total stratigraphic range is less important than first occurrences. Usually it is argued that since the fossil record is incomplete, *total* stratigraphic range cannot be known and hence there is a possibility that first occurrences are preserved in the wrong order. From this it is also usually argued that we need not pay too much attention to the fossil record in constructing phylogenies. Nothing could be further from the truth! As I shall show below, while the *possibility* that first occurrences may be in the wrong order cannot always be ruled out, even in the extreme case where we have only one specimen of each species discovered at random, the *probability* that they are preserved in the wrong order with respect to their real first appearance can only exceed 50% under very special circumstances. In other words if we accept the fossil record as gospel we will be right more often than we are wrong, whereas if we reject its evidence we will be wrong more often than we are right. The incompleteness of the fossil record is almost irrelevant to the order in which the first specimens of species are preserved. Indeed if the fossil record consisted of only two specimens, the probabilities are overwhelming that they would be in the correct order. Let us now examine the arguments in more detail.

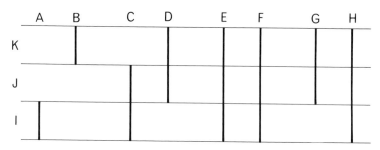

Fig. 14. Hypothetical original ranges of four pairs of species (A–H) which range through three stratigraphical intervals (I–K).

Consider Fig. 14 in which A–H represent the real original ranges of four pairs of species, *not their known stratigraphic ranges*. Clearly in the case of species A and B there is no possibility whatsoever that specimens could be preserved in the wrong order. This is because the older species became extinct before the younger one evolved. (For the present discussion we may ignore derived fossils since they are usually easily recognized as such.) Fossil species can only be preserved in the wrong order with respect to their first appearances if their original ranges overlapped as those of species C–H do in Fig. 14. For this reason, if for no other, most species must be preserved in the correct order since the time range of any one species can only overlap with the ranges of a small proportion of all fossil species.

If we were to discover just one example each of species C and D, what is the probability that they would occur in the correct order, i.e. that the example of species C came from below that of species D? To investigate this, let us assume that the chances of discovering a specimen are equal throughout the ranges of both species. We can return to this assumption later. Now, let the total range of species C be $N_c$ stratigraphic intervals, that of species D $N_d$ intervals and let the zone of overlap be $N_0$ intervals. All intervals are equal. The probability of finding one specimen of species C within the zone of overlap is $N_0/N_c$, and for species D it is $N_0/N_d$. Thus combining, the probability of finding both examples within the zone of overlap is $(N_0)^2/N_cN_d$. Now consider what happens within the overlap zone. We may construct a probability matrix of $N_0$ by $N_0$ intervals (Fig. 15). The black areas represent occurrences in which both specimens came from the same interval and can neither be in the correct

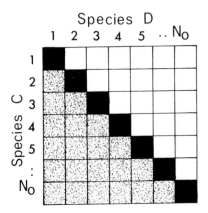

Fig. 15. Probability matrix of $N_0$ by $N_0$ intervals to show all possible outcomes of finding one specimen each of two species with a common range of $N_0$ equal stratigraphical intervals. Solid black squares represent outcomes where both specimens come from the same interval. Shaded squares represent outcomes with the specimens in the wrong stratigraphical order. White squares, outcomes with specimens in the correct order.

nor incorrect sequence. The probability of this happening is $N_0/(N_0)^2$. All positions above the black diagonal represent possible outcomes in the correct order, the stippled areas below the diagonal represent occurrences in the wrong order. These two possibilities are equal and the probability of either is $((N_0)^2 - N_0)/2(N_0)^2$. Now combining the chances of finding both specimens within the zone of overlap with the chances of getting the wrong order within it, we have a total probability of getting the wrong order of:

$$p = \frac{(N_0)^2}{N_c N_d} \times \frac{(N_0)^2 - N_0}{2(N_0)^2} \quad \text{and cancelling}$$

$$p = \frac{(N_0)^2 - N_0}{2 N_c N_d} \tag{5}$$

Figure 16 shows values of $p$ for different values of $N_0$ and $N$. For simplicity it was assumed that $N_c = N_d = N$.

Now it is clear that $p$ will be largest when the overlap is at its maximum. Consider the limiting case of total overlap, e.g. species E and F in Fig. 14. (We must assume that although their ranges are identical, somehow one is older than the other.) Now the value of $p$ reduces to that within the overlap zone, viz:

*C. R. C. Paul*

$$\frac{(N_0)^2 - N_0}{2(N_0)^2} \qquad (6)$$

One specimen of each species chosen randomly

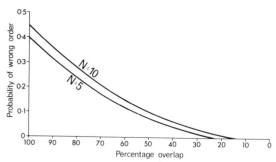

Fig. 16. Graph to show how the probability of finding one specimen each of two species in the wrong stratigraphical order declines with the proportion of overlap of their ranges. Note that the probability of getting the wrong order can never exceed 50% with random discovery.

As $N_0 \to \infty$, $p \to 0.5$. Thus even finding just two examples at random, there is never more than a 50% chance of getting the wrong order. Furthermore, the chances of getting the wrong order only equal the chances of getting the right order when total overlap occurs. In all other cases the chances of getting the correct sequence exceed those of getting the wrong order and, in the vast majority of cases where there was no original overlap, there is no possibility of getting the wrong order.

All the above arguments assume that we only find one example of each species. If we find more than one specimen, the chances of getting the wrong order decline, but it is not easy to describe precisely how. Nevertheless, in the cases of species C and D or G and H in Fig. 14, as soon as we discover a single example of the older species from below the zone of overlap it is no longer possible to get the first appearances in the wrong order. Similarly, within the zone of overlap as soon as we find an example of the older species from the basal interval it is no longer possible to get the wrong order. So we may ask what are the chances of finding a single example of species C or G from the basal zone of overlap or below that level.

This is given by:

$$p = \frac{(N_c - N_0) + 1}{N_c} \tag{7}$$

Now the chances of not finding a specimen of these two species (C or G) from the desired interval are given by $q = (1 - p)$ and in $n$ repeated trials,

$$Q = (1 - p)^n = q^n \tag{8}$$

Since $q$ is always less than 1, $q^n$ decreases as $n$ increases (as we saw in discussing Table I). Thus even with relatively small samples the chances of not getting the right order decline quickly. Bear in mind too that this is a conservative estimate since, even if we do not find an example of the older species below the base of the zone of overlap, in some cases we will still find that the first specimens of the two species are in the correct order.

We may now return to the initial assumption that the chances of finding a specimen are equal throughout the original range. Clearly this is not so. If one accepts the allopatric speciation model, then all new species begin in geographically limited populations and the chances of their being preserved and discovered are initially low. Nevertheless, this should affect all species in a similar manner. Secondly, we know that both the abundance and geographical ranges of species can vary dramatically with time. If species C in Fig. 14 were initially rare and species D initially abundant, this would significantly increase the chances of their first appearances being in the wrong order. However, such a pattern of abundance would correspondingly increase the chances of the first specimens of species C and its ancestor being in the correct order. The same would be true of any comparison between species D and a descendant species derived from it. The maximum distortion to the theoretical probabilities would occur if half of all fossil species had a stratigraphic distribution like species X in Fig. 17 and the other half a distribution like species Y. Even then, only 25% of comparisons between species with overlapping ranges would have their chances of occurring in the wrong order increased. Four types of comparison are possible: comparing type X with type X or type Y with type Y would not alter the probabilities, while comparing type Y with type X or *vice versa* would, but one would increase the chances of getting the right

*C. R. C. Paul*

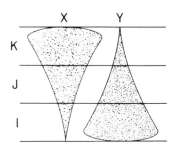

Fig. 17. Hypothetical changes in the abundance of two species (X and Y) which range through three stratigraphical intervals (I—K). Width of shaded areas proportional to abundance.

order. Thus only one of the four possible comparisons would significantly increase the chances of getting the wrong order and this only in comparisons between species with overlapping ranges. Thus no matter how uneven the original abundances of species may have been, there is no way that more than a quarter of comparisons will result in an increase in the chances of getting the wrong order. The same must be true when species migrate into, or out of, an area.

Ultimately even this possibility can be greatly reduced. The real question is what are the chances of getting one specimen of the older species from the early part of its range. It is only necessary to select the value of this probability, and hence determine the sample size necessary, to be, say 99% certain, of finding one specimen using the relationship in equation 2. At the risk of being repetitive, if we have 459 examples of any species we can be 99% certain we have one of the first 1% that have been preserved provided we collect randomly. If a species has been preserved over part of its range, it is akin to special pleading to argue that it has not been preserved from the early part of its range *unless independent evidence suggests that its preservation potential has changed.* Even if this is true, for every comparison where the chances of getting the wrong order are increased there is likely to be another where the chances of getting the right order are increased and two where the chances will not be affected at all.

To summarize, the probability of finding two randomly discovered individuals in the wrong stratigraphic order can only exceed 0.5 in very special circumstances. The probability of getting the wrong order can only equal the probability of getting the right order when total overlap occurs. Normally it is much smaller and in comparisons between species with non-overlapping original ranges it is impossible to get the wrong order. Since most fossil species must have had non-overlapping ranges originally, it follows inevitably that the chance of any two fossil specimens being preserved in the wrong stratigraphic order is very small indeed. Finally, repeated discoveries rapidly decrease the chances of getting the wrong order no matter how small the probability of getting the right order may have originally been with a single trial. So it is exceedingly unlikely that comparisons between the first occurrences of fossil species known from a few hundred specimens could give the wrong stratigraphic order. Once again the incompleteness of the fossil record is largely irrelevant to its adequacy *as a sample* of past life. It does not matter how many species are preserved or known, those that are preserved are likely to be in the correct order. As far as I am aware this idea is new. Since we largely reconstruct phylogenies for known species, recent or fossil, the fossil record is fundamental to phylogenetic reconstruction because it gives us the sequence of events. This brings me to my last point and the relationship between stratigraphic sequence and phylogenetic sequence.

## 3. Sequence in Phylogenetic Reconstruction

We have seen that the fossil record preserves species in the correct sequence; what does this mean for phylogeny? There are two cases to consider. First, if on morphological grounds the species are considered to be related directly, as for example in sequential species of the same genus, then the stratigraphic order is the only possible ancestor–descendant relationship. Consider, for example, Fig. 18 which gives the stratigraphic ranges of species of *Infulaster* and *Hagenowia* in the British Upper Chalk. The data are taken from Gale and Smith (1982) who on morphological grounds deduced an evolutionary sequence involving just two lineages which correspond to their definitions of the two genera. Nowhere are the ranges of species in either of these two genera known to overlap and the stratigraphic sequence coincides exactly with the phylogenetic sequence deduced on morphology.

However, direct relationship cannot always be assumed and a different phylogenetic pattern must obtain where species coexist. In the

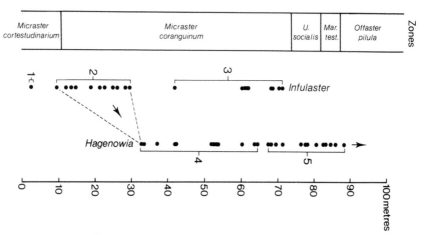

Fig. 18. Ranges of species of *Infulaster* and *Hagenowia* in the Upper Chalk of
south-east England. 1, *I. excentricus*; 2, *I. turberculatus*; 3, *I. infula-
steroides*; 4, *H. rostrata*; 5, *H. anterior*. The *Hagenowia* lineage continues to higher levels in the continental chalk. Data from Gale and
Smith (1982, fig. 1). Symbols indicate occurrence of species not individual specimens

latter case some branching (i.e. speciation) must have occurred. In
Stitt's (1977) work on Late Cambrian to basal Ordovician trilobites in
Oklahoma, for example, there are 16 genera represented by more
than one species which in turn are known from more than isolated
occurrences. In seven genera all or some of the species have overlapping ranges. In seven more genera the species are sequential with a
small interval separating the last occurrence of one species from the
first occurrence of the next. In the two remaining genera the species
are sequential but with enormous gaps between them implying that
something is missing and hence that we cannot argue for direct
relationship. From this limited example it would seem that in considering detailed stratigraphy at specific level the two alternatives
(evolution within lineages or branching evolution) are about equally
common.

Consider Fig. 19A in which there are three species with overlapping
ranges. If we assume they are not directly related, we may arrive at
two incontrovertible phylogenetic conclusions from their known
stratigraphic ranges. First, the latest common ancestor of species 1
and 2 (or 1 and 3, or all three species) must be older than (i.e.
evolved before) the first occurrence of species 1. Secondly, the latest

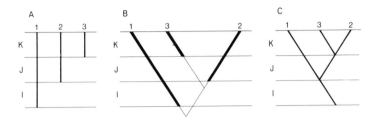

Fig. 19. (A) real original ranges of three species of fossils. (B) cladogram derived
from minimum ages of latest common ancestors of these species. Note
large intervals unrepresented by fossils. (C) the most parsimonious
cladogram. Note that this assumes direct relationship between all three
species.

common ancestor of species 2 and 3 must be older than species 2.
This suggests the pattern in Fig. 19B which is based solely on the
known minimum ages of latest common ancestors. Note that this
pattern cannot be the most parsimonious (Fig. 19C) because that
involves direct relationship. The statements about latest common
ancestors are open-ended. The latest common ancestor of species 2
and 3 *must* be older than species 2, but it could be older than
species 1 or even older than the latest common ancestor of 1 and 2.
Here we are back in a very similar situation to the one we considered
when calculating the probabilities of known specific ranges being in
the wrong order. All the arguments applied to the first occurrences
of species may be applied to the first occurrences of latest common
ancestors. The question is simply what are the chances of not finding
an example of species 2, or the latest common ancestor of species 2
and 3 from the interval between their original evolution and their
first appearances in the fossil record, if they occur at all?

Figure 20 shows three possible patterns of branching between
three species. Clearly with the pattern in Fig. 20A the chances of
finding the first specimens of the three species in the wrong order are
much higher than in either of the other two patterns. For this reason
the experiments conducted by Fortey and Jefferies and described in
this volume cannot yield definitive answers since the probabilities of
getting the wrong order depend to a very large extent on the precise
phylogenetic pattern defined at the outset. If all the important
branching in the phylogeny occurs at the same level (e.g. Fig. 20A),
the chances of the first occurrences of species occurring in the wrong

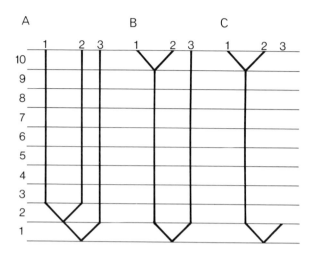

Fig. 20. Three hypothetical phylogenetic relationships between three species. Note the probability of the stratigraphic occurrence of species reflecting their evolutionary relationships is low in A, high in B, while in C there is no possibility of deriving the wrong relationship.

order are relatively high. If some or all the main branches are widely spaced (e.g. Figs 20B, C), the chances of finding species in the wrong order decline significantly. Even so, this simple stratigraphic argument produces the correct branching pattern for the three species chosen by Fortey and Jefferies in their Fig. 3.

So far I have confined my arguments to the stratigraphic ranges and relationships of *species*. It is often argued that with higher taxa the fossil record becomes less reliable. However, just as with fossil species it is the first known individuals which are important, with higher taxa the first known species are critical. By and large the same probabilistic arguments apply. Why should one major preservable taxon evolve several early species which are not preserved when another has its first preserved immediately? Unless it can be argued, as Kier (1977) has done about regular and irregular echinoids, that the preservation potential changed dramatically at some point within its known range, then the fossil record is likely to represent the true order of higher taxa and simple stratigraphic arguments about latest common ancestors apply.

CONCLUSIONS

The fossil record preserves species (and probably higher taxa too) in their correct historical sequence. We may therefore use it in two ways, as an independent test of phylogenies defined on morphology, or as a contributing source of information in phylogenetic reconstruction. If it is to be an independent test, ideally we must define our phylogenies in the total absence of stratigraphic knowledge, because only then can the test be truly independent. I hope I have shown that used this way the fossil record is a powerful test. On the other hand, if we use it as a contributing source of information there is a danger of circular arguments. (This species is primitive because it occurs first. This species is ancestral because it is primitive. This species occurs first because it is ancestral.) In this case we must fall back on the most parsimonious solution that satisfies both morphological and stratigraphic evidence. Again I would stress that the stratigraphic evidence is at least equal to the morphological evidence. No one can know *a priori*, the morphological relationships between organisms and, in practice, the recognition of synapomorphies is beset with problems. If it were not, this symposium would be unnecessary.

The fossil record can only be useless in phylogeny when no fossils are known. It will not be very much help when all fossils appear at exactly the same horizon, although I believe this may well reflect what actually happened. In all other circumstances we ignore the fossil record at our peril. From the consideration of branching patterns above, we may generalize that radiations will pose more problems in phylogenetic reconstruction than widely spaced sequential events. It is partly for this reason that no-one seriously doubts the fossil records of higher plants or vertebrates. The appearance of new classes is widely spaced out sequentially. The origin of the angiosperms may still be something of a problem, but nobody doubts that they were the latest plant group to evolve. The other seductive aspect of the fossil records of these two groups is that they show what we *expect* to see. We must learn to accept the fossil record at face value and construct our theories around it, not the other way round. Too often we have endeavoured to force it into a particular mould or to ignore awkward facts contained in it. For example, the idea of a radiation was known to Darwin. It was this as much as anything else that made him refer to the origin of the angiosperms as an abominable mystery, but it was not until 1944 that this pattern was accepted as a reflection of what actually happened and dignified with a name by Simpson. We still have a long way to go before we look at the fossil record for what it is and not for what we would like it to be. Historically, from Lyell and Darwin onwards, people have looked at the fossil record with a particular pattern in mind. They have failed to find the pattern they sought and have appealed to the incompleteness of the fossil record to explain away this anomaly. We are still doing this, and until we stop it is unlikely that palaeontology will make a significant contribution to phylogenetic reconstruction, or to evolutionary theory in general. This is very regrettable because I believe it has much to offer.

ACKNOWLEDGEMENTS

This chapter has been greatly improved by discussion with several colleagues. I wish particularly to acknowledge the benefit derived from participating in a workshop on species as particles in space and time organized by Professors T. J. M. Schopf and D. M. Raup in Washington, D. C. in June 1978 and from many discussions with Dr R. P. S. Jefferies. I am also grateful to Professor A. J. Cain for reading and commenting on an earlier draft of the manuscript and to Andrew Gale and Andrew Smith for permission to use unpublished data on the stratigraphic distribution of chalk sea urchins.

REFERENCES

Bather, F. A. (1900). The echinoderms. *In* "A Treatise on Zoology" (E. R. Lankester, ed.), Vol. 3. A & C Black, London.

Cain, S. A. (1938). The species-area curve. *Am. Mid. Nat.* **19**, 573–581.

Carroll, R. L. (1977). Patterns of amphibian evolution: an extended example of the incompleteness of the fossil record. *In* "Patterns of evolution as illustrated by the fossil record" (A. Hallam, ed.), pp. 405–437. Elsevier, Amsterdam.

Durham, J. W. (1967). The incompleteness of our knowledge of the fossil record. *J. Paleont.* **41**, 559–565.

Gale, A. S. and Smith, A. B. (1982). The palaeobiology of the Cretaceous irregular echinoids *Infulaster* and *Hagenowia*. *Palaeontology* **25**, 11–42.

Kier, P. M. (1977). The poor fossil record of the regular echinoid. *Paleobiology* **3**, 168–174.

Nakazawa, K. & Runnegar, B. (1973). The Permian-Triassic boundary: a crisis for bivalves? *In* "The Permian and Triassic Systems and their mutual boundary" (A. Logan and L. V. Hills, eds.), pp. 608–621. Canadian Society of Petroleum Geologists, Calgary.

Newell, N. D. (1959). Adequacy of the fossil record. *J. Paleont.* 33: 488–499.

Paul, C. R. C. (1969). *Thomacystis,* a unique new hemicosmitid cystoid from Wales. *Geol. Mag.* 106, 190–196.

Paul, C. R. C. (1973). British Ordovician cystoids. *Palaeontogr. Soc.* [*Monogr.*] (1), 1–64.

Paul, C. R. C. (1976). Palaeogeography of primitive echinoderms in the Ordovician. *In* "The Ordovician System: proceedings of a Palaeontological Association symposium, Birmingham, September, 1974" (M. G. Bassett, ed.), pp. 553–574. Univ. Wales Press and Nat. Mus. Wales, Cardiff.

Paul, C. R. C. (1980). "The natural history of fossils". Weidenfeld and Nicholson, London.

Rushton, A. W. A. (1978). Fossils from the Middle-Upper Cambrian transition in the Nuneaton District. *Palaeontology* 21, 245–283.

Shaw, A. B. (1964). "Time in Stratigraphy". McGraw-Hill, New York.

Smith, A. B. & Paul, C. R. C. (1982). Revision of the class Cyclocystoidea (Echinodermata). *Phil. Trans. R. Soc. Lond.* B 296, 577–679.

Stitt, J. H. (1977). Late Cambrian and earliest Ordovician trilobites, Wichita Mountains area, Oklahoma. *Bull. Okla geol. Surv.* 124, 79pp.

Stokes, R. B. (1977). The echinoids *Micraster* and *Epiaster* from the Turonian and Senonian of England. *Palaeontology* 20, 805–821.

# 4 | Neontological Analysis Versus Palaeontological Stories

## P. L. FOREY

*Department of Palaeontology, British Museum (Natural History), London, England*

Abstract: The arguments between the two schools of phylogeny reconstruction —
cladistics and evolutionary taxonomy — are most obvious in the comparison of
cladograms and trees. Cladograms are the province of neontological analysis
whereas trees have become the province of palaeontological synthesis. Both
schools start with the premise that life is organized in a pattern but here
agreement ends. Cladograms are simple constructs, independent of evolutionary
theory and are concerned with discovering groups. Trees are complicated
statements of characters, the justification of which is to be found in evolutionary theory, and are consequently one step further removed from reality.
Palaeontology has adopted a special role in the construction of trees and has
attempted to vindicate what Darwin saw as one of the major objections to his
theory of evolution. In this chapter several trees are examined and one is found
to be intricately linked with a higher level of abstraction, the scenario.

## INTRODUCTION

Our present theories of phylogeny reconstruction fall into two
schools, phylogenetic systematics (cladistics) and evolutionary
systematics (eclecticism, traditional systematics/taxonomy, evolutionary taxonomy).[1] * Dialogue between the two has been, at
best, rhetorical and at worst polemical. To some people, including
myself, reconciliation between them seems as far off now as it did

---

Systematics Association Special Volume No. 21, "Problems of Phylogenetic Reconstruction", edited by K. A. Joysey and A. E. Friday, 1982, pp. 119–157, Academic Press,
London and New York.

*Superscript numbers in square brackets refer to numbered notes at the end of the
chapter.

fifteen years ago, the publication date of an English translation of Hennig's own manuscript revision (1960) of his book (1950) "Grundzüge einer Theorie der phylogenetischen Systematics". Hennig's (1966) book provided an analytical method for reconstructing phylogeny and the method is based on neontological data. Fossils are not introduced until the last quarter of his book and when they are, they are dealt with in an untraditional fashion (see also Hennig 1965). Hennig's ideas were quickly adopted by other entomologists (e.g. Brundin 1966, 1968; Dupuis 1979 gives a good account of the spread of cladism) and by those vertebrate zoologists faced with problems of reconstructing phylogenies in groups with very large numbers of Recent species and complex classifications (Nelson 1969a).

Palaeontologists were, on the other hand, largely unreceptive to Hennig's book and preferred instead to stand by the synthetic[2] approach as outlined by Simpson (1961) and Mayr (1969). In the euphoria of centenary celebrations commemorating the publication of "The Origin" Newell (1959, p. 275) wrote

> Thus, hypothetical phylogenies based solely on living genera and species cannot express the true relationships. In order to understand the ancestry of, and connections between living genera and families it is necessary to *know* the fossil record (my italics).

To know the fossil record means to find "numberless intermediate varieties, linking closely together all the species of the same group" (Darwin 1859). Palaeontologists took the opportunity to vindicate Darwin who thought that the absence of intermediates was due to imperfections in the geological record but who recognized that the lack of such intermediates "is probably the gravest and most obvious of all the many objections which may be urged against my views" (Darwin 1859, p. 299).

Many fossils have been collected since 1859, tons of them, yet the impact they have had on our understanding of the relationships between living organisms is barely perceptible. For instance, a recent symposium was held by this Society under the heading "The origin of major invertebrate groups". Most of the papers delivered at that meeting had a strong palaeontological bias yet ideas of "relationships" of major invertebrate groups seem very unclear (Manton and Anderson 1979, fig. 20; Yochelson 1979, fig. 2). In fact, I do not think it unfair to say that fossils, or at least the traditional interpret-ation of fossils, have clouded rather than clarified our attempts to reconstruct phylogeny. Compare for instance Boudreaux's neon-tological approach (1979, fig. 33) to arthropod phylogeny with Whittington's palaeontological approach (1979, fig. 2). The preface to that symposium (House 1979, p. vii) poses the question "How far . . . is the pursuit of certainty in matters relating to the origin of major invertebrate groups merely the pursuit of a will-o'-th-wisp? The chase, after all, has been going on for long enough." Perhaps the chase has been misdirected and based on methods of phylogeny reconstruction which have grown up with palaeontology and the theory of evolution, and which never can yield the required answers. An example may illustrate what can happen when the fossils and evolutionary theory take over.

The history of thought concerning the relationships of lungfishes provides such an example of different styles of approach to the same problem (a full history of the lungfish problem is given in Rosen *et al.* 1981). When the Recent lungfishes *Lepidosiren* and *Protopterus* were first examined the method used to classify them (to discover their relationships) was to compare them, feature by feature, with fishes and with reptiles (then including amphibians), two groups thought to be quite distinct, and to assign them to one or the other. Opinion was divided, but the point is that these early workers were trying to solve a three-taxon problem with each of the taxa (lungfishes, fishes and reptiles) regarded as terminal taxa.[3] This is what I here choose to call the neontological approach. The attempt was abandoned when the theory of descent with modifi-cation was almost universally accepted and the spotlight shifted from trying to identify groups to trying to indentify intermediate (ancestral) forms. Lungfishes held such an ancestral position (Haeckel 1866) for a short while. So, the method had changed from trying to characterize groups to attempts to bridge the morphological gaps. But lungfishes were known to be peculiarly specialized and hence unsuitable as ancestors. The insertion of crossopterygians (Cope 1892) as ancestors of tetrapods or common ancestors of lungfishes and tetrapods (Dollo 1896) provided the answer since they were sufficiently primitive; that is, they had neither the characters of lungfishes nor of tetrapods. Since both tetrapods and lungfishes were known to have Palaeozoic representatives then the Devonian crossopterygians helped to provide the links that Darwin predicted would be there were the fossil record more complete and carefully examined. (Of course, the fossils had to be interpreted; mostly on a stegocephalian model, the latter then interpreted in the light of a crossopterygian.) We arrive therefore at the text-book narrative (Romer 1966). But we also arrive at the conclusion that it becomes impossible to characterize a cross-opterygian except in terms of what it did not have, and this is meaningless. Another stage came later, when the theory of cross-opterygian except in terms of what it does not have, and this is ontologists, imposed a causal veneer (a scenario) by explaining how and why the crossopterygian evolved into an amphibian some-time in the Devonian (e.g. Szarski 1962; Olsen 1971). It is scenarios such as these which make evolutionists so vulnerable to anti-evolutionists.

In this, admittedly sketchy, outline of the history of lungfish research I identify three facets of phylogeny reconstruction. The first is to construct hypotheses by trying to discover unique attributes of groups. It was most closely approached by the early workers adopting a purely neontological approach and most closely resembles cladogram construction. The second is the construction of the tree in which ancestors are designated and lineages "worked out" in accordance with evolutionary theory. The third is the scenario, which assumes a given tree and attempts to explain how the tree grew like it did. The second and third are most often practised by palaeontologists since the construction of the tree is intimately connected with fossils and purports to document actual historical events backed up with causal explanations, hence the term "palaeontological stories".[4] Palaeontology has adopted a special role in the formulation of trees since there is a conviction (Simpson 1961) that only trees have a time dimension, controlled by direct historical evidence (fossils), necessary for the documen-tation of actual events.

The distinction between trees and cladograms has been made on several occasions (Tattersall and Eldredge 1977; Platnick 1977; Cracraft 1979; Eldredge 1979a; Gaffney 1979) and was first pointed out by Nelson (unpublished). Briefly, the distinction between trees and cladograms is illustrated in Fig. 1. The cladogram of three taxa (which is one of four possible) can be expressed as six trees

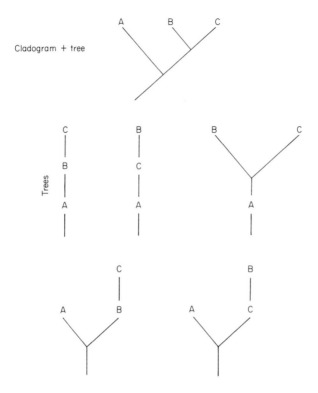

Fig. 1. Diagram to illustrate the topology of a cladogram (one of four possible) of three taxa and their equivalent trees. One tree, that with the same topology of the cladogram, is the only one which can be tested with synapomorphy.

depending on whether ancestors are recognized or not. The problem for the evolutionary taxonomist is which one; that is which, if any, of the branches will be shortened. Platnick (1977) has shown that if one uses the criterion of testing used by cladists (p. 136) then there is only one choice and this is the tree which has the same shape as the cladogram and in which no ancestors are recognized. However, tree construction aims specifically at identifying ancestors and the aim of evolutionary taxonomy would be to choose among the five trees shown below the cladogram. The objective of this chapter is to point out that recognition of ancestors relies on the acceptance of complicated conventions. My reason for raising the issue again is that much of the argument between cladists and evolutionary taxonomists centres on a still prevalent confusion between the two, or on ignorance that there is a difference.

Cladists and evolutionists agree that nature is organized in a pattern and that this pattern is due to historical process, but this seems to be the limit of common ground. A cladist's aim is to discover what he believes to be an hierarchical pattern by identifying nested sets of synapomorphies (Platnick 1980) or ordering characters according to their level of universality. What caused that pattern is outside the scope of cladism and, in particular, has no part to play in the construction of the cladogram. In other words, evolutionary theory is unnecessary for the construction of the cladogram.

Evolutionary taxonomy, on the other hand, *assumes* that the pattern is due to evolutionary theory and so the construction of trees must, of necessity, have regard to evolutionary theory. Hence, trees contain concepts of ancestry, morphological divergence and more complicated evolutionary concepts of paraphyly, polyphyly, phyletic gradualism (gradual transmutation of one species to another), punctuated equilibria, reversals etc. Trees are therefore more complicated statements as I hope to show by examples. Trees are also very closely associated with scenarios and there is often a great deal of cross fertilization as illustrated by the brachiopod example (p. 125). The hope of many of the authors of trees is that the synthetic approach (Bock 1974) will combine the analysis of pattern with the theories of the population biologists/geneticists in mutual support. The problem is that both are ultimately derived from the theory of evolution: the fossil record being interpreted in the light of population biology theory and expressed as ancestor–descendent relationships (Simpson 1953). One might remark that, far from being mutually supportive, this is merely "the blind leading the blind".

A discussion about trees is primarily a discussion about evolutionary theory (Cracraft 1979; Wiley 1979a; Gingerich 1979) and the palaeontological method. I am prepared to discuss the latter but not, at this time, the former. Suffice it to say that there have been recent views (Løtrup 1977; Macbeth 1973; Rosen 1978) expressing serious doubts about the theory of natural selection, or at least the ways in which it may be tested and placed on a scientific plane. I confess sympathy with these views.

I intend to set about the comparison between cladograms and trees chiefly by examples of trees. This is because the rules for the construction of cladograms, their justification and their characteristics have been spelled out on so many occasions (e.g. Hennig 1965, 1966; Brundin 1966, 1968, 1972; Bonde 1977; Griffiths 1972; Kavanaugh 1975; Ross 1974; Schlee 1971) that repetition would be superfluous. But no set of rules exists for tree reconstruction and, in fact, it appears that trees are very much "one-off" jobs, requiring individual attention. As a cladist, working with vertebrates, I was interested in the claims of invertebrate palaeontologists such as Bretsky (1979), Boucot (1979) and Campbell (1975) that the invertebrate fossil record is far more amenable to tree construction than the vertebrate record. The implication here is that invertebrate trees are somehow based on a more solid foundation. In consequence I chose to look at one such invertebrate group – the brachiopods – with the object of trying to analyse a tree. From this a number of generalizations could be made and the question of supraspecific ancestors, widely used in trees, is raised. From here, species as ancestors, are discussed using Gingerich's work on early primates as an example, before leading to general discussion.

## THE BRACHIOPOD EXAMPLE

From invertebrate ranks I singled out brachiopods as study "material" for a number of reasons. First, of the 2500 or so genera all but approximately 70 are extinct. Consequently, ideas of brachiopod phylogeny have been left largely in the hands of the palaeontologist and, with one exception (Hennig 1966, fig. 47) interrelationships have been expressed as trees with scenario elements in some schemes. Secondly, some consensus of opinion exists between those who have studied the phylum as a whole over the shape of the tree (Rudwick 1970) and the methods used have, to some extent, been explained (Williams and Rowell 1965a, b). Thirdly, brachiopods, unlike vertebrates, are represented in the fossil record by large numbers of "complete" specimens in marine sediments and so, according to evolutionary taxonomists such as Bretsky (1979), form ideal candidates for the "palaeontological method" of tree construction. Large collections of brachiopods from well-studied sections have been subjected to careful study and, as such, they would be expected to "give rise to reliable taxonomies and these in turn provide a sound basis for family tree speculations" (Boucot 1979, p. 205). Fourthly, the phylum Brachiopoda is sufficiently small (cf. Arthropoda or Mollusca) that several specialists have been able to look at the phylum as a whole with the result that one can obtain a balanced view (e.g. Williams and Hurst 1977; Rudwick 1970). Fifthly, and finally, I had the ear of a willing, if not entirely sympathetic colleague and brachiopod specialist, Howard C. Brunton.

For the purposes of this chapter I wish to concentrate on two schemes; that of Williams and Hurst/Williams and Rowell/Rudwick, which may be taken as the most widely adopted and that more recently proposed by Wright (1979). I will also concentrate on one area in particular – the relationship between inarticulates and articulates. This may seem over-restrictive but most of the points I wish to raise about trees are exemplified there. Rudwick's scheme (1970, fig. 99) is the most complete and explicit brachiopod tree and would be most suitable for my purposes but he admits that "justification for this particular scheme cannot be adequately discussed within the limits of this work". To my knowledge he has not followed this up. However, Rudwick's diagram is so similar to that of Williams and Hurst (1977, fig. 17; this being an updated version of one published by Williams and Rowell 1965a, fig. 147), reproduced here as Fig. 2, that the two may be considered as one.

Neither Williams and Hurst nor Williams and Rowell state explicitly how the tree was formulated but it was clearly undertaken using the guidelines of evolutionary taxonomy (Simpson 1961; Mayr 1965, 1969; Ghiselin 1972; Bock 1974). It is not altogether surprising that a clear exposition of methodology is not given since, in my opinion, none exists (notwithstanding the excellent attempt by Bock 1974). As Bock says, the aim of evolutionary taxonomy is to incorporate at least two types of relationship reflecting the evolutionary process: genetic relationships (Mayr 1965), which broadly correspond to Simpson's (1961) horizontal relationships and are measured by phenotypic similarity; and genealogical similarity (Mayr 1965), equivalent to Simpson's vertical relationships and expressed as ancestor–descendent sequences and sequences of branching (Nelson 1972 points out that correspondence between

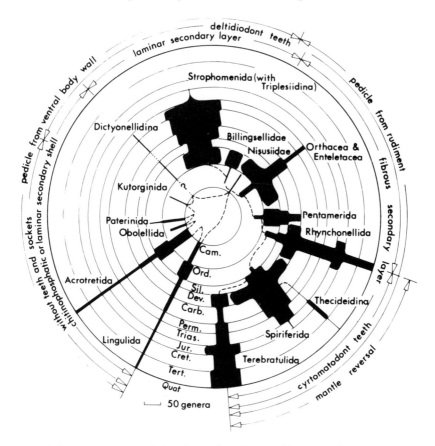

Fig. 2. Phylogenetic tree of brachiopods (from Williams and Hurst 1977). Reproduced with permission from Elsevier Scientific.

Mayrian and Simpsonian relationships is not exact). For the formulation of the brachiopod tree there is, indeed, a very close link with evolutionary theory, specifically with the notion that evolution proceeds by gradual morphological modification through time away from an ancestral species. Morphological gaps "must ultimately be due to breaks in the geological record" (Williams and Rowell 1965a, p. H165). But these morphological gaps are extremely important because they help define the borders of the lineages (clusters of species), and it is with lineages that we are concerned.

The lineages are identified by searching for continuity of morphological change combined with stratigraphic occurrence and are characterized by unique *combinations* of features; "no one character is either exclusive to an order or invariably exhibited by all its members" (Ager *et al.* 1965, p. H299). The particular unique combination of features (the "modal morphology", Williams and Rowell 1965b, p. H228) held to distinguish one lineage from another are those exhibited by a small sample of brachiopods and this sample automatically includes the living species since these are the end products of evolution. If lineages are thus recognized by "modal morphology" then they will only remain recognized if (a) the "modal morphology" of one does not seriously over-lap that of another (where it does then the explanation offered is that this phenomenon demonstrates "the replicating processes in evolution" (Williams and Rowell 1965b, p. H228)) or (b) if the lineage can be traced back through geological strata and still remains distinct. Genera of brachiopods (the unit of study of this group for reasons explained by Williams and Hurst 1977); exhibiting a particular modal morphology, are thus clustered together within lineages using an essentially stratophenetic approach (Gingerich 1976, and see below, p. 137) and expressed as ancestor–descendent relationships (e.g. Ziegler 1966; Smirnova and Pajaud 1968; Walmsley and Boucot 1971).

The relationship between lineages is, apparently (Williams and Rowell 1965a, p. H164), established by "an inductive evaluation of the disparity between unrelated contemporary stocks". These relationships are also expressed as ancestor–descendent sequences but, because of the larger morphological gaps (taxonomically this occurs at the superfamily/subordinal level), this can usually only be done with less confidence (hence dotted lines). The ancestral group is recognized by (a) a best candidate is chosen from among several lineages with representatives at or prior to the appearance of the group in question or (b) the evolutionary trend of one lineage converging backwards in time towards the condition in another. Once again, stratigraphic occurrence is of utmost importance as is illustrated by the following remarks concerning four orders of inarticulate brachiopods:

> Hence it is not known whether they were all independently derived from a remote common ancestor or whether there is a fundamental regularity in the succession of their appearance with one order arising from another. In the absence of chronological information, one is forced

to rely solely on morphological comparison to assess affinities (Williams and Rowell 1965a, p. H167).

True, but the dilemma caused by being forced to rely on morphological criteria is particularly interesting in the case of the inarticulates.

The inarticulates have long been distinguished from the articulates and the high taxonomic rank accorded these groups is presumed to reflect the importance of the characters which distinguish them (the taxonomic importance of a character being measured by the number of species to which it became common, Williams and Rowell 1965a, p. H164). The problem arises when one considers a few Cambrian and Ordovician genera which conform to neither the inarticulate nor the articulate "modal form". The Kutorginida and, to a lesser degree, the Paterinida, show certain character combinations of both. Evolutionary theory predicts that such inter-mediates might exist, yet in the absence of a stratigraphic sequence, their relationships cannot be specified and the kutorginids, at least, are regarded as being derived independently of the Inarticulata and the Articulata (Williams and Rowell 1965a, p. H196). Thus, although it is often stated that phylogeny reconstruction and classification are independent of one another (Mayr 1974; Ashlock 1980) or are "redundant images of each other" (Bock 1974, p. 391) they are, in fact, closely connected in evolutionary taxonomy (Cracraft 1974); the inability to classify a group has led to a theory of relationship (or rather a theory of non-relationship).

As far as brachiopods are concerned this action is carried to an extreme by Wright (1979) who claims that, because several Cambrian and Ordovician groups ("stocks" of Wright) cannot be assigned to either the inarticulates or the articulates, the brachiopods arose polyphyletically from an assemblage of "tubiculous brachiophorates" (hypothetical animals) (Fig. 3). Wright accepts that the sudden appearance of morphologically diverse inarticulate groups at the base of the Cambrian is a real phenomenon and

suggestive of a substantial Precambrian ancestry, possibly as brachiopods with an organic, non-preservable skeleton or alternatively as diversified ancestral lophophorate stocks, whose establishment in the epifauna was simultaneous with the development of a mineralized exoskeleton (Wright 1979, pp. 236–237).

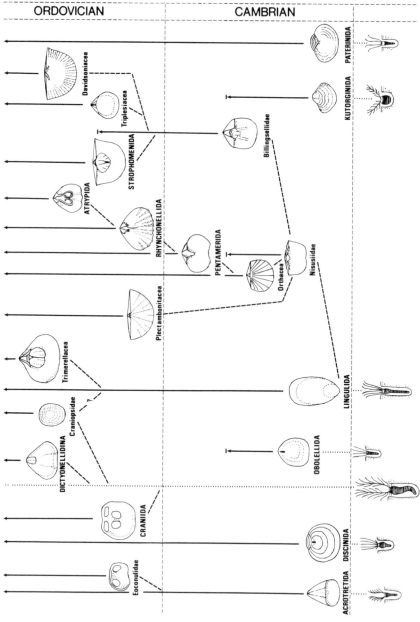

Fig. 3. Phylogenetic tree of brachiopods (from Wright 1979). Reproduced with permission from The Systematics Association.

Wright opts for the latter alternative which was suggested by Valentine (1973, 1975) who was similarly impressed by the sudden appearance of brachiopod stocks and proposed a "new adaptive model of their origin" (1975) — a scenario. The scenario suggested that soft-bodied, infaunal lophophorates moved to an epifaunal position and became anchored by a pedicle. They did this to take advantage of the cleaner water. Once here the lophophore needed protection from predators — hence the shell. Some came right out, lost the pedicle and became cemented. A few dived back into the substrate.

In itself the scenario is harmless, but Valentine suggested that it might be fed into tree construction;

> A functional ecological approach to the origin of higher taxa produces models of adaptive pathways that have much explanatory power and . . . may yet serve to constrain the possibilities so that a well supported model of phylogeny will emerge (1973, p. 101).

Wright appears to have adopted this explanation of why "epifaunal characters" should have developed and gone further by suggesting that they could have been developed on numerous occasions because, apart from those characters (shell, pedicle or cementation) "necessary" for an epifaunal existence, the Lower Cambrian groups display such a mosaic of characters that any coherent pattern seems absent. I have no doubt that, were a stratigraphic record available, the solution may well have been different.

In the absence of a stratigraphic sequence the dilemma we face here is similar to that posed by the origin and interrelationships of arthropods (another assemblage of "stocks" appearing in the Lower Cambrian), also thought to be polyphyletic (Manton and Andersen 1979; Whittington 1979; Bergstrom 1978). This problem was addressed by Patterson who pointed out that there is a confusion between characters and groups; "polyphyly is a term we apply to groups, not to characters" (1978a, p. 101). If brachiopods are polyphyletic then the relationships of each or some of the seven "stocks" lie with two or more named non-brachiopod groups. I do not think that this is what is meant in this case. Rather, Wright appears to be following Valentine (1973, p. 100): "the nearest common ancestor of these various brachiopod stocks was not a brachiopod". Valentine recommends that we raise the taxonomic rank of these various stocks to phylum, thereby making the taxa "monophyletic". By doing this it is assumed that the problem disappears! It does nothing of the sort. What it does do is to remove the problem to the next higher taxonomic rank (in this case the superphylum), and to attribute absolute rank to characters; shells and pedicles are phylum level characters. This decision can only be arbitrary. Muir-Wood (1955) speaks of characters of generic/specific value (shell ornament, convexity of the shell) and those of super-family/order value (nature and position of the pedicle opening, shell attachment).

We see here a common feature of tree construction; a confusion, or at least a lack of clear distinction, between classification (a convention concerned with ranks and with descriptively charac-terizing groups) and phylogeny reconstruction (the establishment of natural groups which, in terms of evolutionary theory, means the clustering of species into lineages). For evolutionary taxonomy the classification is a description which "must include all relevant features of the group to aid in its identification" (Ashlock 1980, p. 448). All relevant features include primitive as well as derived characters and results in each of the brachiopod "stocks" being described by particular "modal morphologies" separated from one another by morphological gaps larger than those occurring within each of the groups. As gaps between these stocks are recognized as being of equal, or non-assessable, value then each group is equally related (or unrelated) to all the others. Relationship, in this case, means genetic relationship, measured by phenetic similarity. To place this in an evolutionary context means that each of the "stocks" has diverged from the common ancestor by the same degree (measured by the assignment of the same rank) but in different directions. Approaches to phylogeny reconstruction such as those of Wright (1979) avoid attempts to discover genealogical relationships because they seek to distinguish more and more groups which could not have evolved from one another because acceptable ancestors have not yet been found. Wright's diagram (Fig. 3) says only that there are seven taxa (Acrotretida, Discinida, Craniida, Dictyonellida, Lingulida + articulates, Kutorginida, Paterinida). Unfortunately, this approach is all too common (e.g. Whittington 1979; Bergstrom 1978; Jarvik 1960; Clemens and Kielan-Jawarowska 1979) and to provide causal explanations, such as those of Valentine

(1973, 1975) and Gutman *et al.* (1978) does not solve the problem. The problem lies elsewhere, with the interrelated concepts of ancestors, relationship and monophyly outlined in the following sections.

TREES AND SUPRASPECIFIC ANCESTORS

From this very brief survey of the ideas on the relationships of early brachiopods I would like to abstract certain features of trees constructed using the guidelines of evolutionary taxonomy. The most general statement about trees is that their justification lies in evolutionary theory and this permits ancestors to be recognized. In the brachiopods it is supraspecific ancestors which are most often recognized (Williams and Hurst 1977). Evolutionary theory does not allow for supraspecific ancestors: a genus does not evolve from a genus, a family from a family and so on. So, in trees, some convention or special meaning must be implied in the use of supraspecific ancestry.

Bretsky (1979, p. 150) defends the use of supraspecific ancestors by saying that they are designed for cases "when the stratigraphic evidence is inadequate for establishing the exact network of connections between species included in the two taxa". But what is really meant is morphological gaps which may be caused by gaps in the stratigraphic record. Bretsky sees no real difference between species and higher taxa, except that the higher the rank the broader the diagnosis, and agrees with Harper (1976) that the degree of confidence one may have in the hypothesis of supraspecific ancestry is directly related to the rank at which the hypothesis is proposed. That is, one may have greater confidence in a class to class relationship (e.g. inarticulates to articulates or reptiles to mammals) than in a genus to genus relationship. This is a curious statement when placed alongside Bretsky's criterion of relative rank (1979, p. 151): "ranking in the taxonomic hierarchy has traditionally been used as a tacit assumption of the level of variability within a higher taxon as well as the *degree of morphological divergence from other taxa*" (my italics). We may, apparently, have greater confidence in proposing ancestor–descendent relationships across larger than smaller gaps.

A convention which has been widely adopted by palaeontologists to recognize a supraspecific ancestor is Simpson's definition of minimal monophyly: a taxon is monophyletic if it can be shown to be derived, through one or more lineages, from a group of equal or lower rank. Cladists have critized this on so many occasions that I need only repeat that this definition depends on some criterion of ranking. Ranks are remarkably mutable concepts (e.g. Valentine's suggestions that we simply raise the rank to solve a problem). Once again we are concerned here with very complicated and subjective assessments of characters, their variability and assignment to rank, and a continuing confusion between classification and phylogeny. The groups, existing in nature, have somehow got lost.

Cladists argue that supraspecific ancestors are not monophyletic (monophyly for a cladist has nothing to do with classification; a monophyletic group is one that contains the latest common "ancestor" plus all descendents). To hive off the ancestor into another group is to leave an incomplete descendent group and a paraphyletic "ancestral group". Paraphyletic groups are not groups — they are taxonomic artifacts. They have no individuality or history of their own and cannot be defined, except by employing a totality of characters primitive with respect to the descendent group. For instance, the statement that the articulates were derived from the inarticulates simply means that inarticulates have primitive characters. But so have a host of other organisms. If some inarticulates are more closely related to articulates then clearly inarticulates are not monophyletic in either a cladistic sense or in the sense of genea-logical relationship of evolutionary taxonomy.

In the above paragraph I have taken the view that supraspecific ancestors are taxonomic artifacts brought about by a confusion between classification and phylogeny which, in turn, results from the lack of a clear statement of what relationship means in evol-utionary taxonomy. But supraspecific ancestors may, in fact, be no more than hypotheses, hypotheses about primitive characters. Hull (1980) has pointed out that when palaeontologists propose supra-specific ancestors what they mean is that, were the ancestor known, it would be classified in the designated ancestral taxon. For instance, the tetrapod ancestor would be classified as a crossopterygian. In other words, the ancestral morphotype conforms phenetically most closely to a crossopterygian (it is the best candidate) and this crossopterygian is one stripped of all its peculiarities. In brachiopods

Williams and Rowell (1965a, p. H195) write: "living articulate species represent the culmination of divergence from ancestors close to primitive inarticulates".

To summarize so far, I have tried to show that supraspecific ancestors of evolutionary taxonomy, although often designated, are taxonomic artifacts. They are not groups and, in fact, they bar any attempt to discover a natural group. They are a convention and do not represent an evolutionary phenomenon. They have no part to play in phylogeny reconstruction — recognizing pattern.

To return to brachiopods finally: Williams and Rowell write: "Few palaeontologists would now dispute the validity of the Inarticulata and Articulata as classes within the phylum" (1965b, p. H227). Maybe so — but they are different; one (the Articulata) is probably a group; the Inarticulata certainly is not since it is based on primitive characters (it lacks what the other has) and is paraphyletic, consisting of a series of nested groups. In Fig. 4 I have attempted to construct a cladogram using information from the Treatise, Rudwick (1970), Williams and Hurst (1977) and Wright (1979). It is incomplete to the extent that the acrotretids, a large assemblage, are not included as I failed to find evidence that they are monophyletic (that is, they have no synapomorphy unique to them and they cannot, as presently recognized, be regarded as a terminal taxon). Some additional information relating to the construction of the diagram is to be found in the legend.

TREES, SPECIES AS ANCESTORS AND DISCUSSION

The preceding section attempted to show that supraspecific ancestors can only be regarded as a convention, either designed to "plug the gaps" in the fossil record or to stand in place of hypotheses of primitive characters. They have no place in phylogeny reconstruction, if only because they are inconsistent with evolutionary theory. Species as ancestors are, however, different; they are allowed by evolutionary theory and consequently should be valid components of phylogenetic trees. Indeed, species trees are held in esteem by their proponents because it is here that actual evolutionary events are specified (Eldredge 1979a; Szalay 1977). But species as ancestors raise several questions which I would like to address: how do we recognize them, what are the consequences for recognizing ancestors,

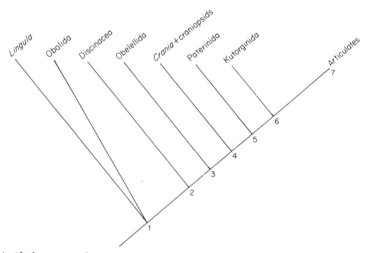

Fig. 4. Cladogram (character dendrogram) of brachiopods constructed using cladistic principles. The taxa, treated as terminal taxa, have been arranged within nested sets to maximize congruence of characters throughout the terminal taxa, with characters being expressed at different levels of universality. The characters at each branch point are: 1, Dorsal and ventral valves secreted by a mantle; 2, Pedicle marking ventral valve, valves unequal; 3, Laminar secondary layer, delthyrium; 4, Posterior margin of pedicle valve straight; 5, Subapical foramen or open gap (Rudwick 1970); Pedicle restricted by deltidium (= homeodeltidium of paterinids); 6, Narrow elongate muscle scars near the mid-line; 7, Synapomorphies of articulates such as teeth and sockets along the hinge-line and characters specified by Hennig (1966, p. 152). The diagram is predictive in the sense that the distribution of other morphological characters, when added, would be expected to remain congruent with the specified groups. Non-congruent distributions provide the test and may mean that the original hypothesis is incorrect (unparsimonious) or that the characters are mistaken (Platnick 1980); that is, they do not specify a group. Absence characters are not considered. The shell structure of two of the terminal taxa (paterinids and kutorginids) is very poorly known (Williams and Hurst 1977; Brunton personal communication) but the diagram does make some predictions about the structure. The cladogram is capable of being criticized (and no doubt it will be). The most obvious incongruence in the diagram is the chitino-phosphatic shell of the paterinids; but placing this group to the left of the obelellids is unparsimonious and suggests that a division of brachio-pods into non-calcareous and calcareous may be over simplistic (see also Williams and Hurst 1977). The basal trichotomy can mean two things in a cladogram, conflicting synapomorphies or, as in this case, a lack of characters of one of the taxa (obolids). For methods of classifying schemes such as this see Patterson and Rosen (1977).

how might we test such hypotheses of relationship and what (if anything) is to be gained by doing so? (see also Eldredge 1979a)

As with the previous consideration of supraspecific ancestors I will begin with an example by choosing the now classic trees produced by Gingerich (1976) and Gingerich and Simons (1977). Unlike many authors Gingerich (1976, 1979, 1980) tries to state clearly how the tree is constructed and he describes the method as the stratophenetic method. This can be briefly outlined as follows. Samples are collected from different horizons (ideally from the same locality but certainly from within the same depositional basin). Samples from each horizon are arranged according to phenetic clustering to determine the number of species present in some predetermined group. The samples are then ordered into a stratigraphic sequence: this must be done independently of the group concerned (a principle not apparently completely followed for the plesiadapid tree reproduced here as Fig. 5, McKenna *et al.* 1977, p. 237). Once this is done, a chosen species from one horizon is linked to species in adjacent levels (principle of minimal stratigraphic gaps; Harper 1976) on the basis of overall similarity (principle of minimal morphological gaps; Harper 1976) and ancestor–descendent relationships are proposed on the basis of stratigraphic occurrence. The result is a minimum spanning tree (Fig. 5). The number of species recognized for each horizon is important in as much as it enables one to recognize extinctions and bifurcations (cladogenesis). Where there is only one species in each successive stratum then the phylogeny is viewed as a single lineage of ancestor–descendent relationships displaying phyletic gradualism. Gingerich's methodology differs little from that practiced by a large number of palaeontologists and, as he says (1980, p. 453) "principles derived from the study of a group with a good historical (i.e. fossil) record should be applicable to less well known groups". He has tried (1979) to apply these principles to less "well known" groups such as the reptiles and the birds.

The first observation on Gingerich's scheme, and species trees in general, is that it represents a factual statement that, for instance, *P. abditus* is ancestral to both *P. frugivorus* and *P. jarrovii*. It is not an approximation as are supraspecific. The second observation is that we must assume that we have, in the sample before us, all of the taxa of that group which lived between the times specified and

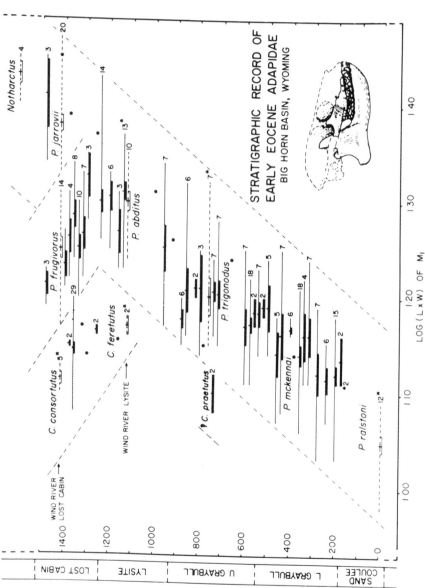

Fig. 5. Phylogeny of the primates *Pelycodus* and *Copelemur* of the Lower Eocene of the Big Horn Basin, Wyoming, showing continuous variation in crown area of the first lower molar. For details see Gingerich and Simons (1977). Reproduced with permission from University of Michigan.

within the area specified in the diagram. In other words, there are no intervening taxa. If one is not willing to accept this, then statements about ancestry have to be translated into best candidate statements. These can range from 1–99% certainty but, more importantly, we need some additional criteria of estimating certainty.

Best candidates are often chosen. Szalay (1977) advocates choosing a species from an assemblage preceding (in time) the taxa under consideration which matches the morphotype (a collection of primitive characters) most closely. Bretsky (1979, p. 145) also suggests that we choose a species which is suitably positioned stratigraphically and which approximates phenetically to the morphotype. She goes on to explain that observed deviation from the ancestral morphotype may be accounted for by inferred individual variation and that "one component of scientific creativity consists in knowing what discrepancies to ignore". But, by what yardstick do we measure discrepancy, and how much can we afford to ignore before rejecting a species as ancestor? Answers to both of these questions surely demand some knowledge of the nature and completeness of the sample under study. Suppose that we have three samples (Fig. 6) representing three species distributed in strata such that either A or B is the ancestor of C. The problem is, which one

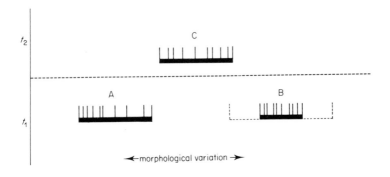

Fig. 6. Diagram to illustrate that sample bias can influence the choice of ancestor using the stratophenetic method. A, B and C represent species each known from the same number of specimens (vertical lines) and assumed, by definition, to have have similar limits of variability (horizontal lines). Species B is represented by a clustered sample so that the actual variability may lie anywhere along the dotted line.

is the ancestor and which becomes extinct. A is closer to C simply because the sample is more representative and thus might be chosen. But, if species are defined as having roughly equivalent ranges of variability (a procedure advocated by Simpson 1961) then B might be the more likely candidate and is only rejected because of a default of sampling.

Harper suggests (1976, principle 1c) that a decision might be reached by looking for trends in other "related" groups and extrapolating back from C to either A or B. Here, however, we are surely using assumed parallelism to determine relationship and such a process can only be self-fulfilling. Of course, most palaeontologists would hope to add more specimens to increase sample size or add a fourth taxon occurring stratigraphically above, below or in between. More specimens may or may not solve the issue but adding a fourth taxon will only change the nature of the problem (see also Platnick 1977).

It is important to realize that when Gingerich places species in ancestor–descendent relationships this is done on purely stratigraphic grounds (1979, p. 58). This means that there is an uneasy compromise adopted between relationships at the family level, based on morphology, while relationships below family level are based on stratigraphic occurrence. McKenna *et al.* (1977), in their review of Gingerich's work, rightly question why morphology is good at and above the family level but not below? The power of stratigraphy over morphology can lead to some rather awkward results as illustrated in Fig. 7. Here, the postulated relationships suggest that in one lineage (*Nannodectes*), a reversal took place in the crown area of the first lower molar so that one of the later species (*N. gidleyi*) is indistinguishable from stratigraphically earlier members of the *Plesiadapis* lineage. In fact, according to the diagnoses provided by Gingerich (1976), other morphological characters are also the same. The conclusion that a reversal has taken place is arrived at because of a literal reading of the stratigraphic record, and the acceptance of this evolutionary phenomenon is widespread among mammal workers (Gingerich 1977). If, however, we accept that ancestral characters might reappear at any time we might as well give up attempts to reconstruct phylogeny from morphology and concentrate instead on getting the stratigraphic sequence correct. The justification for allowing reversals (and

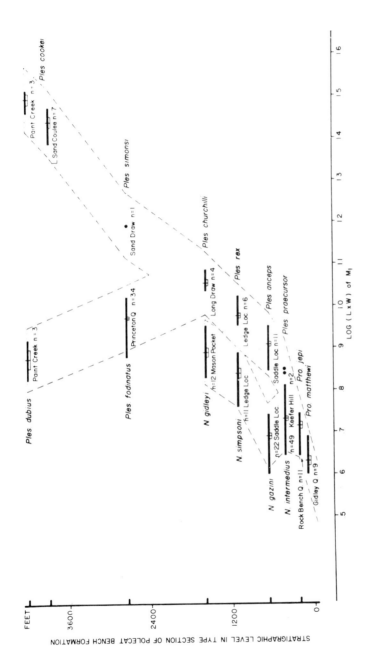

Fig. 7. Phylogeny of the North American species of plesiadapid primates showing reversals and parallelisms in crown area of the first lower molar. For details see Gingerich (1976). Reproduced with permission from University of Michigan.

parallelisms for that matter) lies in evolutionary theory so the formulation of trees is not inconsistent on this point. The problem lies in our inability to choose between competing trees once reversal is allowed (see below, p. 144).

The examples above were drawn up on the premise that all the necessary data to enable one to reconstruct actual historical events are entombed within a dense and continuous fossil record. This immediately raises the question of "how good" is the fossil record in reflecting both species diversity and providing us with an accurate reflection of the temporal distribution of any one species. In other words, are there parameters which would enable a choice to be made on when to use or to reject the stratigraphic record for establishing relationships? Bretsky (1979) does set some limits to the usefulness of stratigraphy by accepting that ancestor–descendent relationships can only be proposed when the stratigraphic gap separating putative ancestor and descendent is less than the expected duration of the species in question.

The first question about species diversity is important because a tree designating ancestors is an absolute statement in contrast to cladograms designating comparative statements: "Hypotheses of ancestor–descendent relationships permit us to view organisms as the result of a specific evolutionary transition" (Szalay 1977, p. 14). There have been numerous attempts to estimate the completeness of species diversity in the fossil record, either for groups (Paul, Chapter 3 this volume; Kier 1977; Bretsky 1979; Raup 1979) or for time periods (Raup 1976, 1979; Valentine 1969; Bambach 1977; Sepkowski 1978). Many of these estimates depend on assumptions as to whether the total species diversity has remained constant throughout Phanerozoic time or whether there has been an increase, and if so, what shape of curve this increase has followed. These assumptions, in turn, depend on models of population ecology (area/species curves, species packing, island equilibrium, competition and so on) and theories of earth history (e.g. amounts of coastline available for colonization). I see no point in dwelling on this issue except to emphasize that estimates of completeness are riddled with assumption and, so far as I know, no one has proposed a 100% record, perhaps because we never know when we have one. In its absence then, we return to best candidate statements.

The second question of the accuracy of the fossil record in

documenting species time range should, theoretically, have no impact on the stratophenetic method as outlined by Gingerich. This is because species in any area are defined by arbitrary sections of time, or falling within sections of time, and these time-defined species must be linked (phenetically) to adjacent time-defined species (or left to go extinct) no matter what reversals occur. But for most palaeontologists species are not defined purely within a standard time framework but are free to wander within the limits of morphological variability through time.[5] In these instances the expected temporal range of a species is not always matched by the fossil record. In as much as stratigraphic position is used (Harper 1976, principle 1a) as a guide to morphological primitiveness and hence an indication of ancestral status then an incomplete, possibly misleading, temporal representation becomes important. Cracraft (1974, fig. 2) and Schaeffer *et al.* (1972) have discussed the problems posed by "misleading" stratigraphic occurrence and these have been acknowledged by evolutionary palaeontologists such as Bretsky (1975, 1979) and Harper (1976). Harper, for instance, acknowledges that the ancestral taxon may appear in the fossil record after its descendent, leading Patterson (1978a) to argue that the logical consequence of this observation is that "Recent species may be regarded as ancestral to other Recent species and even to fossil species" and that "stratigraphic sequence can never be a trustworthy guide to phylogeny" (p. 625).

Despite these arguments there is still a belief in the relevance of stratigraphy for identifying ancestors and ancestral characters which seems to be summed up by Bretsky (1975, p. 114): " The fossil record does not give us the revealed truth about evolution, but neither do rules about determining the direction of evolutionary change without recourse to stratigraphic data." So, even if we cannot recognize ancestors on stratigraphic grounds, can we recognize primitive character states? The most penetrating discussion of the use of stratigraphy for revealing polarity of character-state change is given by Jefferies (1979) who, like most cladists, does not recognize ancestors but, unlike most, tries to bridge an ever increasing gulf between the schools. Jefferies proposes a method-ological rule: "the stratigraphical criterion of primitiveness is applicable when the relevant homologies (synapomorphies) cannot certainly be recognized without it" (p. 454). Jefferies arrives at this conclusion through a consideration of Riedl's (1979) concept of burden (the burden of a homologue (synapomorphy) is the number of other homologues dependent upon it). In this sense, Jefferies provides a relative *a priori* ranking of synapomorphy weight (see also Hecht and Edwards 1976). The lowest weight is no synapomorphy (or no recognizable synapomorphy). Stratigraphy therefore becomes important when no synapomorphies (evolutionary novelties of evolutionary theory) can be identified but here the relationship being specified is purely that of stratigraphic position. If evolutionary theory is concerned with changing adaptations through time then the absence of change (no evolutionary novelties) can have no bearing on the claim that the fossil record demonstrates evolution by way of revealing ancestors or their character-states. To summarize, I find the "rules" for interpreting the stratigraphic record alone, or linked with morphological primitiveness, insufficiently structured to be useful in attempts to reconstruct phylogeny. The number of groups with a sufficiently good fossil record by anyone's criteria is so small that it is doubtful if the method, such as it is, can have wide applicability.

A different, but often associated, criterion for recognizing ancestors is morphological primitiveness and has been briefly referred to above. Ancestors are, by definition, wholly primitive with respect to their descendents and can therefore only be recognized on this criterion. This is true unless we allow reversals, as Gingerich does, but then we are recognizing ancestors on some other criterion (e.g. stratigraphic or geographic distribution). If these wholly primitive organisms occur later than the presumed descendent then it is assumed that we are sampling a persistent population of the ancestor. Ancestors can only be recognized by the absence of characters and justified on negative evidence (Nelson 1969b). This applies to species as ancestors as well as to species groups as ancestors and the discussion can now be widened to include both. For instance, the reptiles, long held to be the ancestors of both the birds and mammals are recognized by the absence of bird and mammal characteristics (Ashlock 1980). Primitive characters need not necessarily imply an ancestor–descendent relationship, they could just as easily imply a sister-group relationship and there appears to be no way of distinguishing between the two (Hull 1980; Platnick 1977).

Some of the consequences of attempting to recognize ancestors (real or hypothetical) may be seen more clearly when placed alongside reasons for recognizing them in the first place. Ancestor recognition is a logical corollary of evolutionary theory and is considered desirable if we are to learn more about the theory. For instance, Gingerich (1979, p. 58) considers that "One of the most important reasons for studying phylogeny at the species level is to gain a better understanding of the dynamics of speciation." (Phylogeny here means ancestor–descendent sequences.) Szalay (1977, p. 14) writes "it is these hypotheses [ancestor–descendent relationships] which may clarify the phylogenetic or functional significance of characters". Both workers are respectful of the continuity of the fossil record and convinced that palaeontology, through its time dimension, can contribute uniquely to the formulation or the justification of the theory. The special contribution in this case is the concept of phyletic gradualism (gradual transformation of one species into another, or others) which is nowadays a palaeontological concept. Phyletic gradualism has justifiably come under attack from Eldredge and Gould (1972) and Gould and Eldredge (1977) who point out that supposed evidence can equally be interpreted in a different light, more in accord with the currently favoured theory of allopatric speciation (Cracraft 1979). Sympatric speciation, implicit in phyletic gradualism, is a theory which needs to be more precisely formulated before evolutionary biologists studying Recent organisms would be willing to accept its usefulness to explain the natural world. But most would accept polyploidy, or hybridization together with polyploidy, as sympatric speciation in some sense. These phenomena, however, are rarely (if ever) characterized by gradual transformation. It would be unwise to push this point too far because the neontological concepts of allopatric and sympatric speciation are themselves theories based on their own premises. But the hope that palaeontological ancestor–descendent sequences will yield ideas of the dynamics of speciation begs the question of by what theory are the fossils to be interpreted (see also Cracraft 1979; Eldredge 1979b)? Faced with the conflict referred to above we might do well to divorce problems of relationship from speciation theory.

Nelson (1970) suggests that the search for ancestors may be an attempt to test a morphotype (this test can never be carried out except by rejecting candidates, see below). This point leads on to the more serious consequences of accepting ancestor–descendent relationships; that is, they deny the possibility of objective testing. Many authors have written on the subject of testing hypotheses of relationship (Wilson 1965; Bock 1974; Cracraft 1974, 1978; Wiley 1975; Platnick 1977; Harper 1976; Kitts 1977; Szalay 1977; Engelmann and Wiley 1977; Patterson 1977, 1978b, this volume; Nelson 1978; Bonde 1977; Gaffney 1979) and clearly there is a diversity of opinion about what constitutes a valid test. One criterion, parsimony, is not a test as is often assumed but is rather an axiom used to choose between two or more alternative hypotheses (Patterson 1978b). On two points there seems to be some agreement: first, attempts to verify do not constitute a rigorous test; secondly, testing consists of, at least, the addition of more characters (finding a fossil out of expected sequence is not a test of ancestor–descendent relationships but may be a test of the fossil record). The addition of characters may do one of two things: nothing, if one is willing to assume reversals or if the characters were deemed primitive *a priori*; or remove the ancestor from that position if the character were an autapomorphy. Bonde (1977), Szalay (1977) and Engelmann and Wiley (1977) have suggested that because the hypothesis of ancestry is more easily refuted (by an autapomorphy) then it is a bolder hypothesis and to be preferred. But this line of argument breaks down as Engelmann and Wiley (1977) go on to demonstrate, because the failure to falsify the ancestor (i.e. to corroborate it) can only be done by applying plesiomorphic characters.[6] Expressed another way, ancestors can only be falsified because no support can ever be found (Nelson 1969b). An ancestor "corroborated" in this way could equally well be an ancestor of some other taxon. An ancestor refuted by an autapomorphy can only result in a reformulation of the problem, that it is the sister-group of its presumed descendent. It then becomes subject to the testing criteria used in cladistic analysis. Platnick (1977) has shown that in a three taxon problem the application of the test (a synapomorphy of two of the three taxa) rejects all possible trees except those with the topology of cladograms where no ancestors are designated. One wonders therefore, whether the proposition of an ancestor is worthwhile in the first place.

One final point on the testability of the ancestor–descendent relationship has been raised by Szalay (1977). If hypotheses of ancestry are formulated by means of and with the purpose of expressing particular events then "Can singular historical events dependent on unique ancestries and circumstances of selection be 'tested'?" (p. 16). The answer is clearly no because there are no principles of generality involved and ancestor–descendent relationships must be viewed in the same light as theories of dispersal biogeography, as one-off explanations for random events.

My last criticism of ancestor–descendent relationships centres on the consequences for the concept of relationship and the discovery of groups. It has already been pointed out that ancestors are devoid of characters or a history of their own; that is, they are not individuals (Patterson 1978b). In cladistic analysis the branching points are recognized by the attributes of the terminal taxa (Nelson 1979) with two (or more) taxa defining the branching point and recognized by synapomorphy. Each of these terminal taxa will be monophyletic and relationship is expressed as nested sets of monophyletic groups. Suppose that we attempt to recognize an ancestor (A) at a branch point leading to the terminal taxa (B and C). We could only do so if A had a synapomorphy of B and C (otherwise there would be no reason to favour A over any organism with more universal attributes than B and C) but lacked an autapomorphy (which would remove it as a possible ancestor). To be ancestral to both it cannot possess the autapomorphous attributes of either B or C: it must be plesiomorphous in all respects. Assuming this to be possible then A, recognized on the absence of characters, would be incapable of being described since it has no characters of its own (it is impossible to describe a group reptiles without reference to birds and mammals; although it is possible to describe turtles, lizards, crocodiles and so on). But let us assume that we allow A to stand. If so, then A in this ancestral position destroys the monophyly of both B and C, if monophyly is defined in cladistic terms (a monophyletic group (holophyletic group of Ashlock 1971) is one containing the latest common ancestor plus *all* descendents). If B must include A, then so must C and we could never recognize a group B + A or C + A because both would remain incomplete. No matter how many additional synapomorphies were discovered common to B + C or A + B + C the original hypothesis would be immune to criticism and hence outside science. To get around the problem evolutionary taxonomists would invoke some classificatory device at this point, either by defining monophyly in terms of taxonomic rank (p. 134) or weighting the autapomorphy of B or C and ascribing some greater evolutionary importance to the feature and observing that there are two kinds of relationship to be expressed (Nelson 1972). These, however, are conventions to deal with a situation caused by a shaky premise – that ancestors can be recognized and constitute meaningful statements. A related problem is associated with the classification of ancestors since they are coextensive with their descendents. This lies outside the scope of this chapter but see Patterson (1977), Patterson and Rosen (1977) and Wiley (1979b).

In summary, I note that trees, being distinguished by the recognition of ancestors, are one step further from reality than cladograms (Eldredge 1979a) and are burdened with assumptions, some of which appear to be in conflict with one another and with the stated aims of those who propose them. The justification for ancestors lies in evolutionary theory and their recognition is based on the premises that ancestors are more primitive (true) and are found earlier (true) but the practice of recognizing these ancestors does not necessarily hold to these premises. There are no clear determinants as to when and by how much we can violate these premises, although there are many explanations offered as to why one or the other should be temporarily set aside. Those explanations depend on assumptions of evolutionary theory or theories of earth history. When the observed differs from the expected (morphotype) then the ancestors being proposed are best candidates but, for species trees at least, this is at odds with the aims of specifying actual and unique historical events in order to demonstrate or contribute to the theory of evolution ("to provide fuel to the evolutionary fire" – Eldredge 1979a). If this is the aim then we must be certain that we have the correct answer (an unattainable objective). If palaeontology in general and tree construction in particular has made contributions to evolutionary theory then they are not apparent to me (see also Eldredge 1979b). One contribution to which palaeontology, through ancestor recognition, has laid claim is the concept of phyletic gradualism but this is inconsistent with the currently favoured view of allopatric speciation.

Ancestors can only be designated at the expense of the ability to recognize groups because ancestors have no individuality which can be justified by a unique history identified by unique characters. They can never be shown to be monophyletic except by the addition of conventions (morphological conventions where characters are tied to absolute rank or stratigraphic conventions) and objective attempts to test hypotheses of ancestor–descendent relationships result only in a reformulation of the problem. In criticizing aspects of evolutionary taxonomy, trees and palaeontology I confess that this chapter is decidedly negative in style. But, in my opinion, this has been the characteristic feature of the palaeontological/evolutionary approach to phylogeny reconstruction. Far from simplifying our view of the living world, tree construction and palaeontology have added confusing elements which have to be explained away. Comparative physiologists, working with living Recent vertebrates, for example, find repeated similarities between groups considered to be far apart by palaeontologists (e.g. lungfishes and tetrapods or birds and mammals). The former, heeding Newell's words (p. 120), generally concede on points of phylogeny by accepting parallelisms, convergences and so on. But, like the comparative physiologist's method, phylogeny reconstruction is no more than an exercise in identifying pattern and testing these patterns with further information. The cause of that pattern if it is considered desirable to know it, is the subject of further analysis. For the recognition of that pattern it is unnecessary to know the cause. However, evolutionary taxonomy and resulting classification by means of tree construction, assumes a causal process which is then used to build the tree.

A common claim (e.g. Bock 1974) is that evolutionary taxonomy and resulting classifications incorporate two "semi-independent" variables of the evolutionary process, genetic similarity and phylogenetic sequence and is therefore the preferable approach. But there seems to be a lack of understanding that one of these variables is included within the other. That is, genealogy is the more inclusive and if there is difficulty with incorporating both into a single system we should opt for the more inclusive. All that is needed for a general reference system is a method to identify and test a pattern and this, I submit, is best met by the cladistic method which has nothing to do with identification of ancestors or of causality. In this, cladistics is similar to phenetics.

Those who attended the symposium or read the resulting papers may have been struck by the contrast between the logic of cladistics and phenetics on the one hand and the apparent absence of logic of evolutionary taxonomy on the other. Evolutionary taxonomy and the traditional approach to palaeontology seems, in this respect, to be in senility; if it is not to be left to go extinct then it needs revitalization through reformulation.

### ACKNOWLEDGEMENTS

I thank Colin Patterson for many hours of discussion and for reading this manu-script, Howard Brunton for his advice on brachiopods (he is, of course absolved from any conclusions expressed here), Mss V. T. Young and A. E. Longbottom for assistance with the illustrations and Ken Joysey and Adrian Friday for the opportunity to participate in this symposium.

### NOTES

[1] I omit phenetics from present discussion only because, in its purest form (Sokal and Sneath 1963) it does not purport to reconstruct phylogeny. There are, however, very close parallels between modern cladistics and phenetics in that both are searching for pattern independent of any con-sideration of the process which caused that pattern.

[2] The term synthesis is used here in a sense defined by "Chambers Twentieth Century Dictionary" as the explanation of certain phenomena by means of principles (in this case, evolutionary principles) which are, for this purpose, assumed as established.

[3] Terminal taxa are those recognized by unique attributes. In attempting to ally lungfishes with either reptiles or fishes the investigators were trying to reduce the three terminal taxa to two (lungfishes + fishes and reptiles or lungfishes + reptiles and fishes). We would not, today, recognize either fishes or reptiles as a terminal taxon.

[4] The use of the word "story" is not meant to be abusive. It is not my intention to imply that palaeontologists tell lies: after all, who knows what the actual events were? But I choose the word because Simpson (1980, p. 240), long an advocate of the palaeontological method, equates story with history.

[5] There has been much discussion over the nature of the species in palae-ontology. Many take the view that because the biological species definition (which itself has been questioned by Scudder (1974), Sokal and Crovello (1970), on phenetic grounds and Ehrlich and Raven (1969), on genetic

grounds) cannot be applied to fossils then they are different. But, oper-ationally, there seems little to choose between the two. Hull (1980) has pointed out that even within Recent populations we may be able to note which individuals mated with which to produce viable offspring but we cannot predict which might have done so but did not. We are perfectly entitled to recognize species in the fossil record, ideally by some unique attribute or perhaps by an ability to repeat the original distinction when confronted with more specimens, and as long as they are treated as terminal taxa. The problem with the palaeontological species is that it is rarely treated as such but rather as a segment of an evolving lineage, either defined morphologically or temporally. Westoll's (1956) holomorphospecies is a wholly typological construct. Simpson (1961) suggests that we divide up the lineage into segments exhibiting the same morphological variability as that shown by modern species related to the group in question. This is similar to the species concept of Raup and Stanley (1971) who suggest that within a chronologically and morphologically continuous series of populations there will be points at which accumulated differences are such that later populations would be reproductively isolated from the initial population were they living at the same time. In other words there is an inherent belief that genetic distance equals phenetic distance (sibling species might present a problem). Another widely held view (Sylvester-Bradley 1956) is that species are limited by stratigraphic gaps. The important point is that for most palaeontologists the species is defined and not discovered. Wiley (1979a) suggests that we use the evolutionary species (a modification of Simpson's evolutionary species) recognizable as segments between branch points. Wiley goes on to say that, for the purposes of analysis these species are treated as terminal taxa in cladistic analysis.

[6] As Nelson (1970) pointed out, the association of plesiomorphy with ancestry can result in characters being assessed as primitive *a priori* when a group or species has already been designated as an ancestor. Ancestors must, by definition, be primitive in all respects. The study of vertebrate phylogeny is replete with examples of this practice: one, sufficient to illustrate the point, is discussed by Butler (1972).

## REFERENCES

Ager, D. V., Amsden, T. W., Biernat, G., Boucot, A. J., Elliot, G. F., Grant, R. E., Hatai, K., Johnson, J. G., McLaren, D. J., Muir-Wood, H. M., Pitrat, C. W., Rowell, A. J., Schmidt, H., Staton, R. D., Stehli, F. G., Williams, A. and Wright, A. D. (1965). Articulata. *In* "Treatise on Invertebrate Paleontology, Part H, Brachiopoda" (R. C. Moore, ed.), pp. H.297–927. University of Kansas Press and The Geological Society of America, Lawrence.

Ashlock, P. D. (1971). Monophly and associated terms. *Syst. Zool.* 20, 63–69.

152          *P. L. Forey*

Ashlock, P. D. (1980). An evolutionary systematist's view of classification. *Syst. Zool.* **28**, 441–450.

Bambach, R. K. (1977). Species richness in marine benthic habitats through the Phanerozoic. *Paleobiology* **3**, 152–167.

Bergstrom, J. (1978). Morphology of fossil arthropods as a guide to phylogenetic relationships. *In* "Arthropod Phylogeny" (A. P. Gupta, ed.), pp. 3–56. Van Nostrand Rheinhold, New York.

Bock, W. J. (1974). Philosophical foundations of classical evolutionary classification. *Syst. Zool.* **22**, 375–392.

Bonde, N. (1977). Cladistic classification as applied to vertebrates. *In* "Major patterns in Vertebrate Evolution" (M. K. Hecht, P. C. Goody and B. M. Hecht, eds), pp. 741–804. Plenum Press, New York.

Boucot, A. J. (1979). Cladistics: is it really different from classical taxonomy? *In* "Phylogenetic Analysis and Palaeontology" (J. Cracraft and N. Eldredge, eds), pp. 199–210. Columbia University Press, New York.

Boudreaux, H. B. (1979). "Arthropod Phylogeny with Special Reference to Insects". John Wiley, New York.

Bretsky, S. S. (1975). Allopatry and ancestors: a response to Cracraft. *Syst. Zool.* **24**, 113–119.

Bretsky, S. S. (1979). Recognition of ancestor-descendant relationships in invertebrate paleontology. *In* "Phylogenetic Analysis and Paleontology" (J. Cracraft and N. Eldredge, eds), pp. 113–163. Columbia University Press, New York.

Brundin, L. (1966). Transantarctic relationships and their significance, as evidenced by chironomid midges. *K. Svenska VetenskAkad. Handl.* **11**, 1–472.

Brundin, L. (1968). Application of phylogenetic principles in systematics and evolutionary theory. *In* "Nobel Symposium 4. Current Problems of lower Vertebrate Phylogeny". (T. Ørvig, ed.), pp. 473–495. Almqvist and Wiksell, Stockholm.

Brundin, L. (1972). Evolution, causal biology, and classification. *Zool. Scripta* **1**, 107–120.

Butler, P. M. (1972). The problem of insectivore classification. *In* "Studies in Vertebrate Evolution" (K. A. Joysey and T. S. Kemp, eds), pp. 253–265. Oliver and Boyd, Edinburgh.

Campbell, K. S. W. (1975). Cladism and phacopid trilobites. *Alcheringia* **1**, 87–96.

Clemens, W. A. and Kielan-Jaworowska, Z. (1979). Multituberculata. *In* "Mesozoic Mammals" (J. A. Lillegraven, Z. Kielan-Jaworowska and W. A. Clemens, eds), pp. 99–149. University of California Press, Berkeley.

Cope, E. D. (1892). On the phylogeny of the Vertebrata. *Proc. Am. Phil. Soc.* **30**, 278–281.

Cracraft, J. (1974). Phylogenetic models and classification. *Syst. Zool.* **23**, 71–90.

Cracraft, J. (1978). Science, philosophy, and systematics. *Syst. Zool.* **27**, 213–216.

Cracraft, J. (1979). Phylogenetic analysis, evolutionary models and paleontology. *In* "Phylogenetic Analysis and Paleontology" (J. Cracraft and N. Eldredge, eds), pp. 7–39. Columbia University Press, New York.

Darwin, C. (1859). "On the origin of species by means of natural selection or the preservation of favoured races in the struggle for life". John Murray, London.

Dollo, L. (1896). Sur la phylogenie des dipneustes. *Bull. Soc. Belge. Geol., Paleont., Hydrologie* 9, 79–128.

Dupuis, C. (1979). Permanence et actualité de la systematique. La 'Systematique Phylogenetique' de W. Hennig. *Cah. Nat.* 34, 1–69.

Ehrlich, P. R. and Raven, P. H. (1969). Differentiation of populations. *Science* 165, 1228–1232.

Eldredge, N. (1979a). Cladism and common sense. *In* "Phylogenetic Analysis and Paleontology" (J. Cracraft and N. Eldredge, eds), pp. 165–198. Columbia University Press, New York.

Eldredge, N. (1979b). Alternative approaches to evolutionary theory. *Bull. Carnegie Mus. nat. Hist.* 13, 7–19.

Eldredge, N. and Gould, S. J. (1972). Punctuated equilibria: An alternative to phyletic gradualism. *In* "Models in Paleobiology" (T. J. M. Schopf, ed.), pp. 82–115. Freeman, Cooper & Co, San Francisco.

Engelmann, G. F. and Wiley, E. O. (1977). The place of ancestor-descendant relationships in phylogeny. *Syst. Zool.* 26, 1–11.

Gaffney, E. S. (1979). Introduction to the logic of phylogeny reconstruction. *In* "Phylogenetic Analysis and Paleontology" (J. Cracraft and N. Eldredge, eds), pp. 79–111. Columbia University Press, New York.

Ghiselin, M. T. (1972). Models in Phylogeny. *In* "Models in Paleobiology". (T. J. M. Schopf, ed.), pp. 130–145. Freeman, Cooper & Co., San Francisco.

Gingerich, P. D. (1976). Cranial anatomy and evolution of early Tertiary Plesiadapidae (Mammalia, Primates). *Univ. Mich. Pap. Paleontol.* 15, 1–140.

Gingerich, P. D. (1977). Patterns of evolution in the mammalian fossil record. *In* "Patterns of Evolution" (A. Hallam, ed.), pp. 469–500. Elsevier Scientific, Amsterdam.

Gingerich, P. D. (1979). The stratophenetic approach to phylogeny reconstruction in vertebrate paleontology, *In* "Phylogenetic Analysis and Paleontology" (J. Cracraft and N. Eldredge, eds), pp. 41–77. Columbia University Press, New York.

Gingerich, P. D. (1980). Paleontology, phylogeny, and classification: an example from the mammalian fossil record. *Syst. Zool.* 28, 451–464.

Gingerich, P. D. and Simons, E. L. (1977). Systematics, phylogeny, and evolution of early Eocene Adapidae (Mammalia: Primates) in North America. *Contr. Mus. Paleont. Univ. Mich.* 24, 245–279.

Gould, S. J. and Eldredge, N. (1977). Punctuated equilibria: the tempo and mode of evolution reconsidered. *Paleobiology* 3, 115–151.

Griffiths, G. C. D. (1972). The phylogenetic classification of the Diptera, Cyclorrhapha, with special reference to the postabdomen. *Series Ent.* 8, 1–340.

Gutmann, W. F., Vogel, K. and Zorn, H. (1978). Brachiopods: biomechanical interdependences governing their origin and phylogeny. *Science* **199**, 890–893.

Haeckel, E. (1866). "Generelle Morphologie der Organismen". G. Reimer, Berlin.

Harper, C. W. (1976). Phylogenetic inference in paleontology. *J. Paleont.* **50**, 180–193.

Hecht, M. K. and Edwards, J. L. (1976). The determination of parallel or monophyletic relationships: the proteid salamanders – a test case. *Am Nat.* **110**, 653–677.

Hennig, W. (1950). "Grundzüge einer Theorie der phylogenetischen Systematik" Deutsche Zentralverlag, Berlin.

Hennig, W. (1965). Phylogenetic systematics. *Ann Rev. Entolmol.* **10**, 97–116.

Hennig, W. (1966). "Phylogenetic Systematics". University of Illinois Press, Urbana.

House, M. R. (1979). Preface. *In* "The Origin of Major Invertebrate Groups" (M. R. House, ed.), pp. v–viii. Academic Press, London and New York.

Hull, D. L. (1980). The limits of cladism. *Syst. Zool.* **28**, 416–440.

Jarvik, E. (1960). Theories de l'evolution des vertebrés. Masson, Paris.

Jefferies, R. P. S. (1979). The origin of chordates – a methodological essay. *In* "The origin of Major Invertebrate Groups" (M. R. House, ed.), pp. 443–477. Academic Press, London and New York.

Kavanaugh, D. H. (1975). Hennigian phylogenetics in contemporary systematics: principles, methods, and uses. *Beltsville Symposia in Agricultural Research* **2**, 139–150.

Kier, P. M. (1977). The poor fossil record of the regular echinoid. *Paleobiology* **3**, 168–174.

Kitts, D. B. (1977). Karl Popper, verifiability, and systematic zoology. *Syst. Zool.* **26**, 185–194.

Løvtrup, S. (1977). "Phylogeny of the Vertebrata". Wiley, London.

Macbeth, N. (1973). "Darwin Retried". Dell Publishing Co., Boston.

McKenna, M. C., Engelmann, F. and Barghoorn, S. F. (1977). Cranial anatomy and evolution of early Tertiary Plesiadapidae (Mammalia, Primates). Review. *Syst. Zool.* **26**, 233–238.

Manton, S. M. and Andersen, D. T. (1979). Polyphyly and the evolution of arthropods. *In* "The Origin of Major Invertebrate Groups" (M. R. House, ed.), pp. 269–321. Academic Press, London and New York.

Mayr, E. (1965). Classification and phylogeny. *Am. Zool.* **5**, 165–174.

Mayr, E. (1969). "Principles of Systematic Zoology". McGraw Hill, New York.

Mayr, E. (1974). Cladistic analysis or cladistic classification? *Z. Zool. Syst. Evol. Forsch.* **12**, 94–128.

Muir-Wood, H. M. (1955). "A History of the Classification of the Phylum Brachiopoda". British Museum (Natural History), London.

Nelson, G. J. (1969a). Gill arches and phylogeny of fishes, with notes on the classification of vertebrates. *Bull Am. Mus. nat. Hist.* **141**, 475–552.

Nelson, G. J. (1969b). The origin and diversification of teleostean fishes. *Ann. New York Acad. Sci.* **167**, 18–30.

Nelson, G. J. (1970). Outline of a theory of comparative biology. *Syst. Zool.* **19**, 373–384.

Nelson, G. J. (1972). Comments on Hennig's "Phylogenetic Systematics" and its influence on ichthylogy. *Syst. Zool.* **21**, 364–374.

Nelson, G. J. (1978). Classification and prediction: A reply to Kitts. *Syst. Zool.* **27**, 216–218.

Nelson, G. J. (1979). Cladistic analysis and synthesis: principles and definitions, with a historical note on Adanson's Familles des Plantes (1763–1764). *Syst. Zool.* **28**, 1–21.

Newell, N. D. (1959). The nature of the fossil record. *Proc. Am. Phil. Soc.* **103**, 264–285.

Olsen, E. C. (1971). "Vertebrate Paleozoology". John Wiley, New York.

Patterson, C. (1977). The contribution of paleontology to teleostean phylogeny. *In* "Major Patterns of Vertebrate Evolution" (M. K. Hecht, P. C. Goody and B. M. Hecht, eds), pp. 579–643. Plenum Press, New York.

Patterson, C. (1978a). Arthropods and ancestors. *Antenna, Bull. R. Ent. Soc., London* **2**, 99–103.

Patterson, C. (1978b). Verifiability in systematics. *Syst. Zool.* **27**, 218–222.

Patterson, C. and Rosen, D. E. (1977). Review of ichthyodectiform and other Mesozoic teleost fishes and the theory and practice of classifying fossils. *Bull. Am. Mus. nat. Hist.* **158**, 81–172.

Platnick, N. (1977). Cladograms, phylogenetic trees and hypothesis testing. *Syst. Zool.* **26**, 438–442.

Platnick, N. (1980). Philosophy and the transformation of cladistics. *Syst. Zool.* **28**, 537–546.

Raup, D. M. (1976). Species diversity in the Phanerozoic. *Paleobiology* **2**, 289–297.

Raup, D. M. (1979). Biases in the fossil record of species and genera. *Bull. Carnegie Mus. nat. Hist.* **13**, 85–91.

Raup, D. M. and Stanley, S. M. (1971). "Principles of Paleontology". Freeman & Co., San Francisco.

Riedl, R. (1979). "Order in Living organisms". J. Wiley, Chichester.

Romer, A. S. (1966). "Vertebrate Paleontology". 3rd ed. Chicago University Press, Chicago.

Rosen, D. E. (1978). Darwin's Demon. *Syst. Zool.* **27**, 370–373.

Rosen, D. E., Forey, P. L., Gardiner, B. G. and Patterson, C. (1981). Lungfishes, tetrapods, paleontology and plesiomorphy. *Bull. Am. Mus. nat. Hist.* **167**, 163–275.

Ross, H. H. (1974). "Biological Systematics". Addison-Wesley, Reading, Mass.

Rudwick, M. (1970). "Living and Fossil Brachiopods". Hutchinson University Library, London.

Schaeffer, B., Hecht, M. K. and Eldredge, N. (1972). Phylogeny and paleontology. *Evolut. Biol.* **6**, 31–46.

Schlee, D. (1971). Die Rekonstruction der Phylogenese mit Hennig's Princip. *Aufsätze Reden Senckenberg. Naturf. Ges.* 20, 1–62.

Scudder, G. G. E. (1974). Species concepts and speciation. *Can. J. Zool.* 52, 1121–1134.

Sepkowski, J. J. (1978). A kinetic model of Phanerozoic taxonomic diversity I. Analysis of marine orders. *Paleobiology* 4, 223–251.

Simpson, G. G. (1953). "The Major Patterns of Evolution". Columbia University Press, New York.

Simpson, G. G. (1961). "Principles of Animal Taxonomy" Columbia University Press, New York.

Simpson, G. G. (1980). "Splendid Isolation". Yale University Press, New Haven.

Smirnova, T. N. and Pajaud, D. (1968). Contibution à la connaissance des Thécidées (Brachiopoda) du Cretace d'Europe. *Bull. Soc. geol. France* (7th series) 10, 138–147.

Sokal, R. R. and Crovello, T. J. (1970). The biological species concept: a critical evaluation. *Am. Nat.* 104, 127–153.

Sokal, R. R. and Sneath, P. H. A. (1963). "Principles of Numerical Taxonomy". Freeman, San Francisco.

Sylvester-Bradley, P. C. (1956). The new palaeontology. *In* "The Species Concept in Palaeontology" (P. C. Sylvester-Bradley, ed.), pp. 1–8. Academic Press, London and New York.

Szalay, F. S. (1977). Ancestors, descendants, sister groups and testing of phylogenetic hypotheses. *Syst. Zool.* 26, 12–18.

Szarski, H. (1962). The origin of the Amphibia. *Q. Rev. Biol.* 37, 189–241.

Tattersall, I. and Eldredge, N. (1977). Fact, theory, and fantasy in human paleontology. *Am. Scient.* 65, 204–211.

Valentine, J. W. (1969). Patterns of taxonomic and ecological structure of the shelf benthos during Phanerozoic time. *Palaeontology* 12, 684–709.

Valentine, J. W. (1973). Coelomate superphyla. *Syst. Zool.* 22, 97–102.

Valentine, J. W. (1975). Adaptive strategy and the origin of grades and ground plans. *Am. Zool.* 15, 391–404.

Walmsley, V. G. and Boucot, A. J. (1971). The Resserellinae – a new subfamily of late Ordovician to early Devonian dalmanellid brachiopods. *Paleontology* 14, 487–531.

Westoll, T. S. (1956). The nature of the fossil species. *In* "The Species Concept in Palaeontology" (P. C. Sylvester-Bradley, ed.), pp. 53–62. The Systematics Association, London.

Whittington, H. B. (1979). Early arthropods, their appendages and relationships. *In* "The Origin of Major Invertebrate Groups" (M. R. House, ed.), pp. 253–268. Academic Press, London and New York.

Wiley, E. O. (1975). Karl R. Popper, systematics, and classification: a reply to Walter Bock and other evolutionary taxonomists. *Syst. Zool.* 24, 233–243.

Wiley, E. O. (1979a). Ancestors, species, and cladograms – remarks on the symposium. *In* "Phylogenetic Analysis and Paleontology". (J. Cracraft and N. Eldredge, eds), pp. 211–225. Columbia University Press, New York.

Wiley, E. O. (1979b). An annotated Linnaean hierarchy, with comments on natural taxa and competing systems. *Syst. Zool.* 28, 308–337.

Williams, A. and Hurst, J. M. (1977). Brachiopod evolution. *In* "Patterns of Evolution" (A. Hallam, ed.), pp. 79–121. Elsevier Scientific, Amsterdam.

Williams, A. and Rowell, A. J. (1965a). Evolution and phylogeny. *In* "Treatise on Invertebrate Paleontology. Part H. Brachiopoda" (R. C. Moore, ed.), pp. H.164–199. University of Kansas Press and The Geological Society of America, Kansas.

Williams, A. and Rowell, A. J. (1965b). Classification. *In* "Treatise on Invertebrate Paleontology, Part H, Brachiopoda" (R. C. Moore, ed.), pp. H214–237.

Wilson, E. O. (1965). A consistency test for phylogenies based on contemporaneous species. *Syst. Zool.* 14, 214–220.

Wright, A. (1979). Brachiopod radiation. *In* "The Origin of Major Invertebrate Groups" (M. R. House, ed.), pp. 235–252. Academic Press, London and New York.

Yochelson, E. L. (1979). Early radiation of Mollusca and mollusc-like groups. *In* "The Origin of Major Invertebrate Groups" (M. R. House, ed.), pp. 323–358. Academic Press, London and New York.

Ziegler, A. M. (1966). The Silurian brachiopod *Eocoelia hemisphaeria* (J. de C. Sowerby) and related species. *Palaeontology* 9, 523–543.

# 5 | Evolutionary Trends and the Phylogeny of the Agnatha

## L. B. HALSTEAD

*Departments of Geology and Zoology, University of Reading,*
*Reading, England*

Abstract: The controversy between cladists and evolutionary systematists is essentially a re-run of the arguments put forward in the last century by Engels and Darwin. Cladistics is claimed by its adherents to constitute part of science as defined by Popper, whereas evolutionary theory and palaeontology are beyond the boundary of science. This latter is merely a reflection of Popper's irrational attitude towards the historical sciences.

The difficulty in recognizing primitive and advanced features which is the prerequisite for constructing cladograms is illustrated by a consideration of the evolutionary trends in agnathan dermal armour. It is demonstrated that trends towards fusion as well as breakdown both took place at the same time among contemporary groups. The cladogram of agnathan relationships presented by Janvier and Blieck (1979) which purports to show that heterostracans were more closely related to the hagfish than to the gnathostomes is discussed in detail and the alternative view that the heterostracans were more closely related to the gnathostomes is considered to be better supported by the available data. An alternative cladogram is constructed. It is demonstrated that as more data becomes available, the cladogram becomes more complex and unwieldy. It is concluded that the construction of family trees is a more valuable method of portraying suggested relationships than cladograms.

### INTRODUCTION

In recent years there has been a revival of interest in phylogenetic reconstruction which can be attributed in large measure to the

Systematics Association Special Volume No. 21, "Problems of Phylogenetic Reconstruction", edited by K. A. Joysey and A. E. Friday, 1982, pp. 159–196, Academic Press, London and New York.

publication of Hennig's "Phylogenetic Systematics" (1966). Although designed for living animals, insects especially, Hennig's cladistic methodology has been applied to the fossil record, one of the more notable contributions being that of Patterson (1977) on teleosts; he stressed that a living model was the prime prerequisite before any real advance could be achieved. It was assumed that cladistics could hardly be applied to groups with no living representatives. However, Miles (1973, 1977) showed how the approach could be used with placoderms and acanthodians.

One of the great barriers to understanding cladistics was the terminology of numerous neologisms with which it was expressed. This has been overcome to some extent in the major work "Dinosaurs and Their Living Relatives" (Anon. 1979) (see also Halstead 1980b). Cladistics is being adopted by a number of palaeontologists, although it is firmly rejected by others (see Halstead 1978b; Gardiner *et al.* 1979; Halstead *et al.* 1979; Panchen 1979; Fink and Wiley 1979). Cladists claim that their methodology is to be preferred because it is consistent with the hypothetico-deductive method propounded by Karl Popper and that, furthermore, cladistics falls within Popper's definition of science unlike the practices of "evolutionary systematics". Popper seemed to provide the intellectual justification for the general approach to phylogeny, systematics and even the policy of the form of display in the public galleries in the Natural History Museums which takes an avowedly cladistic stand (Halstead 1978a).

It seemed to me that cladistics was wrong in its approach and if this was being justified on the ground that it was consistent with Popper then there were only two possible explanations; either the cladists had misunderstood Popper or Popper was in error. By far the easiest introduction to Popper is Magee (1973). It was evident that Popper had something to say and that since it appeared so influential among palaeontologists then there was clearly no alternative but to read Popper ("The Logic of Scientific Discovery" (1959), "The Poverty of Historicism" (1957), "Objective Knowledge" (1972) and "Conjectures and Refutations" (1963) as well as his autobiography "Unended Quest" (1976). Popper is a philosopher and his attitude towards his own discipline is instructive. If one changes "philosophy" for "science" one gets the following:

> There is no such thing as an essence of science, to be distilled and con-
> densed into a definition. A definition of the word "science" can only

have the character of a convention, of an agreement; and I, at any rate see no merit in the arbitrary proposal to define the word "science" in a way that may well prevent a student of science from trying to contribute, *qua* scientist, to the advancement of our knowledge of the world (Popper 1959).

Yet when it comes to science he proposes a definition which by its very nature automatically excludes the historical sciences. Popper (1976) reached "the conclusion that Darwinism is not a testable scientific theory but a metaphysical research programme" (see Halstead 1980a).

The notion that the theory of evolution is not scientific but is metaphysical is being promoted apparently on the grounds that such a view is consistent with Popper (Patterson 1978b). On the other hand, Hennig provides a method of producing phylogenetic reconstructions which can be "tested" according to the precepts of Popper's scientific hypothetico-deductive method. This is spelt out most clearly by Gaffney (1979), and in the summaries of Popper's precepts by Platnick and Gaffney (1977).

Hennig thus allows a minor aspect of palaeontology to return to the fold of science having passed the critical test of falsifiability which is Popper's line of demarcation between science and non-science. Bonde (1975) and Patterson (1978a) almost alone among the cladists have recognized that the sort of testing by comparing cladograms is not what Popper had in mind.

Popper's deep emotional antipathy to the historic sciences stems from his disenchantment with communism and in his rejection of both the practice and theory of Marxism or historical materialism which is dialectical materialism applied to human society. He felt compelled by the same token to exclude from science all the disciplines with a time component. If all the historical sciences were to be excluded by definition from science, then there could be no way for dialectical materialism to claim its methods were scientific and hence Marxism would be unable to claim that it was based on the scientific study of history. This, I believe, was the underlying motive of Popper's attitude. But here we come to the irony of the situation, we seem to be acquiring a new dogma and one moreover which seems to be letting in dialectical materialism by the back door, in the guise of cladism.

The principle source of discord between cladists and evolutionary systematists is the nature of change. G. G. Simpson appears to be the main bogeyman of the cladists because he firmly upholds the notion of gradualism in the evolutionary process. The theory of punctuated equilibria expounded by Eldredge and Gould (1972) is much more consistent with cladistic analysis (Cracraft 1979). The two underlying concepts were propounded with great clarity by Frederick Engels in "Anti-Dühring" (1878) and "The Dialectics of Nature" (1925). He recognized the great value of the contributions made by geology in establishing that there was constant movement and change in nature and the significance of Darwin's demonstration that this applied also to the organic world. The main tenets of dialectical materialism are not only unexceptional, their application can be of tremendous value in the natural sciences. The theory states three fundamental premises: (1) In order to understand a phenomenon it must be considered in its relationship to the world around, i.e. it cannot be understood in isolation; (2) furthermore, it should be seen in its stage of development, whether it is something arising or dying away; (3) finally, it must be recognized that there is a development from imperceptible quantitative changes to qualitative changes. The crux of the entire theoretical framework, however, is in the nature of qualitative changes. This is also spelt out by Engels in "The Dialectics of Nature" — "a development in which the qualitative changes occur not gradually but rapidly and abruptly, taking the form of a leap from one state to another". In "Anti-Dühring" Engels states "at definite nodal points, the purely quantitative increase or decrease gives rise to a qualitative leap".

Darwin (1859) had a different view of the process of evolution, he considered that it was a gradual process and the gaps in the fossil record and the absence of missing links he attributed to the imperfections of the fossil record. In any event the fossil record did sketch a broad canvas of what had happened in the past and in a general way the sequence of changes that had occurred. Even Engels (1925) felt constrained to admit that

> hard and fast lines are incompatible with the theory of evolution. Even the borderline between vertebrates and invertebrates is no longer rigid, just as little is that between fishes and amphibians, while that between birds and reptiles dwindles more and more every day. Between *Compsognathus* and *Archaeopteryx* only a few intermediate links are wanting.

His deeply held conviction that there were sudden qualitative changes found little support in the evidence of evolution and his attempt out of this dilemma does not sound at all convincing. "Intermediate links prove only that there are no leaps in nature, *precisely because* nature is composed entirely of leaps" (Engels 1925).

The fact that quantitative changes can be shown to lead to qualitative changes cannot be denied. The question at issue is whether the process is gradual and hence the point of transition is in the final analysis an arbitrary one drawn across what is in reality a continuum or whether there really is a sudden qualitative leap. Whether the history of life is fundamentally evolutionary or revolutionary, whether the evidence supports gradualism or punctuated equilibria.

I would contend that whenever sufficient evidence is available (see for example Gingerich 1979) the pattern that emerges is a gradualist one. The main arguments that devolve around cladistics are a re-run of the fundamental dichotomy between Darwin (and Simpson) on the one hand and Engels (and Hennig) on the other.

The position of Popper in all this is one of supreme irony. By virtue of his blind antagonism towards Marxism and his complete lack of understanding of the historical sciences, he has unwittingly ended up providing intellectual respectability for what is in essence a brand of dialectical materialism within the natural sciences. The underlying theoretical framework of cladistics (as well as punctuated equilibria) is certainly unsound. Nevertheless, the advent of prosyletizing cladists has shaken many out of their complacency and in this sense has provided a useful service obliging many of us to re-examine the basic assumptions of our disciplines. The currently fashionable worship of Popper is singularly inappropriate for anyone remotely connected with the historical sciences and is perhaps the most serious aspect of the current anti-Darwin and anti-evolution craze that has infected certain areas of palaeontology.

In spite of the underlying philosophical considerations the cladistic approach is capable of providing the occasional insight into particular problems. In order to construct cladograms only advanced or derived character states (synapomorphies) can be employed. In order to do this, it is first necessary to be able to distinguish advanced from primitive features. I will discuss the example of evolutionary trends in the dermal armour of agnathans to illustrate the very real difficulty of determining the nature of such characters in actual practice.

The living Agnatha, the cyclostomes, are naked with unpleasant habits, being either scavengers or parasites, whereas the fossil forms, the ostracoderms, lived a life without blemish and were covered in a bony armour. It is the study primarily of this armour from which most of our knowledge has been gained. It is possible to divide the different types of fossil into clearly defined groups and it has been recognized from very early on that some bony carapaces are composed of a solid, single bony covering and in others of numerous separate elements or simply a few separate elements. There has been a controversy for many years as to whether the evolutionary trend was towards the break-up of a simple carapace into separate units or the opposite, whether the trend was towards fusion or breakdown. Rohon (1892) at first favoured the idea of fusion but later (1893) changed his mind and plumped for breakdown. This simply illustrated that the evidence could be viewed in two different ways even by the same man within a year. Traquair (1899) believed that the trend was from small denticles which fused to form small plates and these subsequently fused to form large plates, the alternative view he considered seemed to be "rather like putting the cart before the horse". Jaekel (1911) was the next to postulate that from a simple carapace there was a subsequent breakdown but this view received little support until it was put forward by Stensiö (1927) and such was the power of his advocacy that this view came to be the received wisdom for a generation.

As an undergraduate my major interest was in Palaeozoic Agnatha and, as my main training was in geology, when it came to considering this particular topic I began by considering the stratigraphic sequence of forms as a working hypothesis. The nature of the Ordovician genera *Astraspis* and *Eriptychius* seemed to me to provide convincing support for the view that the main evolutionary trend was towards fusion. It suggested that the development of the bony plates was towards their fusion and not their division.

Obruchev (1945) considered that the zones of tesserae in the drepanaspids and psammosteids were not the surviving remnants of a primitive condition but were a neoformation. Obruchev had already suggested that the trend was towards fusion from his study. Later when I met Stensiö in 1956 he delightedly regaled me with the fact that he had changed his mind with regard to the main evolutionary trend and that now he was firmly convinced that the trend was towards fusion and apparent simplicity. Stensiö (1958) recognized that there were three basic kinds of armour that could be distinguished and that these represented three contrasting types of growth. There were carapaces in which the plates were formed from initial primordia around which further elements were added by accretion of cyclomorial growth, forms in which the plates appeared simultaneously or synchronomorially, in a regular pattern, and finally plates which showed evidence of cyclomorial or areal growth. It was on the basis of these types of growth that he distinguished different groups of heterostracans.

As a result of studying the heterostracan *Corvaspis* I arranged the variation in the ornamentation in a sequence of different types (Tarlo 1960a). In some specimens the plates were divided up into separate units that must have formed synchronomorially, in other parts they were separated by zones of small tubercles that looked as if they had formed after the main units and were an indication of subsequent cyclomorial or areal growth. Occasionally there were specimens in which the ornamentation ran along the length of the plate with no hint of any superficial subdivisions. In all cases there was a narrow zone of cyclomorial growth around the entire plate. From these observations of variation all within a single species from a single locality I produced a hypothesis suggesting that the three main types of growth and ornamentation recognized by Stensiö could be shown to be related (Tarlo 1960a). This was confirmed by the finding of small tesserae, as well as a complete plate, of *Kallostrakon macanuffi* which occurred in the underlying geological formation and showed cyclomorial growth of the small elements. The theory I propounded was that the primitive condition was where the armour began with numerous isolated units (primordia) around which areal growth of further units developed until they met to form a terrazzo. This pattern then appeared synchronomorially as for example in *Corvaspis* and in time produced large synchronomorial plates. The pattern of these plates, which must have been formed when the animals reached their definitive size, was subsequently acquired earlier and earlier in development so that although the pattern of plates was retained these increased in size as the animal grew by normal areal growth, that is cyclomorially. When the psammosteids which were the dominant agnathans of the Middle and Upper Devonian were studied, it could be seen that a different trend could be followed whereby plates which were formed by normal cyclomorial growth began to acquire a surrounding zone of synchronomorial units, which became more extensive until the entire surface of the median plates was so formed, and eventually these individual small units were developed as small cyclomorial units, beginning with a central primordium around which subsequent units were added on by areal growth cyclomorially until they abutted against adjacent tesserae to form a terrazzo. This pattern on the outer surface seemed not to differ in any way from the condition first observed in the Ordovician genera *Astraspis* and *Eriptychius*. There was, however, a fundamental difference. In the case of the latter psammosteids, the small tesserae formed a layer superficial to the underlying bony plate, which itself showed quite clearly evidence of overall cyclomorial growth. As the armour matured the superficial tesserae fused to the underlying plates.

The interpretation of these trends is perhaps little more than a redescription of observation. The theory which I developed was that of progressive fusion followed by subsequent breakdown at least as far as the superficial elements of the armour were concerned (Tarlo 1962, 1967).

The relationship of superficial tesserae to the underlying pattern of plates was dealt with by Westoll (1967) in the discussion of the tesserate condition in such arthrodires as *Radotina* as well as in the psammosteids. In contrast to my theory he suggested it was a phenomenon illustrating Holmgren's principle of delamination which was developed by Jarvik (1959). It is surely legitimate to attempt to synthesize both theories as the data on which both are based can be confirmed. The two views need not be incompatible. The fusion of small elements to form a pattern of plates and the subsequent sinking of these plates to deeper levels so that, in the case described by Westoll (1967), they end up as cartilaginous plates, is in complete accord with the principle of delamination as a developmental process. The redevelopment of superficial tesserae does present problems. These structures can be traced in certain groups up to their elimination and then there is evidence that they reappear again going through the same essential stages in which they were eliminated. It is difficult to imagine that their apparent identity does not imply some kind of ultimate genetic connection.

There is general agreement that the psammosteids, especially the well known *Drepanaspis*, has much the same pattern of plates as *Pteraspis* and one can envisage that there must be some kind of relationship. The key difference is that between the lateral and median plates there is a zone of small tesserae. This is certainly the case in the young stages of *Traquairaspis* and *Phialaspis* and these later in ontogeney become incorporated into the main plates. In *Drepanaspis* and the psammosteids these intervening fields of tesserae are retained. The origin of these fields can be explained in two ways: either this is a new invention or it is a surviving remnant of a primitive condition. The evidence is conflicting. Kutcher (1933) described a young *Drepanaspis* in which there were no tesserae and this has been used as evidence of the drepanaspids having been derived from a pteraspid-like ancestor. The immature psammosteids thus showed the evidence of their pteraspid ancestry. An alternative interpretation is that the young drepanaspid was an example of paedomorphosis and that the original tesserate condition was suppressed and only came back later in development. The situation in the basal Devonian *Weigeltaspis* where the carapace comprises a single dorsal plate, lateral branchials separated by a field of tesserae and an entirely tessellated ventral region supports the hypothesis that such primitive fields of tesserae are a retention of a primitive condition. There is certainly a relationship that links the psammosteids (including *Drepanaspis*), the traquairaspids and pteraspids as well as *Weigeltaspis*. The question is what exactly is this relationship.

On the basis of my analysis of the heterostracans, I continued to maintain the postulate that the cyathaspids represented a separate evolutionary lineage to that of the pteraspids and psammosteids (1973).

The simple type of carapace still appeared to me to represent the end line of a trend towards fusion of small units, although the amphiaspids could well have been derived from them by still further fusion. In spite of, for example, White's (1935) and Denison's

(1970) view that the cyathaspids and pteraspids were very close and that the latter could be derived from the former by break-up of the simple carapace, I did not accept this, as all the evidence as I saw it pointed to the trend towards fusion and not breakdown. The redevelopment of tesserae in the psammosteids was on top of and extra to an established pattern of plates and in no sense did it constitute any evidence for break-up of large simple plates into more numerous ones. If it could be demonstrated that the *Drepanaspis– Pteraspis* link was one in which the pteraspid pattern gave rise to the drepanaspid then it would put a different complexion on matters. Part of this particular problem has been effectively resolved by Dave Elliott who has described early pteraspids in which the pattern of plates is essentially cyathaspid but with regions clearly demarcated foreshadowing the pattern of pteraspid plates. To my mind Elliott has established beyond any shadow of doubt the close affinity between the pteraspids and cyathaspids. His evidence gives a clear indication that there must have existed an evolutionary trend towards breakdown. This does not in any way invalidate the evidence of fusion in other groups but certainly it suggests a greater degree of complexity than was formerly suspected.

The next significant discovery was of the Australian genera described by Ritchie and Gilbert-Tomlinson (1977) such as *Arandaspis*. Here was Ordovician material, contemporary if not older than the American *Astraspis* and *Eriptychius*, in which a simple cyathaspid-type carapace was developed and not one composed of tesserae. Here was evidence that the simple type of carapace as postulated by Jaekel was the starting-off point and from which by progressive break-up, which according to White (1935) would have conferred the advantage of added flexibility, could have given rise to the pteraspids and later still the psammosteids.

There is now clear evidence of both fusion and break-up and both processes can be documented in different groups. There is the distinct possibility that a tessellated carapace and a simple cyathaspid-type carapace may be equally primitive. I do not think there is sufficient evidence to be able to choose between fusion or breakdown as the basic pattern. Indeed, it is possible to put forward the hypothesis that both processes occurred simultaneously, albeit in different evolutionary lines. There is certainly evidence to support both hypotheses regarding evolutionary trends so that rather than trying to choose between one hypothesis and the other, why not accept both? This would imply that there were two types of carapace formation which occurred simultaneously.

The formation of a tessellated carapace, a partially tessellated or a "simple" one were equally possible and are equally likely to have occurred. In the laying down of a hard tissue there are a limited number of geometries possible and the likelihood is that none was especially favoured. Among the various early vertebrates all possibilities could have developed. This would seem to be the hypothesis which covers all the evidence which is presently available. The two conflicting hypotheses can be satisfactorily combined, both evolutionary trends can be recognized and their progress documented. An understanding of the process of formation of mesodermal hard tissues and their subsequent growth simply suggests that wherever there is more than one possible line of development open then all are likely to be taken, if they are at all viable. The conclusion to be drawn from all this is that it is not simple to identify the direction of a single evolutionary trend. The recognition of primitive or plesiomorphic conditions and advanced derived synapomorphies is not always as straightforward as one might hope.

### CYCLOSTOME NAKEDNESS – PRIMITIVE OR DERIVED

The relationship between the extinct ostracoderms and the cyclostomes, the hagfish and lampreys, has been, and still is, a subject of controversy. One fact over which there is complete agreement is that ostracoderms had a bony armour and the cyclostomes are naked. It is generally agreed that the vertebrates arose from some kind of organism not too dissimilar from the living cephalochordates. By outgroup comparisons they are considered to be the sister group of the vertebrates or craniates. The feature in this context which is important is that the cephalochordates do not possess any bony tissue. The acquisition of bone is taken as a major step in vertebrate development, yet the two groups acknowledged as the most primitive of living vertebrates are devoid of such a tissue.

Before it was realized that the ostracoderms were agnathans there was no problem. The cyclostomes were simply taken as examples of primitive vertebrates and their lack of any kind of armour was attributed to their being surviving remnants of the pre-bony vertebrate condition, in which the internal skeleton was still cartilaginous.

The discovery that the earliest vertebrates in the fossil record were heavily armoured led to a reappraisal of the condition of the living agnathans and the idea that they had secondarily lost their original armour became accepted as the received opinion that was rarely challenged. An analogy was drawn with the history of actinopterygian fishes where the most primitive living representatives demonstrate a secondary reduction of bone from their fossil precursors. Another aspect of the living forms was a consideration of their respective modes of life. The hagfishes are burrowing scavengers being part of the infauna of the bottom sediments in the oceans. The lampreys in the main are ectoparasites. An unarmoured condition, especially the ability to secrete copious amounts of mucus, would seem to be more advantageous for both these particular modes of life. The loss of both armour and paired fins was compensated for by the elongation of the body to an eel- or worm-like shape, so that control of movement could be accomplished by both vertical as well as lateral flexing of the body.

The general view, therefore, was that the living forms had lost the attributes from their armoured precursors as a consequence of taking up specialized modes of life and in fact their survival to the present-day was a direct result of them occupying these two particular ecological niches – scavengers or parasites, both extremely successful ways of obtaining a living. As White (1946) remarked, the vertebrates did not "spring forth like Pallas Athenae fully armed onto an unsuspecting world", there must have been unarmoured vertebrates before there were armoured ones. The Euphanerida were established as just such a group based on the original interpretation of Jamoytius (White 1946). This genus was assigned to the anaspid ostracoderms by Stensiö (1958) with little comment but when I re-examined the original material I was able to confirm that Jamoytius was in fact covered by well-developed ornamented scales (Tarlo 1960b). Subsequent discoveries by Ritchie (1960, 1968) confirmed this but also revealed the existence of a cartilaginous branchial basket of essentially lamprey-type. Furthermore, the arrangement of the gill openings was not compressed as in the anaspids but more akin to the situation of the lampreys. This led to the view that *Jamoytius* could well have been close to the origin of both the typical anaspids as well as the lampreys. Obruchev's (1945) view deriving the anaspids from the lampreys was always difficult to envisage in view of the concentration of the gills in the anaspids, but the discovery of the gill arrangement in *Jamoytius* made the origin of the lampreys from an unspecialized anaspid in the form of *Jamoytius* much more feasible, especially in view of the circular mouth of *Jamoytius*. The only outstanding problem was the loss of the paired fins, which were well developed in *Jamoytius* but of which there is not a sign in the lampreys. The discovery of Carboniferous lampreys (Bardack and Zangerl 1968, 1971) duly unarmoured but without apparently parasitic specializations, fitted in the general scenario. The evolutionary pattern may be thought to have been confirmed; at least it was not falsified. The evidence is incontrovertible, *Jamoytius* exists, so too do the anaspids and so do the Carboniferous lampreys, finally the living lampreys exist. What must also be patently obvious is that they are all related.

The evidence of nakedness and armouredness is not quite as easy to assess as the general opinion in the literature implies. It may well be that the primitive anaspids developed armour and therefore, evolved in two directions one leading to the typical anaspids and the other by the secondary loss of armour and paired fins to the lampreys. An alternative view which involves a greater economy of hypothesis is that the lamprey had never developed either armour or paired fins in the first place and hence never had the bother of getting rid of them. In this sense the lamprey is a surviving representative of a primitive type of vertebrate. *Jamoytius*, on this view, would have been a subsequent evolutionary development in which armour and paired fins developed and which became still further modified, including the structure of the mouth in the later typical anaspids. This hypothesis is certainly the more parsimonious, it involves less complications and obviates postulating the acquisition of structures only immediately to lose them again.

The purpose of this discussion is to demonstrate that what has been accepted fairly uncritically by one and all, including myself, is not necessarily justified.

## PHYLOGENY OF THE AGNATHA

Janvier and Blieck (1979) published a cladistic analysis of the phylogeny of the craniates claiming that the phylogeny they proposed was the most parsimonious one with respect to the available data on both extant and fossil material. Now it is frequently stated (for example, Fink and Wiley 1979) that "cladograms are in fact hypotheses about relative recency of common ancestry"; and Eldredge (1979) that "a cladogram is nothing more than a branching diagram depicting the nested pattern of synapomorphies [or shared derived characters] among the taxa under study".

So we have cladograms with their clearly defined data and clearly stated hypotheses, and since the rationale of cladograms is for them to be subjected to test then it is necessary to examine their construct in some detail.

Before performing this exercise, it is perhaps worth beginning with the agreed data. There are two groups of living agnatha: the lampreys and hagfish that share a number of features in common. The nature of the similarities may indicate merely the retention of primitive structures from an ultimate common ancestor. The differences, which may be attributed to later specializations or may be original, suggest that the hags and lampreys should be grouped as two independent groups. The fossil lampreys occasion no real difficulty and can satisfactorily be encompassed in the same overall major taxon as the living genera. Among the armoured fossil forms there is also general agreement on the basis of the dorsal nasohypophyseal and pineal openings and separate gill openings and in the case of the cephalaspids the structure of the brain and ear region. The cephalaspids and anaspids can be grouped with the lampreys but not the hagfish. The Chinese galeaspids seem to be best brigaded with the lampreys and cephalaspids in some way or other. Finally, the heterostracans and thelodonts are recognized as distinct from the lamprey group but their relationship to the hagfish on the one hand and the later jawed vertebrates on the other are very much a matter of controversy and much of this is emphasized by the application of cladistic methods in evaluating or recognizing synapomorphies.

There is one further issue on which there is some agreement, it is that we can gain important information with regard to the early possible development of the vertebrates by studying the non-vertebrate chordates, especially the cephalochordates, by outgroup comparison as these represent the "putative sister group" of the vertebrates.

In the following section the sequence of character states as presented by Janvier and Blieck (1979) is discussed in the order given by these authors (Fig. 1).

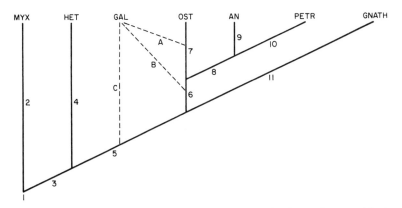

Fig. 1. Diagram illustrating opinion of Janvier and Blieck on the possible inter-relationships of fossil and living aganthans. AN, Anaspida; GAL, Galeaspida; GNATH, Gnathostomata; HET, Heterostraci; MYX, Myxinoidea; OST, Osteostraci; PETR, Petromyzonida; A, B, C, possible affinities of the Galeaspida. For explanation of numerals and discussion thereof see text below. (After Janvier and Blieck 1979.)

## 1. Janvier-Blieck Agnathan Cladogram

1: Primitive craniote condition (cephalization, terminal nasohypophysial opening, jaws absent, endodermal gill lamellae,? one semicircular canal). canal).

The acquisition of bilateral symmetry evident in the cephalochor-dates suggests the development of sense organs in the first ver-ebrates or craniates, i.e. olfactory, optic and otic senses, can be visualized as having taken place in different ways. All craniates have paired olfactory lobes and the regions of the brain concerned with sight and the sense of balance and hearing are similarly paired. When it comes to the sense organs themselves the matter is not quite as clear cut. The organs concerned with the detection of light include paired eyes together with an unpaired organ the pineal, the ear apparatus, the acousticolateralis system of lower vertebrates, is paired but again as in the cephalaspids can have an unpaired median element. The olfactory organ is even more difficult. Taking the simplest type of chemical-testing sense organ among the chordates with a further outgroup comparison, the neural gland of the ascidian tunicates is an unpaired structure. The hypophysis and the pituitary are similarly unpaired structures. The nasohypophyseal structures of the living agnathans could, from this standpoint, be considered as simply primitive or plesiomorphic features that have been retained from a pre-craniate condition. This feature could therefore be included as an integral part of the basic craniate pattern. There is of course the possibility that the initial craniate condition comprised paired olfactory organs, but this involves postulating the derivation of paired organs from a primitive unpaired condition and subsequently the secondary development of fused unpaired organs and according to Janvier and Blieck's hypothesis in doing this on two separate occasions. This is a rather involved procedure and it seems much more economic to postulate that the monorhinal condition of both lampreys and hagfish is the retention of a primitive pattern.

> 2: Backward migration of the branchial apparatus and concentration of the external branchial openings. "Monorhinal" condition of the olfactory organ. Slime glands. Elongate prenasal sinus.

The cephalochordates and presumably the primitive vertebrates are likely to have had a branchial apparatus that extended far posteriorly but the myxinoid backward migration is certainly a specialization. The concentration of the branchial openings may well be simply an adaptation for a burrowing organism and may not be of any particular phylogenetic significance. There is no evidence that the monorhinal condition arose *de novo* to produce the myxinoid condition. The formation of an elongated prenasal sinus as a new feature is postulated simply because it is assumed that in the primitive craniate condition the nasohypophyseal opening was terminal. There is in fact no evidence that this was necessarily the case. The myxinoids have a single semicircular canal in the ear region and this was included as a possible feature of the primitive craniate condition. In fact there are two ampullae and the general view is that the myxinoid condition is a derived one (see Halstead 1979).

> 3: Ossified exoskeleton. Two vertical semicircular canals. Vertebral elements? Lateral sensory line system.

The ossification of the exoskeleton was certainly a major innovation. The assumption here is that the type of mineralization was common to all the craniates. It is established that the deposition of crystallites of hydroxyapatite can take place on a variety of organic matrices such as glycosaminoglycans (acid mucopolysaccharides) and keratin as well as collagen (Halstead 1974). The occurrence of globular calcified cartilage as well as bone, dentine and aspidin suggests that there were different ways by which this was achieved. The heterostracans and thelodonts, the latter not figuring in the cladogram under discussion, are characterized by an acellular bone-like tissue, aspidin, together with a superficial tissue, dentine. In the cephalaspids the connective tissue within the cranium was mineralized but the nature of this is not entirely clear, although cellular bone seems to be an important constituent of the armour. The superficial ornamentation is composed of a specialized bone-like material to which the name mesodentine has been applied. However, it is not dentine in the same sense as the tissue found in the heterostracans, thelodonts and gnathostomes. The evidence at present suggests two different routes of hard tissue development. The distinctive feature attributed to the Heterostraci, that of a single common branchial external opening does not seem to be an especially significant feature. The thelodonts, which can best be brigaded with the heterostracans, seem to have had separate gill openings.

> 4: One single common external branchial opening on each side. Exoskeleton of the branchiocephalic region fused into large plates. Exoskeleton cancellar.

The characterization of the Heterostraci, having the exoskeleton of the branchiocephalic region fused into large plates is a feature that applied only to certain heterostracans, it is not a diagnostic feature of the group nor is the microstructure invariably cancellar.

> 5: True paired fins (with musculature and endoskeleton).? Photosensory pineal organ.? Extrinsic eye muscles. Loss of the preanal fold in adults.

The appearance of true paired fins is the next attribute given. These were not possessed by the galeaspids and were not present in such cephalaspids as *Tremataspis*. Furthermore, there is no convincing evidence of which I am aware to show that the cephalaspid flaps of say *Ateleaspis* and *Aceraspis* were homologous with the lateral fin-fold of the primitive anaspid *Jamoytius*. The possible photo-sensory pineal organ was not present in the galeaspids since in all known forms it was covered by the bony armour.

> 6: Dorsal migration of the naso-hypophysial opening. "Monorhinal" condition.

It seems difficult to claim the dorsal migration of the nasohypophyseal opening since it has not been established that it was not primitively in this position anyhow. To further postulate the development of the monorhinal condition at this stage implies that from a primitive monorhinal condition there arose in the first craniates the diplorhine condition from which the hagfish single nasal apparatus was formed, then subsequently this situation was repeated again and there was a second reversal to the monorhinal condition. This is a situation where Occam's Razor can be used to advantage.

By far the simplest hypothesis is to start with the primitive monorhinal condition and from it evolve the double structure once. On this interpretation, by far the most parsimonious, it is only necessary to invoke one change instead of three.

> 7: Forward migration of the branchial apparatus. Ossified endoskeleton forming a shield in the head region. Sensory or electric fields on the dorsal face of the shield.

The synapomorphies which distinguish the cephalaspids include the forward migration of the branchial apparatus and the formation of the lateral and dorsal sensory fields. With the diagnostic features of the cephalaspids there seems to be no disagreement. The ossification of the endoskeleton applies equally to the galeaspids and hence, although evidence of a relationship between these two groups exists, it cannot be used as a distinguishing feature of the cephalaspids alone.

8: Backward migration of the branchial apparatus. Branchial compartments developed into pouches in adults (but unlike those of the Myxinoidea). Hypocercal tail.

The backward migration of the branchial apparatus seems to be an entirely redundant character to postulate. The basic craniate condition was one in which numerous branchial elements extended some distance posteriorly. The most economical theory is simply to accept that at this level of the proceedings the branchial apparatus still retained its basic distribution. The development of gill pouches seems to have been already established in both the galeaspids and cephalaspids and so it seems a little strange to wait till this point to introduce this character state.

9: Paired fins extending far back, into a ventrolateral fin (also possible primitive myopterygian condition). Numerous external branchial openings (up to 15).

The paired fins, in this case the lateral fin-fold, extend backwards and there are numerous gill openings. This implies an increase yet such a number is more likely to have been the primitive condition. It seems unnecessary to postulate the reduction of branchial elements from say the cephalochordate condition only to have to postulate a further increase.

10: Loss of the paired fins. Development of a sucking oral disk. Loss of the exoskeleton.

The lamprey condition is now supposedly arrived at by the loss of the paired fins that have only lately developed and also by the loss of the exoskeleton. The number of branchial openings, the absence of a bony armour and paired fins could just as well be interpreted as original features and there has been no evidence put forward to support the contention that these structures have been secondarily lost. To fall back yet again on the principle of parsimony, it seems more economic a hypothesis to never have had structures, than to have acquired them only immediately to lose them, especially as there is no evidence forthcoming to support such a notion. The one feature, however, that does characterize the modern lampreys is the development of a sucking oral disk.

> 11: True jaws (formed by the premandibular and mandibular arches). Three semicircular canals. Ectodermal gill lamellae.

Finally the development of true jaws from premandibular and mandibular arches. This character state which is the common derived character of all the jawed vertebrates is certainly agreed but there is no evidence that this was the stage at which there developed three semicircular canals in the ear. Just as there is evidence that the hyoidean gill or spiracular gill was eliminated in certain agnathans there is no evidence that for example the Heterostraci had only two semicircular canals in the ear. That they had two is not in doubt, but the third or horizontal canal would not have been able to leave any impression or hint of its former presence on the dorsal carapace, hence there is no way in which this condition can ever be decided one way or the other. In consequence, it is simply not possible for any categoric statement to be made on the number of the canals in the heterostracan ear apparatus beyond that they had at least two.

The cladistic analysis presented by Janvier and Blieck (1979) has now been subjected to the sort of scrutiny which their presentation was intended to provoke.

The construction of the Janvier and Blieck cladogram or indeed any other is dependent entirely on the subjective assessment of certain characters being truly shared derived characters. As I have demonstrated, these are virtually all contentious and depending on one's personal assessment whether characters are derived or primitive, entirely different pictures will emerge. The main objection to the construction of such cladograms is the spurious impression of exactitude that is given.

## 2. Evolutionary Systematist Approach to the Agnatha

An alternative approach to the sudden appearance of new character states is that put forward by Strahan (1958) in what I consider to be one of the major contributions towards an understanding of the mutual relationships of the Agnatha. Strahan employed co-ordinate deformations for comparing the different agnathans. D'Arcy Thompson (1916) used the deformation of Cartesian co-ordinates to suggest the sequence of small changes by which one type of skull for example could be transformed into another. This technique is one which can be applied to situations where there is continuous gradual change and it is perhaps most clearly applicable in the consideration of the embryological development of chosen organisms. However, if one takes what Strahan terms the basic agnathan embryo or *bauplan* together with the main groups of Agnatha it is possible to indicate the embryological development from embryo to adult. With the living Agnatha, it is possible to produce an accurate pattern of the embryological development and also produce a hypothetical predictive pattern. The value of this is that it indicates the overall validity of the approach that allows the procedure to be applied to the fossil Agnatha. The oral hood of the ammocoete need not be considered as a hypertrophied larval adaptation, it could just as easily be taken as a normal stage in the development of the adult buccal funnel. When the anaspids are considered, it is possible for there to have been an ammocoete stage in development and Strahan concludes that the lampreys may well have arisen from near the base of the branch which gave rise to the anaspids.

The subsequent discovery of *Jamoytius* or rather its anaspid affinities, non-concentrated gill openings, round mouth and lamprey-like gill basket would support this view. There is no need however for Strahan to have postulated the posterior migration of the branchial apparatus and the loss of exoskeleton for the production of the lamprey line. These conditions could simply be the retention of primitive features. The position of the cephalaspids with their concentrated branchial apparatus and dorsal and lateral fields does not seem to support the developmental pathway in which an ammocoete oral hood was present (Fig. 2). The dorsal position of the nasophypophyseal opening in the cephalaspids is not due to the formation of an oral hood but rather to the anterior expansion and the forward migration of the branchial region, so that the

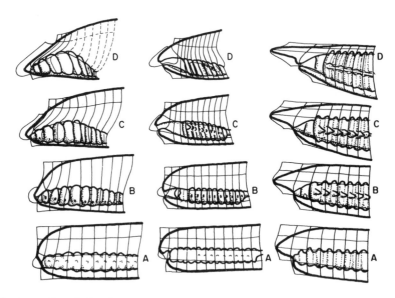

Fig. 2. Predicted development of fossil agnathans. Left to right: Cephalaspid *Kiaeraspis*, Anaspid *Pterolepis*, Heterostracan *Pteraspis*. A–C equidistant intermediate transformations between the basic agnathan embryo or *bauplan* and adult (D). (After Strahan 1958.)

nasohypophyseal complex "was not so much 'displaced' as 'left behind'" (Strahan 1958). The pathway of development would suggest that by the time cephalaspids diverged from the anaspid–lamprey line the ammocoete stage had evolved. Such a hypothesis would involve the formation of a mineralized tissue independently in the cephalaspids and anaspids. The histological studies that have been done show the bony material of the cephalspids and anaspids to be different and hence it is not unreasonable to suggest that they may not be homologous.

The discovery of the galeaspids (see Halstead, Liu and P'an 1979), which underwent an important evolutionary radiation in the Chinese province in Lower Devonian times, has added a new dimension to our understanding of the evolution of the Agnatha if only to demonstrate its complexity. There are two major groups of galeaspid: the Polybranchiaspidiformes and the Eugaleaspidiformes. The main distinction is that the eugaleaspids show an anterior concentration of the branchial apparatus as in the cephalaspids, whereas in the polybranchiaspids the primitive extended arrangement is retained. On this feature the eugaleaspids would be more advanced. On this character one could perhaps envisage the cephalaspid condition having arisen from that of the polybranchiaspids and this could imply that as far as the polybranchiaspids were concerned there could have been an ammocoete development stage during embryological development. The corollary of this is that the ammocoete condition was subsequently suppressed during the evolution of the cephalaspids and for that matter the eugaleaspids.

The overall similarity of cephalaspids and eugaleaspids may be an indication of a close relationship or it may be simply a consequence of similar adaptations to the same basic ecological niche. There is no evidence to decide the matter one way or the other. What is clear is that the mineralization of the connective tissue, the formation of endochondral ossification, is identical in both the polybranchiaspids and the cephalaspids. The possession of two semicircular canals in the ear, the large venous sinuses and the overall morphology of the brain are remarkably close. The cephalaspid specializations not present in the galeaspids are the dorsal and lateral sensory fields, but a new type of galeaspid under description by Pan and Wang shows two regions near the lateral margins of the dorsal carapace which seem to be possible to interpret as lateral sensory fields of cephalaspid type. All the evidence points towards a very close relationship of the galeaspids and cephalaspids.

The most striking feature of the galeaspids is the large median excavation on the dorsal surface, which although originally interpreted as a mouth in the polybranchiaspids is now accepted in both groups to have probably housed a nasohypophyseal organ (Halstead, Liu and P'an 1979). This excavation is extremely large and of variable outline, but by far the most curious aspect of it is that it was covered by a large plate during life. As well as this supposed sensory organ being covered, the pineal organ behind it was also covered by the bony carapace.

In both the cephalaspids and anaspids the nasohypophyseal opening and pineal organ appear to have been uncovered in the same way as both these organs are in the lamprey. There are two possible explanations of this situation: either all were originally covered by some structure which was subsequently lost in the cephalaspids, anaspids and lampreys and retained only in the galeaspids, or it was a specialized feature confined to the galeaspids and related to the environment they inhabited. The initial impression is that it is a derived character which distinguishes the galeaspids. However, there is a possibility that such coverings were present in the ancestral groups and were lost at the point where the cephalaspids as well as the anaspid–lamprey stock diverged from the primitive galeaspids or polybranchiaspids. This seems the less likely hypothesis but the existence of galeaspids with possible lateral sensory fields ensures that such a hypothesis is not entirely dismissed. We are thus left with the galeaspids and cephalaspids closely related to one another, but with no certain way of determining what character states are synapomorphies, at least not with any real confidence.

The heterostracans can be seen to be easily derived from a basic agnathan embryo or *bauplan* with perhaps the least amount of modification. There is certainly no displacement of the hypophysis to a dorsal position, simply the elongation of the rostral region. The olfactory apparatus is generally considered to be double and opened ventrally. There is no evidence whatsoever that the Heterostraci "inherited an inhalent nasohypophyseal duct from the early craniotes" (Janvier and Blieck 1979). There has never been any evidence that they even had a nasohypophyseal duct. What evidence there is suggests that such a structure did not exist. The lateral notches on the carapace in front of the eyes in certain cyathaspids (Kiaer and Heintz 1935), the impressions near the anterior internal margin of the carapace, have been generally interpreted as external nares and olfactory sacs. Stensiö (1958, 1964) with his determination to impose a myxinoid morphology on the heterostracans postulated that the notches were for sensory tentacles as in myxinoids, and the paired impressions were the position of cartilage. Heintz (1962) has demonstrated that even on Stensiö's own reckoning the hypothetical prenasal sinus, which Stensiö invented again to fit in with his notion of myxinoid affinities, was clearly a bilobate structure entirely different from the median prenasal sinus of myxinoids. The paired nasal capsules in many of the heterostracans must have opened into the buccal cavity but there is no necessity to postulate as Janvier and Blieck (1979) do, following Stensiö (1958), that "it is reasonable to assume that the olfactory organ was separated from the oral cavity by a wall or a membrane corresponding to the palatosubnasal lamina of the Myxinoidea".

The other heterostracan feature to which Janvier and Blieck draw attention is the presence of only two semicircular canals in the ear and in this regard they contrast this condition with that in the gnathostomes. This is not a valid point because the third semicircular canal is oriented horizontally and hence would be quite incapable of leaving an impression on the underside of the dorsal carapace. The presence in some of the Siberian amphiasphid heterostracans of a crescentic opening lateral to the orbits on the dorsal carapace has been interpreted as a spiracle (Halstead 1971) and as a prespiracular opening by Novitskaya (1971). On the undersurface there are impressions and depending on how they are interpreted the opening can be associated with a hyoidean or mandibular gill. When the basic embryology of vertebrates is examined the evidence for the existence of a mandibular gill is nowhere convincing and once again using Occam's Razor, the most reasonable interpretation is that the first gill, the hyoidean, has become modified to form a spiracle as a likely specialization for life in a muddy turbid environment. Janvier and Blieck (1979) simply state that this structure

> could be interpreted more parsimoniously. It may possibly represent inhalent openings secondarily derived from the more medial inhalent duct of the other Heterostraci, as a consequence of the flattening of the whole carapace.

To simply state that an opinion is more parsimonious is not sufficient, especially when it entails the migration of a median inhalent duct for which no evidence has been proferred, so that it divides in two and comes to be situated lateral to the orbits. It is not usual in primitive vertebrates for the olfactory organ to be concerned with inspiration, it is merely a water testing or tasting organ. With regard to the interpretation of paired structures on either side of the midline, between impressions left by the dorsal configuration of the central nervous system there are paired oval impressions subdivided longitudinally. Tarlo and Whiting (1965) interpreted these as impressions of somatic muscle blocks situated above gill structures on the basis that such an interpretation was consistent with vertebrate anatomy and embryology, whereas such a dorsal extent of gill structures as postulated by among others Watson (1954) was unknown among the vertebrates. Janvier and Blieck (1979) make some of these structures spinal nerves.

The perforations they describe in the inner surfaces of *Ctenaspis*, a minute cyathaspid, seem to mark the position on the outer surface of the junctions between branching parts of the sensory canal system, and cannot be in any way related visceral arches. There is one way of testing this interpretation: that is to paint the specimen with oil of aniseed so that the bony material will become translucent and the pattern of sensory canals can then be seen.

Regarding the affinities of the heterostracans, it has been recognized for over a century that they differ strikingly from the cephalaspids and their allies. There is one feature that is, however, notable, the overall shape of the carapace of the cyathaspids, polybranchiaspids and the primitive cephalaspid tremataspids are all remarkably similar, suggesting a primitive grade of organization. There are three hypotheses that have been championed; the first being that the heterostracans are not related to anything, and the second that they have myxinoid affinities, a view championed in different guises by Stensiö (1927, 1932, 1958, 1964). This subject has been dealt with in detail elsewhere (Halstead 1973b), suffice it to say that Stensiö's hypothesis became so fixed in his mind that in both the "Traités de Paléontologie" and "Zoologie", he invested the heterostracans with a suite of imaginary structures in their internal anatomy for which there was not one iota of evidence. The restorations were exceedingly beautiful but as T. H. Huxley remarked "one of the tragedies of science is the slaying of beautiful hypotheses by ugly facts". In spite of Stensiö's mistaken endeavour to impose a myxinoid structure on the heterostracans, he was unable to give them either a central nervous system or sense organs that were remotely comparable to myxinoids. There are at least two semicircular canals in the ear, there were two well-formed eyes and there was a prominent pineal organ. The configuration of the brain was comparable to that found in cephalaspids, galeaspids and primitive gnathostomes. There is little that can be said for attempting to establish any affinity between the hagfish and the heterostracans. It is gratifying that Janvier has moved away from this position although the Janvier–Blieck paper indicates that there is still room for further movement.

The third interpretation that has been put forward is that the heterostracans are nearer to the jawed vertebrates than they are to any of the other agnathan groups (Romer 1966, 1968; Halstead 1969, 1973). The double nasal sacs and elongated olfactory tract, Janvier and Blieck interpret as due to convergence. Janvier (1979) rejects double nasal sacs as of no consequence as they are plesiomorphic characters and only shared derived characters or synapomorphies can be used in determining relationships. He does not provide any evidence on which to determine the criteria or suggest any by which it can be established which condition is primitive and which derived. From outgroup comparisons the balance of evidence suggests to me that the double nasal sac is the more advanced condition, simply that a double structure is a more efficient directional aid than a median one and it is more difficult to envisage a reversion from an efficient double system back to a single less efficient one, which Janvier and Blieck postulate having occurred not just once but twice.

The nature of the gills of heterostracans has not been demonstrated to be of the pouch type associated with all the other agnathans and the Y- or V-shaped structures towards the lateral margins of the carapace have been interpreted as having been made by visceral arches of gnathostome pattern. The gnathostome type of gill can be derived directly from the vertebrate *bauplan* but not in any sense from the extant agnathan type. Admittedly the evidence for a gnathostome pattern in the heterostracans is not weighty but evidence for the alternative is non-existent. The formation of spiracles, if this interpretation of the amphiaspids is accepted, is a character state that is only known in heterostracans and gnathostomes. The one morphological feature that is the trademark of the heterostracans, and for that matter the thelodonts and the gnathostomes, is the tissue dentine. This is only found in these two groups and is unknown among any other agnathan group. If one insists in claiming the existence of a synapomorphy to link the heterostracans and gnathostomes, and this is what Janvier (1979) has challenged me to do, then I offer dentine.

The final question that remains in any attempt to determine the interrelationships of the Agnatha is the position of the myxinoids. They can be derived from the basic agnathan embryo but not along the same developmental pathway as that followed by the

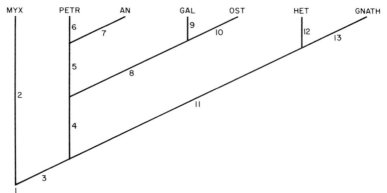

Fig. 3. Diagram illustrating possible interrelationships of fossil and living agnathans. AN, Anaspida; GAL, Galeaspida; GNATH, Gnathostomata; HET, Heterostraci; MYX, Myxinoidea; OST, Osteostraci; PETR, Petromyzonida.

1. Primitive craniate condition (cephalization, single nasal sac, paired eyes, acousticolateralis system, branchial openings extend posteriorly, membranous covering of brain).
2. Concentration of external branchial openings, gill pouches of endodermal origin, nasohypophyseal opening into pharynx, elongate prenasal sinus, pair of longitudinal lingual horny teeth, reduction of eyes, elongation of body.
3. Formation of braincase (cartilaginous), development of brain into fore- (telencephalon and diencephalon), mid- (mesencephalon) and hind- (cerebellum and medulla oblongata) regions, pineal organ (possible formation of calcium phosphate mineralized tissue but see item 6 below).
4. Formation of ectodermal gill pouches, separate gill openings, dorsal nasohypophyseal opening, two semicircular canals in ear.
5. Fusiform shape (possible development of paired fins, but see item 7 below).
6. Development of sucker mouth and rasping tongue (secondary loss of external skeleton, if previously acquired at 3, secondary loss of paired fins, if previously acquired at 5).
7. Development of paired fins, lateral fin fold (unless acquired at 5), external armour of acellular laminated bony plates (unless retained from 3), concentration of branchial apparatus.
8. Acquisition of cellular bone and mesodentine (unless acquired at 3), endochondral ossification.
9. Formation of cover plate over nasohypophyseal opening, pineal covered by armour.
10. Concentration of branchial apparatus, development of dorsal and lateral sensory fields, formation of pectoral flaps.
11. Formation of acellular bony tissue (aspidin and dentine) (unless acquired at 3), development of double nasal sacs.

galeaspid—cephalaspid—anaspid—lamprey route nor along the hetero-stracan—gnathostome line. There are certain features which give a hint of lamprey affinities and perhaps they can best be brigaded as a separate and independent line from a basic craniate stock, which retains more primitive features than the others. The central nervous system is unique among the vertebrates and they are best kept as a separate group with as yet no known relatives among the other Agnatha. Their primitive nature would ally them more with the cephalaspid groups than with the heterostracans. The nature of the configuration of the central nervous system suggests that the cephalaspids and heterostracans are closer to one another than either is to the hagfish. The fact that both the lampreys and hagfish have retained the monorhinal condition indicates that they are among the more primitive groups of the vertebrates, whereas the possession of double nasal sacs as well as dentine, the major con-stituent of vertebrate teeth, suggests that these are advanced features and that the Agnatha include three basic lineages one of which, the heterostracan, is closer to the gnathostomes than are either of the other two lineages.

CONCLUSIONS

In the main part of this chapter I have discussed the problem of working out evolutionary trends and how one can determine the direction of such trends and the problems involved in postulating whether features represent gains or losses. The cladistic methodology has been applied in attempting to establish the mutual relationships of the major agnathan groups. The hypotheses incorporated in the Janvier—Blieck (1979) cladogram do not seem to rest on solid foundations and in consequence I consider that their cladogram can be rejected. Such an exercise does not invalidate the cladistic methodology as such, and in order to facilitate discussion with cladistic colleagues I have presented my views in the form of cladograms (Figs 3—5). The first cladogram (Fig. 3) deals with the

---

12. Common lateral external branchial openings, pineal organ covered by armour (see item 9), main tissue of armour aspidin.

13. Development of jaws from mandibular arch and reduction of hyoidean gill to a spiracle, development of paired fins (see items 7 and 10).

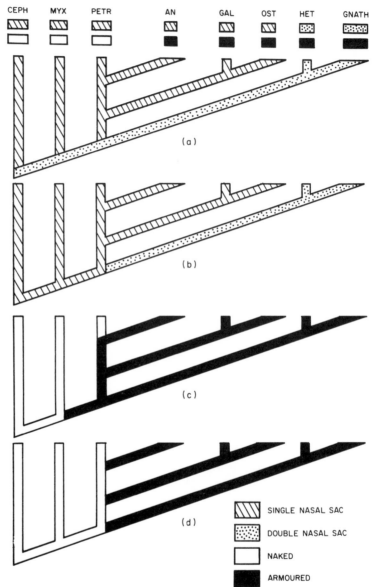

Fig. 4. Cladograms as in Fig. 3, with addition of CEPH Cephalochordata, of two pairs of characters distributed among groups as shown at top of figure. (a) Double nasal sac primitive; (b) double nasal sac derived; (c) armour appears before lampreys and is secondarily lost in lampreys; (d) lamprey retains primitive unarmoured condition, and armour appears independently in three lines.

identical groups of the Janvier–Blieck cladogram and deliberately excludes a number of key forms, which are important in trying to work out the interrelationships of these groups such as, for example *Jamoytius*, which Janvier and Blieck considered to be "more closely related to the Petromyzonida than to the typical Anaspida".

Although it is possible to construct cladograms to present, in a graphic manner, the substance of one's beliefs, much hinges on the recognition of which characters are considered to be plesiomorphies or synapomorphies. Taking the nasal sacs, if double nasal sacs are primitive features as claimed by Janvier and Blieck, then it is necessary to postulate the derivation of the myxinoid and cephalaspidomorph conditions having arisen independently. If the double nasal sac condition is accepted as a derived condition, as I contend, this view can in fact still be accommodated in the same cladogram. In either case the relationships as portrayed on the cladogram can remain the same. The dichotomies are all hypothetical and the evidence can be assessed in different ways within the same framework (Fig. 4).

A slightly more difficult case is seen when it comes to considering the significance of the lack of armour of the living agnathans. The view of Janvier and Blieck is that the hagfish condition is primitive and that the lamprey is secondarily derived from an armoured precursor. The alternative interpretation is that both are primitively naked. The corollary of this view is that it is necessary to postulate the independent origin of a mineralized tissue in three separate lineages. This view seems to be the least parsimonious but there is as yet no evidence of the lampreys having ever possessed an armour nor for that matter paired fins. On the other hand the hard tissues in the three armoured lineages seem to be histologically distinct. In spite of these two fundamentally different interpretations they can both be equally well accommodated in the same cladogram and do not in any way suggest any difference in the mutual relationships of the different groups (Fig. 4). The real difficulty in many cladograms is that they can obscure major areas of controversy. The portrayal of alternative schemes of interpretation as in Fig. 4 can usefully highlight such areas of dispute.

In discussing the possible relationships of major groups as in the Janvier and Blieck cladogram and my restructured version, the

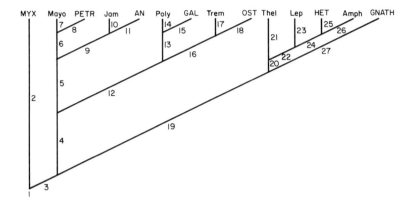

Fig. 5. Cladogram as in Fig. 3 with addition of further groups. AMPH, Amphiaspidiformes; JAM, *Jamoytius*; LEP, *Lepidiaspis*; MAYO, *Mayomyzon*; POLY, Polybranchiaspidiformes; THEL, Thelodonti; TREM, Tremataspidiformes.

1. Primitive craniate condition (cephalization, single nasal sac, paired eyes, acousticolateralis system, branchial openings extend posteriorly, membranous covering of brain).
2. Concentration of external branchial openings, gill pouches of endodermal origin, nasohypophyseal opening into pharynx, elongate prenasal sinus, pair of longitudinal lingual horny teeth, reduction of eyes, elongation of body.
3. Formation of braincase (cartilaginous), development of brain into fore- (telencephalon and diencephalon), mid- (mesencephalon) and hind- (cerebellum and medulla oblongata) regions, pineal organ, (possible formation of calcium phosphate mineralized tissue but see item 6).
4. Formation of ectodermal gill pouches, separate gill openings, dorsal nasohypophyseal opening, two semicircular canals in inner ear.
5. Round mouth, cartilaginous branchial basket.
6. Elongate body, cartilaginous lingual skeleton, annular cartilage around mouth.
7. Cannot be distinguished from 6.
8. Sucker mouth and rasping tongue (see item 2).
9. Armour, paired fins – lateral finfold.
10. Cannot be distinguished from 9.
11. Branchial apparatus compressed, mouth of variable shape – not round.
12. Acquisition of armour of cellular bone and mesodentine (unless acquired at 3), endochondral ossification.
13. Cover plate over nasohypophyseal opening, pineal covered.
14. Cannot be distinguished from 13.
15. Compression of branchial apparatus.

system allows the different underlying assumptions of different workers in the field to be graphically contrasted. However, when the fossil record provides further materials which are relevant in the construction of an overall phylogeny, then cladograms tend to become unwieldy. Each new piece of evidence necessitates a new dichotomy as a new character state is recognized. As the gaps in the fossil record are gradually filled so too do the branches of the cladogram proliferate. The discovery of the Carboniferous lamprey *Mayomyzon* and the Upper Silurian anaspid *Jamoytius*, the recognition of the two groups of galeaspid (the polybranchiaspids and eugaleaspids), the position of the tremataspids as well as the thelodonts, and the genus *Lepidaspis*, all allow the series of character states in the simple cladogram to be more finely separated so that the cladogram is inevitably extended. This may well be thought to be a useful exercise. If the aim is a graphical display of hypotheses then it gives an impression of increased complexity whereas the additional evidence in reality simplifies the overall picture. One of the problems that I encountered in constructing a more detailed cladogram (Fig. 5) was my inability to designate character states to enable dichotomy to be distinguished. For example at the dichotomy leading to the lampreys and anaspids I do not know how it can be possible to distinguish on morphological grounds

---

16. Concentration of branchial apparatus, development of lateral and dorsal sensory apparatus.
17. Cannot be distinguished from 16.
18. Development of pectoral flaps.
19. Formation of dentine (unless acquired at 3), development of double nasal sacs.
20. Formation of acellular bony tissue, aspidin unless acquired at 3.
21. Separate gill openings, paired lateral fins.
22. Dentine fused to aspidin plate.
23. Cannot be distinguished from 22.
24. Common branchial opening, pineal organ covered by armour (see item 13), main tissue of armour aspidin.
25. Cannot be distinguished from 24.
26. Development of spiracle from hyoidean gill.
27. Development of jaws from mandibular arch and reduction of hyoidean gill to a spiracle, development of paired fins.
Note: spiracle developed at both 26 and 27, concentration of branchial apparatus at 11, 15 and 16.

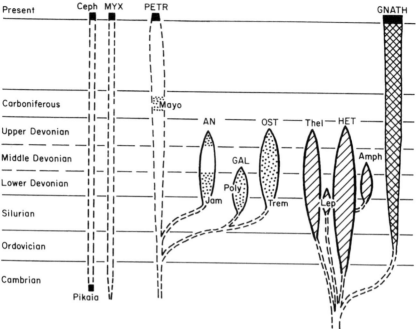

Fig. 6. Hypothetical family trees of agnathans (abbreviations as in Figs 3–5, with addition of possible Middle Cambrian representative *Pikaia*). Possible interrelationships shown by pecked lines. This family tree denotes the same degrees of relationships, as shown in cladograms (Figs 3–5).

the character state from which *Mayomyzon* and the lamprey can be derived and that of *Mayomyzon* itself. The same situation appertains between character states from which *Jamoytius* and the typical anaspids are derived, in what way can this be separated from the character state which distinguishes *Jamoytius*? The same situation occurs with the galeaspids and cephalaspids (osteostracans).

One of the valuable features which emerges from the more detailed cladogram is the recognition that within small segments of the cladogram the same derived character state can be arrived at independently such as the concentration of the branchial apparatus in advanced anaspids, the eugaleaspids and osteostracans.

The inability to distinguish between items 6 and 7, 9 and 10, 13 and 14 and 16 and 17 is a reflection of the underlying reality. From a basic *Mayomyzon* stock there arose the lampreys with their sucker mouth. The *Mayomyzon* stock experienced no morphological change, similarly with *Jamoytius* and with the polybranchiaspids. There is no way of morphologically characterizing the character states before and after the divergence of an advanced group from the main line.

The representation of overall relationships by means of the traditional family tree is not dissimilar from the pattern produced by a simplified cladogram except that the temporal aspect is excluded from the latter. Furthermore the tree emphasizes the difference between fact and conjecture (Fig. 6). The hypothetical links are shown as such and in this regard contrast dramatically with the construct of typical cladograms which give a spurious impression of exactitude to the unwary reader. Theoretically a cladogram is simply one worker's set of ideas or hypotheses and to delineate them so graphically would seem to be a commendable endeavour. Unfortunately such is human nature that such forthright presentations seem destined to inhibit discussion rather than provoke it. If the value of cladograms is to stimulate discussion then at least as far as the agnathans are concerned, cladistics can certainly claim a success on this score. Only by attempting to apply the methodology is it possible to uncover its limitations and rediscover the value of contructing family trees.

ACKNOWLEDGEMENTS

Mr Alan Cross drew the figures and Mrs Irene Gillett typed the text.

REFERENCES

Anon. (1979). "Dinosaurs and Their Living Relatives". British Museum (Natural History) and Cambridge University Press, London and Cambridge.
Bardack, D. and Zangerl, R. (1968). The first fossil lamprey: a record in the Pennsylvanian of Illinois. *Science* **162**, 1265–1267.
Bardack, D. and Zangerl, R. (1971). Lampreys in the fossil record. *In* "The Biology of Lampreys" (M. W. Hardisty and I. C. Potter, eds), pp. 67–84. Academic Press, London and New York.
Bonde, N. (1974). Review Interrelationships of fishes. *Syst. Zool.* **23**, 562–569.
Darwin, C. (1859). "The Origin of Species by Means of Natural Selection or the Preservation of Favoured Races in the Struggle for Life." John Murray, London.
Denison, R. H. (1970). Revised classification of Pteraspidae with description of new forms from Wyoming. *Fieldiana Geol.* **20**, 1–41.

Eldredge, N. (1979). Cladism and common sense. *In* "Phylogenetic Analysis and Paleontology" (J. Cracraft and N. Eldredge, eds), pp. 165–198. Columbia University Press, New York.

Eldredge, N. and Gould, S. J. (1972). Punctuated equilibria: an alternative to phyletic gradualism. *In* "Models in Paleobiology" (T. J. M. Schopf, ed.), pp. 82–115. Freeman, San Francisco.

Engels, F. (1878). "Anti-Dühring". Lawrence and Wishart, 1955, London.

Engels, R. (1925). "Dialectics of Nature". Lawrence and Wishart, 1955, London.

Fink, W. L. and Wiley, E. O. (1979). Cladism defended. *Nature, Lond.* 280, 542.

Gaffney, E. S. (1979). An introduction to the logic of phylogeny reconstruction. *In* "Phylogenetic Analysis and Paleontology" (J. Cracraft and N. Eldredge, eds), pp. 79–111. Columbia University Press, New York.

Gardiner, B. G., Janvier, P., Patterson, C., Forey, P. L., Greenwood, P. H., Miles, R. S. and Jefferies, R. P. S. (1979). The salmon, the lungfish and the cow: a reply. *Nature* 277, 175–176.

Gingerich, P. D. (1979). Stratophenetic approach to phylogenetic reconstruction in vertebrate paleontology. *In* "Phylogenetic Analysis and Paleontology" (J. Cracraft and N. Eldredge, eds). pp. 41–77. Columbia University Press, New York.

Halstead, L. B. (1969). "The Pattern of Vertebrate Evolution". Oliver and Boyd, Edinburgh.

Halstead, L. B. (1971). The presence of a spiracle in the Heterostraci (Agnatha). *J. Linn. Soc. (Zool.)* 50, 195–197.

Halstead, L. B. (1973a). The heterostracan fishes. *Biol. Rev.* 48, 279–332.

Halstead, L. B. (1973b). Affinities of the Heterostraci. *Bio. J. Linn Soc.* 5, 339–349.

Halstead, L. B. (1974). "Vertebrate Hard Tissues". Wykeham Publications, London and Winchester.

Halstead, L. B. (1978a). Whither the Natural History Museum? *Nature* 275, 683.

Halstead, L. B. (1978b). The cladistic revolution – can it make the grade? *Nature* 276, 759–760.

Halstead, L. B. (1979). Internal anatomy of the polybranchiaspids (Agnatha, Galeaspida) *Nature* 282, 833–836.

Halstead, L. B. (1980a). Popper: good philosphy, bad science? *New Scientist* 85, 215–217.

Halstead, L. B. (1980b). Review Dinosaurs and their living relatives. *Pal. Assoc. Circ.* 101, 10.

Halstead, L. B., White, E. I. and Macintyre, G. T. (1979). The salmon, the lungfish and the cow: a reply. *Nature* 277, 176.

Halstead, L. B., Liu, Y-H. and P'an, K. (1979). Agnathans from the Devonian of China. *Nature* 282, 831–833.

Heintz, A. (1962). Les organes olfactifs des Hétérostracés. *Collognes int. Cent. natn. Rech. scient.* 104, 13–29.

Hennig, W. (1966). "Phylogenetic Systematics". Chicago University Press, Chicago.

Jaekel, O. (1911). "Die Wirbeltiere". Berlin.

Jarvik, E. (1959). Dermal fin rays and Holmgren's principle of delamination. *Kugl. svenska Vetensk-Akad. Handl.* 6, 1–51.

Janvier, P. (1979). Cladism defended. *Nature* 280, 542.

Janvier, P. and Blieck, A. (1979). New data on the internal anatomy of the Heterostraci, with general remarks on the phylogeny of the Craniotes. *Zool. Scripta* 8, 287–296.

Kiaer, J. and Heintz, A (1935). The Downtonian and Devonian Vertebrates of Spitzbergen. V. Suborder Cyathaspida, Part 1, Tribe Poraspidei, Family Poraspidae Kiaer. *Skr. Svalb. Ishavet.* 40, 1–138.

Kutscher, F. (1933). Fossilien aus dem Hunsruckschiefer. I. *Jb. preuss. geol. Landesamt. Berg. Akad.* 54, 628–641.

Magee, B. (1973). "Popper". Fontana, Glasgow.

Miles, R. S. (1973). Relationships of acanthodians. *In* "Interrelationships of Fishes" (P. H. Greenwood, R. S. Miles and C. Patterson, eds), pp. 63–103. Academic Press, London and New York.

Miles, R. S. and Young, G. C. (1977). Placoderm interrelationships reconsidered in the light of new ptyctodontids from Gogo, Western Australia. *In* "Problems in Vertebrate Evolution" (S. M. Andrews, R. S. Miles and A. D. Walker, eds), pp. 123–198. Academic Press, London and New York.

Novitskaya, L. (1971). Les amphiaspides (Heterostraci) du Devonien de ia Siberie. *Cahiers Paleont. Paris* 1971, 1–130.

Obruchev, D. (1945). The evolution of the Agnatha. *Zool. Zh. Moscow* 24, 257–272.

Panchen, A. J. (1979). The cladistic debate continued. *Nature* 280, 541.

Patterson, C. (1977). The contribution of paleontology to teleostean phylogeny. *In* "Major Patterns in Vertebrate Evolution" (M. K. Hecht, P. C. Goody and B. M. Hecht, eds), pp. 579–643. Plenum, New York.

Patterson, C. (1978a). Verifiability in Systematics. *Syst. Zool.* 27, 218–222.

Patterson, C. (1978b). "Evolution" British Museum (Natural History), London.

Platnick, N. I. and Gaffney, E. S. (1977). Systematics: a Popperian perspective. *Syst. Zool.* 26, 360–365.

Popper, K. R. (1957). "The Poverty of Historicism". Routledge and Kegan Paul, London and Henley.

Popper, K. R. (1959). "The Logic of Scientific Discovery". Hutchinson, London.

Popper, K. R. (1963). "Conjectures and Refutations: The Growth of Scientific Knowledge". Routledge and Kegan Paul, London and Henley.

Popper, K. R. (1972). "Objective Knowledge: An Evolutionary Approach". Oxford University Press, Oxford.

Popper, K. R. (1976). "Unended Quest: An Intellectual Autobiography". Fontana, Glasgow.

Ritchie, A. (1960). A new interpretation of *Jamoytius kerwoodi* White. *Nature, Lond.* 188, 647–649.

Ritchie, A. (1968). New evidence on *Jamoytius kerwoodi* White, an important ostracoderm from the Silurian of Lanarkshire, Scotland. *Palaeontology* 11, 21–39.

Ritchie, A. and Gilbert-Tomlinson, J. (1977). First Ordovician Vertebrates from the southern hemisphere. *Alcheringa* 1, 351–368.

Rohon, J. V. (1892). Die obersilurischen Fische von Oesel. Thiel I: Thestidae und Tremataspidae. *Zap. imp. Akad. Nauk.* 38, 1–88.

Rohon, J. V. (1893). Die obersilurischen Fische von Oesel. Thiel II: Selachii, Dipnoi, Ganoidei, Pteraspidae und Cephalaspidae. *Zap. imp. Akad. Nauk.* 41, 1–124.

Romer, A. S. (1966). "Vertebrate Paleontology". University of Chicago Press, Chicago.

Romer, A. S. (1968). "Notes and Comments on Vertebrate Paleontology". University of Chicago Press, Chicago.

Stensiö, E. A. (1927). The Downtonian and Devonian Vertebrates of Spitzbergen. 1. Family Cephalaspidae. *Skr. Svalb. Ishavet* 12, 1–391.

Stensiö, E. A. (1932). "The Cephalaspids of Great Britain". British Museum (Natural History), London.

Stensiö, E. A. (1958). Les Cyclostomes Fossiles. In "Traité de Zoologie 13" (P. P. Grasse, ed.), pp. 173–425. Masson, Paris.

Stensiö, E. A. (1964). Les Cyclostomes Fossiles on Ostracodermes. *In* "Traité de Paleontologie 4" (J. Pivetean, ed.), pp. 96–385. Masson, Paris.

Strahan, R. (1958). Speculations on the evolution of the agnathan head. *Proc. Cent. Bicent. Congr. Biol. Singapore*, 83–94.

Tarlo, L. B. (1960a). The Downtonian ostracoderm *Corvaspis kingi* Woodward, with notes on the development of dermal plates in the Heterostraci. *Palaeontology* 3, 217–216.

Tarlo, L. B. (1960b). The invertebrate origin of the vertebrates. *Rept Int. Geol. Cong. XXI Sess. Norden* 22, 113–123.

Tarlo, L. B. (1962). Lignées évolutive chez les ostracodermes hétérostracés. *Colloques int. Cent. natn. Rech. scient.* 104, 31–37.

Tarlo, L. B. (1967). The tessellated pattern of dermal armour in the Heterostraci *J. Linn. Soc. (Zool.)* 47, 45–54.

Tarlo, L. B. H. and Whiting, H. P. (1965). A new interpretation of the internal anatomy of the Heterostraci (Agnatha). *Nature, Lond.* 206, 148–150.

Thompson, D'A. W. (1916). "On Growth and Form". Cambridge University Press, Cambridge.

Traquair, R. H. (1899). Report on fossil fishes collected by the Geological Survey of Scotland in the Silurian rocks of the South of Scotland. *Trans. R. Soc. Edinb.* 39, 827–864.

Watson, D. M. S. (1954). A consideration of ostracoderms. *Phil. Trans. R. Soc. Lond. B*, 238, 1–25.

Westoll, T. S. (1967). *Radotina* and other tesserate fishes. *J. Linn. Soc. (Zool.)* 47, 83–98.

White, E. I. (1935). The ostracoderm *Pteraspis* Kner and the relationships of the agnathous vertebrates. *Phil. Trans R. Soc. Lond.*, B, 225, 381–457.

White, E. I. (1946). *Jamoytius kerwoodi*, a new chordate from the Silurian of Lanarkshire. *Geol. Mag.* 88, 89–97.

NOTE ADDED IN PROOF

Following the publication of my article on Popper (Halstead 1980a), Popper replied (1980, *New Scientist* 85, 611) acknowledging the scientific status of the theory of evolution and the historical sciences.

# 6 | Fossils and Phylogeny - A Compromise Approach

R. A. FORTEY

and

R. P. S. JEFFERIES

*Department of Palaeontology, British Museum (Natural History), London, England*

Abstract: Two opposite schools have recently developed concerning the use of fossils in reconstructing phylogeny. Extreme cladists deny that fossils can ever be useful, while extreme stratopheneticists deny that primitiveness can ever be recognized without consulting stratigraphy. In this chapter we stress the need for compromise. We advocate an eclectic approach, arguing that both schools are right in some circumstances but wrong in others. Stratigraphy is all-important when dealing with differences of low burden (in Riedl's sense) between closely related species and with a good fossil record. On the other hand, stratigraphy can and should be ignored when dealing with differences of high burden between distantly related taxa where the record is poor. Between these two extremes both methods should be applied as seems appropriate. We have investigated the opposing influences of burden and percentage fossil record by using stochastic models.

## INTRODUCTION

Phylogeneticists have recently divided into two camps as concerns their attitude to fossils. On the one hand there are those who pursue the stratophenetic approach, with a conviction that evolution can still be read more or less directly from the rocks on the basis of the

Systematics Association Special Volume No. 21, "Problems of Phylogenetic Reconstruction", edited by K. A. Joysey and A. E. Friday, 1982, pp. 197–234, Academic Press, London and New York.

stratigraphic sequence of taxa found there; and on the other hand there are extreme cladists for whom the stratigraphic record is, at best, a check *a posteriori* on the latest possible date for the origin of synapomorphies. The former claim to recognize actual ancestors; the latter regard the identification of such as an impossibility. The two attitudes have become mutually exclusive in the heads of their protagonists, as a glance at the recently published account of another symposium will show (Cracraft and Eldredge 1979). Extreme cladism has sought scientific respectability by demonstrations of its conformity to the criteria elaborated by Karl Popper on the distinctions of "good" science from metaphysics. There have been attempts at methodological justification of stratigraphy-based lineages (Gingerich 1976; Bretsky 1979), but many invertebrate palaeontologists have been little affected by the debate, although there has been some vigorous, if ill-aimed carping (Campbell 1975). This chapter is about the need for compromise. We believe it can be shown that neither method is inherently better than the other. They are simply appropriate to different circumstances.

### AIMS OF PHYLOGENETIC ANALYSIS

Since evolution has happened, there will be one unique solution to the problem of the history of life. If this history were known there would be no arguments about phylogeny, only about nomenclature. There is only one correct answer concerning the time, place and taxonomic origin of all species, living and extinct. We shall refer to this as the correct tree. Nelson, Eldredge and others have observed that a cladogram is *not* equivalent to the correct tree; any cladogram corresponds to several tree interpretations. Statements about phylogeny are models about part of the correct tree. Since the tree includes (indeed consists largely of) ancestors, palaeontological evidence is essential to produce the complete tree. This is not to say that important phylogenetic statements cannot be made based on any selection of evidence (such as entirely Recent species); it is just that these cannot completely portray the historical course of events. Any phylogenetic statement is capable of being successfully refuted until such time as it acquires a one-to-one correspondence with the correct tree. It is hardly necessary to add that certain parts of the correct tree can never be known — those animals and plants with no fossil record, for example. But in other cases, with phyla such as Mollusca or Brachiopoda, there is no theoretical reason why large parts of the tree should not be observable. The information included in the correct tree exceeds that of any model for the tree (with the identification of ancestors and time of origin of synapomorphies included); hence the portrayal of the correct tree is the ultimate goal of phylogenetic analysis. It includes within it all other forms of phylogenetic presentation, whether cladograms or biostratigraphical trees. The success of any phylogenetic method can be defined by the extent to which it corresponds with the correct tree.

### CLADOGRAMS VERSUS STRATIGRAPHIC SPECIES LINEAGES

A phylogeneticist will prefer the cladogram which portrays the most parsimonious arrangement of shared, derived characters, believing that it is most likely to reflect correctly the evolution of the group under study. Cladograms are objective, if synapomorphies are correctly identified as homologous. The time dimension is represented in terms of recency of common ancestry, i.e. is a relative scale. Unless the cladogram is translated into a biostratigraphical tree (or trees) neither the historical time of origin of synapomorphies nor the actual ancestors can be resolved. Cladograms can relate any taxa (three or more) across any phylogenetic distance. Stratigraphic species lineages, on the other hand, attempt to show ancestor–descendent relationships in correct sequence for continuously related (i.e. taxonomically adjacent) species. The stratigraphic species lineages that are most likely to be adequate are based on continuous sampling from richly fossiliferous sequences deposited continuously. They deal with species-to-species transformations. Stratigraphy has been used in a much broader, laxer sense, however, to distinguish primitive from derived characters; and when the organisms compared are not closely related this is a misapplication, i.e. the phylogenetic relationships that it implies may well approximate less closely to the correct tree than a cladogram that uses morphology but intentionally ignores stratigraphy. This is discussed further below. The main point to establish here is that cladograms, based on derived character complexes, and stratigraphic species series, based on a sequence which must be *consistent with morphology*, can both be valid models for portions of the correct tree, but they will be valid in different circumstances.

Many palaeontological phylogenetic studies are concerned with species–species relationships. The best such studies are upon groups with a good fossil record, which for the moment may be taken to mean that the bulk of species that existed have been sampled, and that continuous stratigraphic sections containing them can be collected (Bretsky 1979). The procedure is generally as follows: presumed natural (i.e. genetically continuous) groups are identified from morphological criteria; the order in which the species appear then indicates an ancestor–descendent series. In some cases the splitting of a species into two can be identified in this manner, but many palaeontologists also recognize a more or less orthogenetic series of species, sometimes with gradual boundaries between species. Although the notion of such gradual change is less respectable today than formerly (Gould and Eldredge 1977) the literature has many examples (notably among graptolites, echinoids, ammonites and foraminifera) where change of this kind does seem to have happened. Regardless of whether the boundaries between species in a presumed phylogenetic–morphological series are saltations or gradual, the *characters* involved are usually of the most superficial kind, relating to such features as rib number, surface ornamentation, overall size, and the like. Palaeontologists concerned with this kind of species-to-species change insist on the primacy of stratigraphy. The usual implication is that there is no *a priori* way to determine the direction in which the characters would trend, i.e. the stratigraphic order, if the geological circumstances are right, is used to determine primitive and advanced character states at the species level. In some cases the stratigraphic sequence is permitted to overturn previously established notions of primitive and advanced character states. In the ammonites, for example, relatively uncoiled forms were always considered advanced (by comparison with the usual tightly coiled forms). But Casey showed that some ammonites could secondarily coil up. This was established from a stratigraphic species-lineage with progressive secondary coiling. We discuss below an example where trilobite genal spines, primitive for the group as a whole, are secondarily re-acquired in a group that otherwise lacks them, and where this secondary, derived, pseudo-primitive condition is betrayed by a series of stratigraphically successive species.

## 1. Geological evidence

There is geological and sedimentological evidence that may be applied to determine whether or not the record is probably complete. This evidence has the advantage that it is logically independent of the palaeontological information. The geological circumstances where a complete record may be anticipated include the following:

(i) Sites where there is reason to suppose that deposition is likely to be conformable and continuous, as, for example, in former shelf-edge to oceanic environments.

(ii) Where there is sedimentological evidence of fast and continuous deposition. There is no reason why this should not apply to certain terrestrial sites. If deposition is *too* fast, however, a thick sequence may not record enough time to pick up species-to-species changes.

(iii) Lack of drastic changes in sedimentary facies within a vertical sequence.

These three circumstances may occur singly or together, and the best case is when they all occur together. There is an obvious corollary to this list of favourable circumstances. They all tend to imply lack of drastic change in environment or facies. Many of the major changes of most concern to phylogeneticists, however, occur at environmental discontinuities; for example, the evolution of the tetrapod limb at the interface between fluviatile and terrestrial habitats. The very term radiation implies non-continuity, invasion of new habitats, and so on, all producing violations of conditions (i)–(iii) above. Certainly, evolution of new morphologies, seminal to major phyletic units, will have proceeded by a series of species-to-species changes, but the circumstances propitious for such changes are exactly those which minimize the geological chances of finding observable species-to-species records. The detailed species-to-species sequences of the stratigraphic palaeontologist seldom help, therefore, in reconstructing the origins of major groups. There are some lucky exceptions, however. The ammonoids are connected by a series of intermediate species with the nautiloids, and the origin of the planktonic graptoloids is similarly documented at species level. Such examples are within the marine habitat where the change in life habits involved does not imply a change in fossilization potential. Species-to-species changes of the kind that may indicate true ancestor–descendant relationships are distributed widely through phyla with good fossil record: Trilobita (Olenidae), echinoderms (*Micraster*), brachiopods (*Eocoelia*), foraminifera (*Orbulina*, *Globigerina*), ammonoidea (*Harpoceras, Amaltheus, Dactylioceras*), graptolites (*Monograptus triangulatus* group), and so on. We shall not pretend to catalogue all these here.

## 2. Repetition of an Observed Lineage from Place to Place

Once the sequence of species in phylogenetic order is established from one section, in groups with a good fossil record there is the expectation that is can be repeated elsewhere indefinitely. So if the proposed phylogeny was: "species A ancestral to B, which was the ancestor of C and D", then in all subsequent occurrences A should occur below B, and B in turn below C and D:

$$
\begin{array}{cc}
\mathrm{C} & \mathrm{D} \\
\hline
\multicolumn{2}{c}{\mathrm{B}} \\
\hline
\multicolumn{2}{c}{\mathrm{A}} \\
\hline
\end{array}
$$

Such sequences naturally become of biostratigraphic significance. Since it is theoretically possible to arrange A, B, C and D in other ways than the observed one, it is possible to regard each subsequent occurrence as falsifying all alternative hypotheses, i.e. it is possible to make these models conform to a Popperian paradigm (though this is an artificial device, in just the same way as writing out all alternative cladograms). In the case of some well known species-to-species lineages, as, e.g. the lineage leading to the spherical Tertiary foraminiferan *Orbulina*, the repeatability extends to dozens of sites, any one of which was a potential falsifier of the phylogeny proposed, and these sites occur over a huge circum-tropical belt. One might add that in this example the oil company palaeontologists who utilize the lineage have no vested interest in retaining the academic status quo!

The crux of the acceptability of species-to-species lineages with actual ancestors is that, once the sequence of species is repeated (or if one prefers, has survived potential falsification on a number of occasions), the chances of that sequence being a part of the correct tree is correspondingly increased. Of course, one can never know for certain whether or not one has *arrived* at the correct tree; the phylogenetic sequence is perpetually "on trial" – but this is all to the good in Popperian terms. In practice there is a point at which palaeontologists tend to accept a given phylogeny of species as corresponding to the correct tree, and this does not correspond to the initial proposition of a phylogenetic hypothesis, but to its repeated occurrence, and preferably oft-repeated in diverse localities. Since the correct tree contains all other formulations of phylogenetic relationships within it, there can be no doubt that the recognition of portions of the tree with increasing confidence is a valid goal of phylogenetic activity. These portions may be of importance, and rightly so, to stratigraphers, but unless they include the origin of a major taxon they are not of interest to neontologists apart from light they may throw on the mechanism of speciation. The chances of finding species-to-species series for the origin of, say, tetrapods is low for purely stochastic reasons, and because, as mentioned above, the stratigraphic record is likely to be interrupted when major evolutionary changes are occurring.

Finally we note that there are important gaps in the fossil record, across which species-to-species lineages are hardly known. The Permian to Triassic interval is one. Obviously the chances of working out stratigraphy-based species-to-species sequences for the major evolutionary novelties of the Mesozoic are minimal as far as marine organisms are concerned.

A THEORETICAL DEMONSTRATION

The relative advantages of a cladistic or a stratigraphic approach in resolving phylogenetic relationships can be demonstrated by means of a simple model (Figs 1, 2) which was suggested initially by a discussion in Schaeffer *et al.* (1972). In the example we assume that the form of the correct tree is as shown. The shape of the correct

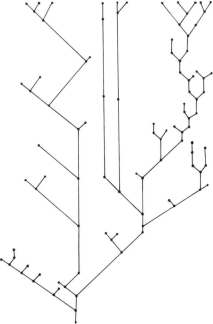

Fig. 1. Hypothetical Correct Tree containing 100 species represented by points which are capable of being sampled by the fossil record. The form of the tree is arbitrary, and the reader is invited to devise one of his own conforming to his preferences for the way his animals speciate. This one is as generalized as possible, incorporating both rapid dichotomies and linear ("orthogenetic") changes. The horizontal scale is any measure of morphologic dissimilarity, the vertical scale is time. For simplicity of representation species on orthogenetic lines are placed vertically above one another (i.e. the morphologic change is not visually recorded here).

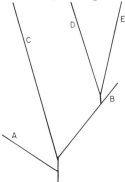

Fig. 2. Diagram to show the bare bones (major evolutionary lines) of the correct tree. The letters identify the main lines referred to on subsequent diagrams.

tree is not of great moment to the exercise, and the reader is invited to try out any tree which he thinks may correspond to the evolutionary behaviour of his own speciality. The hundred points on the diagram represent a hundred taxa (e.g. species) that may be sampled. Those species which are to be sampled can be obtained from a table of random numbers, or from a hat. The object of the game is to compare the closeness to the correct tree of models based on stratigraphic occurrence, or on a cladogram, at various percentage completenesses of random record. We assume that synapomorphies are always correctly identified.

At a low record, say 3%, the species identified by the random process usually have the kind of distribution shown in Fig. 3.

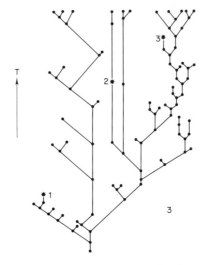

Fig. 3. Selection from the correct tree at 3% record, selected at random.

Drawing three species in sequence on one branch is extremely unlikely. For the example selected, what one might term the "naive stratigraphical" approach, which is simply placing the species in stratigraphic order to approximate phylogeny, yields a simple 1-2-3 model (Fig. 4). This bears virtually no relationship to the correct tree, in fact it cuts through most of the major branches. On the other hand a cladogram relating the same three species *does* produce a replica of the main branching pattern, with the origin of the synapomorphies identified in the correct order. The implication is that

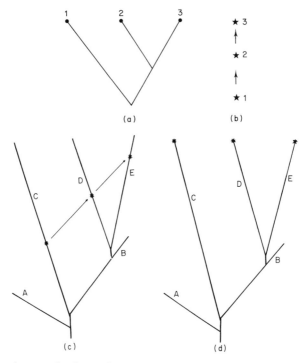

Fig. 4. Hypotheses of relationships about the three species shown in Fig. 3.
(a) Cladogram, (b) simplistic stratigraphic view. (c) Shows how the
simplistic stratigraphic view cuts across the correct tree at this level of
record. (d) Shows how the cladogram correctly identifies the main
relationships of the three species on the correct tree.

at low record the cladogram produces the best approximation to
the correct tree, and the naive stratigraphic approach is liable to
confuse, rather than to make the phylogenetic relationships explicit.

Conversely, at high record, and two-thirds is sufficient (Fig. 5),
there are always a large number of "hits" along any lineage of the
correct tree. For these segments a stratigraphic species-lineage
approximates closely to the correct tree; for the middle segment,
in the example chosen, the stratigraphic series would expose the
actual ancestors, and the succession of the descendant species. A
cladogram of the same portion, with non-recognition of ancestors,
obviously approximates less closely to the correct tree. In "real
life" a good record is also likely to be repeatable, and hence the
reality of that segment is affirmed by subsequent attempts to upset

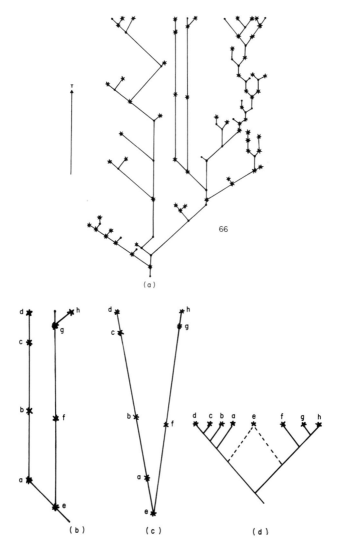

Fig. 5. (a) Two-thirds (66%) record selected from the correct tree randomly. (b–d) To show how one segment of the 66% tree (b) is approximated to more closely by the stratigraphic-based tree (c) than by the cladogram (d) which can neither resolve the problem of the ancestors nor the sequence of appearance of species.

it. At this level the observation of stratigraphic species lineages is the superior method, since it approximates closer to the correct tree.

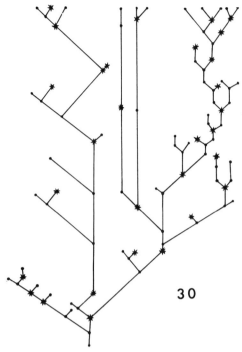

30

Fig. 6. A 30% record selected randomly from the correct tree.

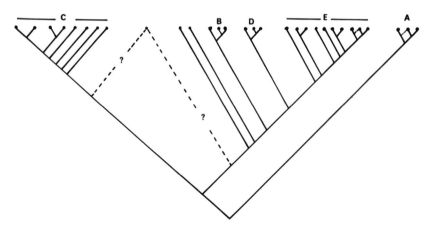

Fig. 7. Cladogram based on Fig. 6. At 30% record the cladogram picks out the main groups as on Fig. 2 very well, but cannot resolve the symplesiomorphous forms so well as the stratigraphically based "tree" in Fig. 8.

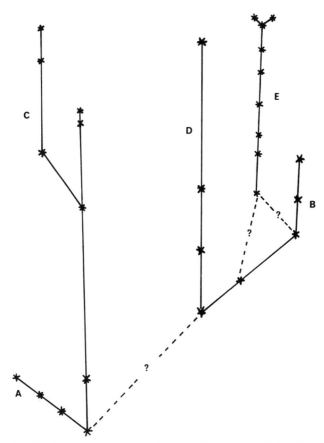

Fig. 8. The kind of "tree" a stratigraphic-based analysis of the data in Fig. 6 might produce. As with the cladogram the main groups are identified, although not as clearly, but the relationships between the groups are unsatisfactory. On the other hand the position of some of the plesiomorphous, ancestral species is more satisfactorily resolved by comparison with the correct tree. Both cladograms and "stratograms" are useful at this percentage record.

The prime purpose of this model is to establish the validity of the different approaches at the extremes. There will obviously be a "middle ground" (Figs 6—8). On this version of the correct tree it falls between 20—50% record, but will probably have different values according to the presumed nature of the tree. For example, on a tree constructed mostly of orthogenetic lines (such as some ammonite workers believe in), the stratigraphic method becomes useful at a lower % record. Most of the methodological fights that occur might be in this middle ground. *Both* methods will here be useful approaches to the correct tree.

The assumption that the record is random is a minimal one, but not generally a correct one. One type of "good" record is the single horizon where most or all of the existent available taxa are sampled (a horizontal line across the correct tree). For this kind of sample the cladogram is applicable for unique complete assemblages of fossil forms (i.e. in the absence of stratigraphical evidence of sequence). A sample at time $T$ on the diagram, if correctly analysed, will reveal the main lineaments of the correct tree by cladistic analysis (Fig. 9).

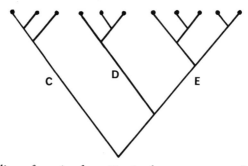

Fig. 9. The top line of species from Fig. 1 taken to represent the Recent. Only a cladistic analysis correctly resolves the relationships.

However, when the complete samples become frequent ($T_1$, $T_2$, $T_3$) segments of the correct tree will again be picked up in the stratigraphic order. Again the extremes are identifiable, the middle ground contentious.

In summary, for poor record, cladogram analysis produces models of the correct tree that are preferable; for good record (perhaps 70% or more of all species), stratigraphic species lineages are likely to approximate more closely to the correct tree. If the percentage record improves, the remodelling of hypotheses of relationships to fit stratigraphical sequence may become the better approach; if new morphological information on previously known species becomes available cladistic refinement is likely to prove the more fruitful.

### THE RECOGNIZABILITY OF HOMOLOGUES

So far we have assumed that homologues can always be correctly identified as such, but this is clearly untrue. Homology can be defined as fundamental structural identity particularly between different species; it is a special case of order and the recognition of order has recently been elucidated from a zoological standpoint by Riedl (1975, 1979). He stresses that repetition is necessary for the finding of order. Thus if a source sends out the sequence of digits: 11086165, then this, so far as the receiver can tell, seems entirely random. But if the same sequence is repeated only once (11086165, 11086165), then everyone would be convinced of the existence of order. We intuitively feel that the "decisions" behind the choice of the first set of digits must somehow be the same as the "decisions" behind the identical second set. In other words, we judge that the chance of accidental repetition of the first sequence is infinitesimal. And we feel this without actually calculating that chance, and without knowing what the decisions are.

The calculation of the accidental probability of this repetition can be approached in at least two ways. First, we could assume that each digit of the eight could be chosen directly from the range 0 to 9 with equal probability. This would make the chance of accidental repetition equal to $1:10^8$. Alternatively we could follow Riedl in regarding all decisions as permutations of yes/no decisions in which case 27 ranks of such binary decisions would be needed for selecting the series of eight digits since $2^{27} = 1.34 \times 10^8$ while $2^{26}$ is only $6.7 \times 10^7$. The probability is effectively zero, however it is calculated and we recognize this fact without calculation. And moreover we do so without knowing the nature of the decisions behind the choice of the particular sequence. In a similar way, the alternation of night and day was recognized as orderly long before even speech was invented. The recognition of this orderliness was independent of, and long preceded, the explanation of it (though once attained the explanation has affected our perception of night and day).

A definition of order follows from these considerations, and also a quantitative measure of it. Thus consider that a transmitter sends out the message:

11086165    11086165    11086165    11086165

We can distinguish here between any one occurrence of the repeated sequence (law content = 1) and the number of instances of the sequence ($a$). The determinacy content ($D$), or order content, of a message can then be expressed by the equation:

$$D = l \cdot a \qquad (1)$$

This can be measured in bits of information and the determinacy content of the above message = $(2^4 \times 8) \times 4 = 512$ bits.

Living order, which is much more complex than non-living order, can only recur when a linking mechanism somehow transmits the decisions behind the first occurrence to the subsequent occurrences. Riedl called this linking mechanism "Tradierung" in the German edition of his book (1975) and for the English edition one of us translated this as "traditive inheritance" (Riedl 1979; translation by Jefferies). It could better be called "transmission". Non-living order does not need transmission; thus crystals of NaCl will form by the mere evaporation of a saturated solution of common salt. By contrast, molecules of insulin will not form from the appropriate mix of amino acids without the instructions for its formation being present (DNA, RNA, ribosomes and so on).

Language is a special case of living order, and has the advantage, for purposes of exposition, that humans understand one language intuitively. The transmission of a language usually depends on the instinctive tendency of a child to imitate the language that it hears around it, but other forms of transmission exist, such as involve reading, printing and writing. We will take a linguistic example to illustrate some of the characteristics of living transmitted order, by quoting the first line of Chaucer's "Canterbury Tales":

Whan that Aprill with his shoures soote . . .

This sequence of seven common Middle English words has a remarkable property. For, whenever we meet it, we feel intuitively certain that it has been transmitted directly, by copying and recopying, from a common source, and we know from other evidence that this source was the creation that sprang into Chaucer's brain about 1380. But our conviction that each quotation of this line is fundamentally identical to all other quotations of it, i.e. has descended by copying from the same original, does not depend on all the details being the same. Thus the line is sometimes quoted, with an extra *e* (and an extra syllable) on the end of Aprill, as follows:

Whan that Aprille with his shoures soote . . .

Nobody would doubt that both these versions originate from the same thought of Chaucer's. Indeed this is even true of a gross modern translation such as:

When April with its sweet showers . . .

Fundamental identity of this sort, which is identity disregarding changeable details, is something that we all recognize and use constantly in thought and in everyday life without ever being able to define it. The recognition of fundamental identity depends on recognizing a gestalt — a particular complex law content with a particular arrangement within itself, and without a gestalt there is no intuitive conviction of fundamental identity. Thus the same seven words of this quotation no doubt exist scattered throughout the near-contemporary "Piers Plowman" of William Langland, but since they do not there form a gestalt, nobody would take this as showing that Langland was indirectly quoting Chaucer, nor Chaucer Langland. Moreover, the *e* on Aprille does not form an essential part of the gestalt of Chaucer's first line which is why it is not immediately obvious whether it was part of his original thought or not.

A complex law content, therefore, needs fewer repetitions to convince us of orderliness than a simple law content needs. Thus, in the tossing of coins, if a player who has bet on tails throws tails once, this will be no grounds for suspicion, since the accidental probability is 1:2. But if he throws tails ten times in succession he is almost certainly cheating since the accidental probability will be $1:2^{10} \sim 0.001$. The parsimony rule, which tells us to assume that like events have like causes, is valid after ten consecutive instances of tails, but not after one only.

Homology, as the word is commonly used in comparative anatomy, can be defined as fundamental structural identity between species, which means a recurrence of the same law content. The recognition of homology does not demand an identical repetition in every detail, which never happens. But if we find complex (and accidentally improbable) structures repeated in two different species, we deduce that they have been transmitted from an identical structure in the latest common ancestor. Thus we assume that the right forelimbs of man and gorilla, because they are so complex and so like each other, must therefore depend on like decisions and must be transmitted from a common original in the same way as "11086165" or "Whan that Aprill . . ."

Complex homologues are hierarchical. Riedl's preferred illustration of this is the mammalian vertebral column, with its constituent regions, vertebrae and parts of vertebrae. It is important to note that the division of a mammalian vertebral column into parts homologous with those of other such columns cannot be continued indefinitely. Within the smallest homologizable parts of the vertebrae we do not find other single homologizable parts. The ventral articular facet of the odontoid process of the axis vertebra, for example, can be recognized in all healthy mammalian vertebral columns. But if we seek to divide it further we find, not single entities, homologizable with similar entities in other vertebral columns, but mutually similar parts present in large numbers, such as Haversian columns. These similar multitudes are called homonoms in the German literature. The ventral articular facet of the odontoid process is called a minimal homologue by Riedl. And the homologues of higher rank that include it and other minimal homologues are called cadre homologues. Lower-rank cadre homologues (such as the axis vertebra) are included in higher-rank cadre homologues (such as the cervical region). And the different ranks of cadre homologue are themselves included in the great hierarchy which is the vertebral column (which itself is a cadre homologue within the skeletal system). Riedl estimates the total number of homologues in the human vertebral column (i.e. its complexity) as 4381. The chance of two human columns sharing them by accident alone, assuming conservatively that each depended only on a yes/no decision, would be $2^{-4381}$, which is negligible as was already obvious by intuition.

This leads to Riedl's concept of burden which can be defined as the number of decisions which depend on a decision, or of features which depend on a feature. The vertebral column is at the top of its particular hierarchy and its constituent cadre and minimal homologues constitute a burden of about 4381 unit homologues, i.e. 4381 homologues depend on it, being included in it. Cadre homologues will have a smaller burden and minimal homologues no burden at all within the vertebral column. The effect of these differences in burden is that low-ranking and minimal homologues will be easier to change than high-ranking ones. Thus the decision (mutation) not to form a vertebral column could only produce a viable mammal if 4381 constituent unit homologues had somehow become unnecessary for life, and it is obvious that such a mutation would kill the animal. A mutation that eliminated the ventral articular facet of the odontoid process, on the other hand, would probably not be fatal but subvital. For it would not in itself affect other constituents of the vertebral column (or not much) and its harmful results would come by disturbing the relations with other systems.

Several forms of hierarchy can be recognized within a homologue. These forms seem partly parallel to each other, but must be conceptually distinguished. The form that has already been discussed consists of the hierarchy of accepted anatomical features within larger anatomical features which has been traditionally recognized by anatomists; for want of a better name we call this anatomical hierarchy. (It was called homological hierarchy in Jefferies (1979), but this was a misnomer.) Another hierarchy is of more fundamental significance, being the features arranged in order of ontogenetic appearance. Thus, as concerns the mammalian vertebral column, the notochord is at the top of the ontogenetic hierarchy; followed by the dorsal nerve cord, neural crest and muscle blocks; followed by the arcualia of the vertebrae; followed by the vertebral centra; and so forth. Another hierarchy is that of position, which might better be called "centrality". In this hierarchy, central (proximal) features have a higher rank than peripheral (distal) ones. It is characteristic of all these hierarchies that the features of low rank (physically smaller, later in ontogeny or distal in position) can be altered more readily than features of higher rank. We can speak in this connection of high or low anatomical, ontogenetic or positional burden. There is a parallel here with human hierarchies in that changes at the top have more effect than changes at lower levels. The death of a pope has more effect than the death of a parish priest.

Ontogenetic burden is the basis of Haeckel's law. A new feature is most likely to be added to the end of ontogeny, because only thus can it be introduced without disturbing the development of features that already exist. And, when it is added, no other feature will depend on it, i.e. it will have no burden and will therefore be easily modifiable or liable to disappear completely. As time goes by, however, it may acquire dependent features, i.e. features for which its presence is a developmental precondition. These new features are presumably useful, or they would not have been favoured by natural selection, and consequently the original feature cannot now be got rid of without losing the useful aspects of these newer dependent features. In other words, complexity will increase with time; novelties will usually be added at the end of development; and previously present features, having acquired burden, will thus become more difficult to eliminate.

Also a feature will become more recognizable as the number of features that developmentally depend on it increases. The feature $a$ could easily have arisen twice in parallel as a minimal homologue; but if we find two organisms where $a$ is present, and also $b$ which depends developmentally on $a$, and $c$ which depends developmentally on $b$, then we can assume with high probability that $a$ is homologous in both organisms; with slightly less probability that $(a + b)$ is homologous; and with slightly less probability again that $(a + b + c)$ is homologous. The complex "$a$ preconditional to $b$, preconditional to $c$" is a gestalt or complex law content like "Whan that Aprill . . ." and the gestalt can be recognized even if details are altered, as, for example, if $c$ were omitted or $d$ substituted for $c$. Thus when the notochord was evolved it was presumably a simple extension of the gut wall and could easily have arisen twice in related animals. This is no longer true of the notochords of two mammals, with tens of thousands of identical features developmentally dependent on them.

A curious paradox, pointed out by Riedl (1979, p. 112 ff.), is that, with mass homologues (homonoms), a uniform change in the members of higher ranks is easier to introduce successfully than a uniform change in the members of lower ranks. An accidental change to the design of all the cilia in a mammal, for example, would seldom be of benefit. This is because mammalian cilia have many different functions, at least one of which would probably be precluded by any change to the principle. The more numerous the mass homologue (i.e. the lower its rank) the more likely it is to be involved in many aspects of life, and the more difficult uniform change in it will be.

However, mass homologues have not been much used in reconstructing phylogeny, so this paradox does not greatly affect the present discussion.

There is therefore a spectrum of inherent recognizability for any given segment in the cladogram of a group (Fig. 10). Imagine three species $a$, $b$ and $c$, in which $b$ and $c$ are the sister group of $a$. There will then be a segment of the cladogram corresponding to the exclusive common ancestry of $[b + c]$ which $a$ does not share. As regards evolution within this segment, four situations are conceivable with differing degrees of inherent recognizability. First, there may be no change in the segment and $b$ and $c$, although forming a strictly monophyletic group, have no synapomorphies and never had any (Fig. 10a). The sister-group relationship of these two species to $a$ could only be recognized, if at all, by using stratigraphical, or perhaps biogeographical, evidence. Secondly, a single simple change may have occurred in the segment (Fig. 10b). It could be recognized as a synapomorphy of $[b + c]$ provided that neither species has lost it (but this could easily happen with a simple change) and provided that it has not evolved in parallel in closely related groups. Thirdly, more than one simple change may have occurred in the segment (Fig. 10c), each change being independent of the others. All such changes will be synapomorphies of $[b + c]$ if they have not been lost subsequently. Such a segment will be more recognizable than in the first or second case, though still not indubitable. Fourthly, several changes may have occurred within the segment, such that each is dependent on the others (Fig. 10d). In the figure, for example, change 2 will be a precondition for 3, and 3 for 4. The result will be a gestalt which can be recognized with something approaching certainty, which will not easily be lost, and which will not easily arise in parallel in related species. Such gestalten will usually arise gradually by a succession of changes separated in time. But sometimes they may happen suddenly as when a mutation eliminates a sub-sequence of amino acid residues (perhaps glycine-alanine-glycine-proline-arginine-) at one blow from the middle of a protein.

The position of the cladogram segment within this spectrum of inherent recognizability therefore depends on the "quantity of evolution" that has happened within the segment. As concerns still-living organisms the "empty" case of least recognizability (Fig. 10a) might

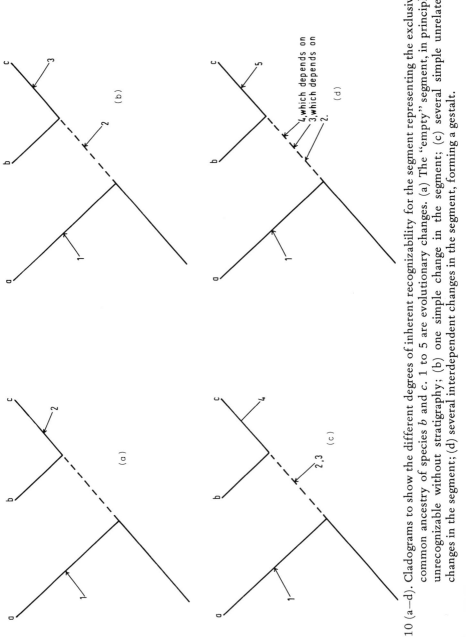

Fig. 10 (a–d). Cladograms to show the different degrees of inherent recognizability for the segment representing the exclusive common ancestry of species *b* and *c*. 1 to 5 are evolutionary changes. (a) The "empty" segment, in principle unrecognizable without stratigraphy; (b) one simple change in the segment; (c) several simple unrelated changes in the segment; (d) several interdependent changes in the segment, forming a gestalt.

correspond to three closely related species within a genus, perhaps the Wood Warbler, Willow Warbler and Chiff Chaff (*Phylloscopus sibilatrix, P. trochilus, P. colybita*). The second case (Fig. 10b) might correspond to three less closely similar species within a genus, perhaps the Song Thrush, Blackbird and Ring Ouzel (*Turdus philomelos, T. merula, T. torquatus*); in this instance blackness would be the synapomorphy corresponding to change 2 in the figure. In the third case the species separated by the segment are still further apart from each other, possibly corresponding to different genera within a family, as perhaps the Mallard, Pochard and Tufted Duck (*Anas platyrhynchos, Aythya ferina, Aythya fuligula*) all Anatidae, with small size and the downward-pointing tail feathers as the unrelated changes 2 and 3. In the fourth case the segment in question may separate very different organisms such as Nile Crocodile (*Crocodilus niloticus*), Song Thrush and Mallard; the most striking gestalt that had arisen in the segment in question would, in this example, be the feather, with 2 as the origin of the feather, 3 the evolution of barbs and shaft, and 4 the evolution of barbules on the barbs. Naturally the examples given are purely for purposes of illustration and are not serious statements about the phylogeny of the animals cited. So far we have spoken of what might be called the inherent recognizability of the segment.

However, whether a segment can be distinguished depends not on inherent recognizability alone, but also on changes with the lapse of time. This is true of all except the empty segment (case 1, Fig. 10a) which, being unrecognizable from the beginning, cannot become more so. In the other cases the segment becomes more difficult to descry from two basic causes: (1) species descended from the segment may lose features evolved within it; (2) related species may acquire features seemingly identical with those evolved in the segment. Roughly speaking, segments of high inherent recognizability will take longer to disappear into the depths of time than segments of low inherent recognizability.

For a segment to become unrecognizable, it is not necessary that all its descendants lose the features evolved in the segment. This is because if many, but not all, of the descendants lose such a feature, anatomists will perhaps assume that the feature is not primitive for all the species descended from the segment. As an example, primitive fossil chordates and echinoderms (living and fossil) shared the common feature of a calcitic mesodermal skeleton in which each plate is a single crystal of calcite. For this and other reasons, echinoderms and chordates should be placed together in the monophyletic group of the Dexiothetica (Jefferies 1979). This skeleton has disappeared among living chordates, however, and the loss has probably happened three times, independently, in the stem groups of acraniates, tunicates and vertebrates. As concerns living animals, therefore, such a skeleton does not appear as a synapomorphy of chordates and echinoderms, but falsely mimics an echinoderm specialization.

Fossils may conceivably play two distinct rôles in helping to reconstruct phylogeny. The first is that they may supply distinct morphologies not known among living organisms; the second is that, because of their stratigraphical sequences, they may tell us which features are primitive for a group, assuming that the oldest species known are more likely to display the primitive condition. The first rôle for fossils, where they are used merely as extinct anatomies without reference to stratigraphy, will only be possible if the fossils in question display synapomorphies which living forms either do not show at all, or else which have become restricted among living forms and are therefore wrongly regarded as specializations of the groups that retain them. The calcite skeleton of calcichordates and echinoderms is an example of the latter; an example of the former condition, of a synapomorphy lost in all living relatives, may be certain common features (dorsal bearing surfaces) of the tail skeleton of stem vertebrates and stem tunicates among calcichordates, which stem acraniates do not share. These features suggest that tunicates and vertebrates together form the sister-group of the acraniates, supporting a conclusion derived from the innervation of the musculature of the living animals (Jefferies and Lewis 1978, p. 290; Jefferies 1973, p. 463). Use of fossils as extinct anatomies, in this manner, is independent of stratigraphical sequence. It does however depend on the disappearance, among some or all of the closest living relatives of the fossil in question, of features of high enough burden to be recognized. Since features of high burden tend to persist, such disappearances will be rare events, only likely to happen with the lapse of long periods of time. This is the main reason why fossils, used as extinct anatomies without consulting stratigraphy, only seldom help in the construction of the cladograms of still living groups, and why, when they do help, they connect together groups of high rank whose phylogenetic connections are ancient (as with chordates and echinoderms). Løvtrup (1977, p. 21) has stated that: "The discovery of a new fossil has no impact on classification." His reasons are that the soft anatomy of fossils is difficult to reconstruct and their biochemistry unknowable. Certainly these two considerations make fossils harder to use in phylogeny but the fact that features of high burden will seldom be lost is a more fundamental difficulty. Palaeontologists should continue to search for such features, however, heartened by the thought that, if in this way they find fresh synapomorphies, the connections so established are likely to be important ones between high-ranking groups.

The second use of fossils is in connection with the stratigraphical criterion of primitiveness. Hennig (1966, p. 95) called this the criterion of geological character precedence and said: "If in a monophyletic group a particular character condition occurs only in older fossils, and another only in younger fossils, then obviously the former is the plesiomorphous and the latter the apomorphous condition of a morphocline a, a′, a″." In fact, however, as already pointed out above, it is quite possible for this criterion of primitiveness to mislead (as with all other such criteria) since the fossil record may be incomplete. The importance of the stratigraphical criterion will decrease gradually down the spectrum of recognizability of a cladogram segment, i.e. Fig. 10a to d. In Fig. 10a, stratigraphy is the only way of establishing the existence of the segment. In Fig. 10b, c, d stratigraphy will be of decreasing value as the features evolved within the segment become more recognizable and less easily repeated in parallel. In the case of complex features forming a gestalt whose homology is certain, the stratigraphical criterion of primitiveness becomes both unnecessary and misleading, for outgroup comparison will supply a more appropriate criterion of primitiveness, not being subject to the vagaries of fossil collection (Jefferies 1979). The interaction of "perfection of record" and "recognizability of segment" is summed up in Table I.

An experimental simulation of the effects of burden in the reconstruction of phylogeny is summarized in Figs 11–16. The simulation assumes a group which evolved by branching occasionally into two, whereby one of the daughter species was always changed from the

Table I. Interaction of "perfection of record" and "recognizability of segment"

| Evolution within the segment | Quality of stratigraphical record | |
| --- | --- | --- |
| | Absent or poor | Good or perfect |
| None (Fig. 10a) | Segment can never be recognized | Segment recognizable only by stratigraphical observation |
| One simple change (Fig. 10b) | Segment not safely recognizable | Segment recognizable by using stratigraphy to exclude loss or parallelism, i.e. by combining the cladistic and the stratigraphical approaches |
| Several simple changes (Fig. 10c) | As above, but segment can be postulated with greater probability | As above, but a less perfect stratigraphical record will be adequate for recognition of the segment |
| Origin of feature of high burden (Fig. 10d) | Segment recognizable with near certainty. Stratigraphical sequence of fossils (if any) should be ignored | Stratigraphy irrelevant to recognition of segment which is virtually certain anyway; stratigraphy may be useful for working out the sequence of changes (each of smaller burden than the total change) within the segment; however this sequence may sometimes be revealed by fossils compared as anatomies, without reference to stratigraphy |

ancestral condition but the other was not. This corresponds to an allopatric model of speciation involving punctuated equilibria (Gould and Eldredge 1977). We have adopted this model, not because we are convinced of its truth, but because it implies a large number of empty segments in the corresponding cladogram (Fig. 10a). It is important not to lose sight of these inconvenient segments which can never leave a synapomorphy behind them. The imaginary species in the simulation were given a morphology. This consisted of an upper and lower appendage (string of digits within the range 1 to 5) which could grow by the addition of digits to the right-hand, distal end, or shorten by the loss of digits from this end.

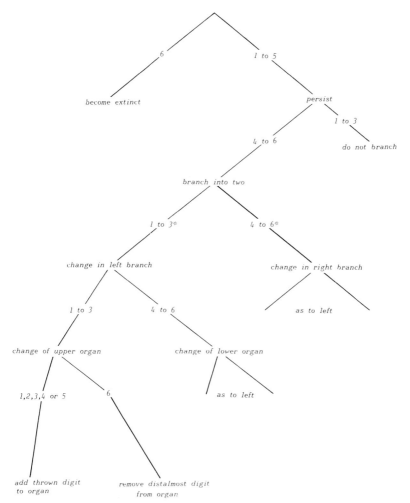

Fig. 11. Playing rules for throwing dice in the construction of the correct tree shown in Fig. 12. The throw * was not necessary for randomness but was included in the game as played.

Capital letters (A, B, C . . .) were used to signify position in the upper organ and lower case letters (a, b, c . . .) were used for this purpose in the lower organ. Thus in the diagrams, E5 means the addition of 5 in position E of the upper organ, while − e3 would mean the loss of 3 from position e of the lower organ. As digits by chance were joined by new digits to the right of them, therefore, they themselves became less likely to be lost, i.e. they acquired a

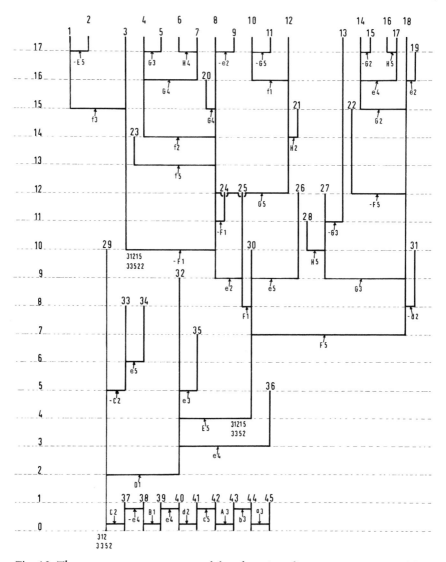

Fig. 12. The correct tree, constructed by throwing dice. 1 to 18 are surviving species; 19 to 45 are extinct species; morphologies are shown for species 29, 30 and 3; e4, − F1 etc. = mutations as explained in text.

burden of features which stabilized them. The evolution of the group was decided by throwing dice according to the playing rules in Fig. 11. Evolution took place at a series of 17 catastrophes at which change or extinction could occur, separated and succeeded

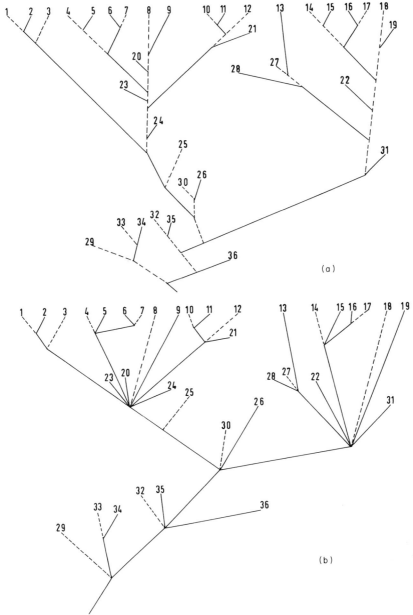

Fig. 13. (a) Complete correct cladogram of the correct tree shown in Fig. 12; non-terminal empty segments are included; empty segments are shown as broken lines. (b) Best possible correct non-stratigraphical cladogram of the correct tree — empty non-terminal segments have been omitted.

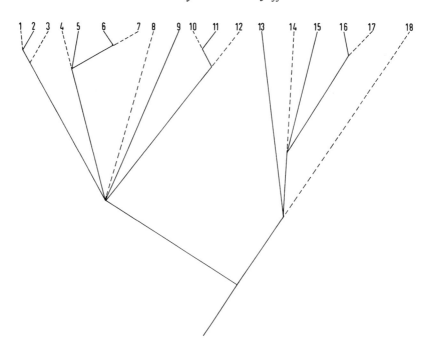

Fig. 14. Best possible correct non-stratigraphical cladogram of living species
only; broken lines = empty segments.

by horizons within which no changes were permitted. Horizon 17.5
was taken as representing the present day. There were nine preliminary catastrophes at which evolution occurred in accordance with
the fall of dice, but with no branching allowed.

Figure 12 presents the correct biostratigraphical tree in which, in
the period when branching was allowed to happen, 36 morphospecies
evolved of which 18 still survive.

Figure 13a shows the complete correct cladogram for this tree,
with all segments included, whether empty or full, terminal or
non-terminal. Figure 13b shows the best possible non-stratigraphical
cladogram for the group, i.e. one in which, because stratigraphy
was not consulted, empty non-terminal segments could not be
detected. The resolution of this cladogram can be expressed as a
percentage since with $n = 36$ species there ought to be $n - 1 = 35$
branch points, but in fact only 16 exist, so that the resolution is

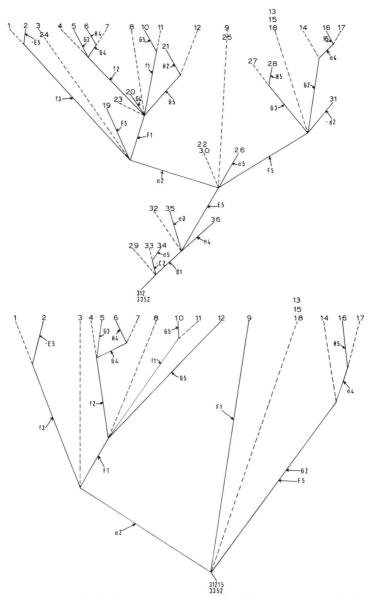

Fig. 15. Reconstructed cladograms. (a) Fossil and living species; (b) living
species only.

$\frac{16}{35} \times 100 = 46\%$. Figure 14 shows a similar cladogram but includes
only the 18 living species. The resolution of this is better at $\frac{11}{17} \times$
$100 = 65\%$.

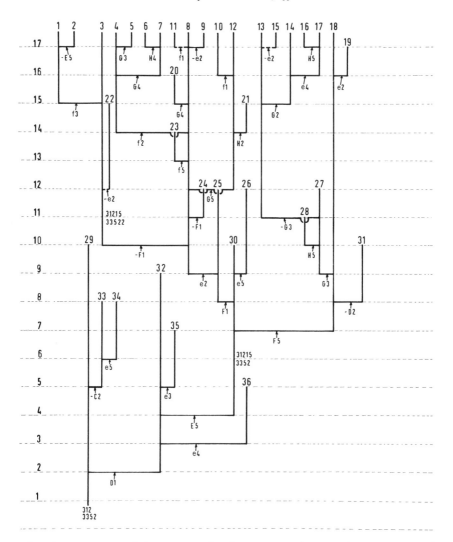

Fig. 16. Reconstructed biostratigraphical tree. Mistakes in the reconstructed origins of the species 3, 11 and 15 are shown as broken lines.

Figure 15b shows a cladogram for the living species as obtained by reconstruction from the known morphologies. For purposes of this reconstruction the morphologies of the 18 living species were each written on the backs of small cards and the cladogram constructed on this information alone. This was done as follows: (1) We reconstructed the morphotype for all living species, i.e. a morphology consisting of the features present in all of them and therefore assumed to have existed in the latest common ancestor. This reconstructed morphology was $\frac{31215}{3352}$ corresponding, correctly as it happened, to the latest common ancestral fossil species (no. 30). (2) We tried to recognize successively smaller monophyletic groups assuming that evolution by gain (of digits 1 to 5) was more probable than evolution by loss – this assumption is parsimonious, but incorrect according to the playing rules (Fig. 11). The major mistake caused by this assumption was that the group [4 to 12 less 9] was wrongly considered as monophyletic with F1 as its synapomorphy, whereas it was in fact paraphyletic having retained F1 when the true monophyletic group [1 to 3] had lost it. The other major reconstructional error is the inevitable lumping of 13, 15 and 18 whose morphologies were identical ($\frac{312155}{3352}$). In actual fact 18 represents the primitive condition for the monophyletic group [13 to 18], whereas 13 had gained and then lost G3, and 15 had gained and then lost G2.

Figure 15a is a cladogram reconstructed like Fig. 15b from the morphologies alone, but in it fossil species are included. These were known to be extinct, but their stratigraphical ranges were otherwise not known during reconstruction. The noteworthy result is that all this extra information did not improve, nor require the alteration of, the previously reconstructed cladogram for the recent species. It merely contributed to the history of the group and increased the number of unanalysable homoeomorphs (3, 24; 22, 30; 9, 25). The likely cause for the unhelpfulness of the fossils was that no complex synapomorphies were lost in the total course of the evolution of the group.

Figure 16 is a reconstructed biostratigraphical tree based on complete knowledge of the morphologies and stratigraphical ranges of all species, living and extinct. The new constraint is that a species can only arise from another species known to exist at the time of origin. This tree achieves much better discrimination than the reconstructed non-stratigraphical cladograms. For example groups [1 to 12] and [13 to 18] are correctly recognized as monophyletic groups among the living species. There are only three mistakes, all of them, coincidentally, as to the origin of terminal species (11, 15, 22). The stratigraphy is mainly useful as a way of distinguishing the origins of homoeomorphs and, in particular, recognizing cases where a feature is gained and then in some descendants lost again. Examples are: 18 + G3 → 27 − G3 → 13 and 30 + F1 → 25 + e2 → 8 − F1 → 3. The latter sequence allows F1 to be recognized as a synapomorphy of group [1 to 12] and − F1 as a synapomorphy of [1 to 3]; it avoids, in other words, the interpretation of F1 as a synapomorphy of a false monophyletic group [4, 5, 6, 7, 8, 10, 11, 12].

This simulated evolution might seem to confirm the common palaeontological belief that the phylogeny of a group of fossils can only be worked out by consulting the stratigraphy. However, all the changes involved were of low burden and the game did not continue long enough to allow features of high burden to be lost. The simulation only suggests, in fact, that stratigraphy is essential to distinguish segments of low inherent recognizability – it concerns Fig. 10a–c, not Fig. 10d.

## A TRILOBITE EXAMPLE

Eldredge (1973, 1979) has used trilobites to exemplify a cladistic approach to phylogeny in these extinct animals. It might be instructive to show how our eclectic approach can be used for phylogenetic analysis and the recognition of parallelism. The three trilobite families Cyclopygidae, Nileidae and Taihungshaniidae have been placed in two different superfamilies in the *Treatise on Invertebrate Palaeontology*. Examination of the most primitive known cyclopygid, which also happens to be the oldest, reveals the primitive glabellar form (in all later cyclopygids the glabella becomes effaced) while having the typical synapomorphy for the group (hypertrophied eyes). Out-group comparison using this, and features of the thorax, shows that the Nileidae are the sister-group of the Cyclopygidae, and that the *Treatise* assignment of the Nileidae to the Asaphacea was incorrect. This comparison was again based on an early nileid, because, like the cyclopygids, later ones are effaced (and hence their homeomorphic resemblance to asaphids). Primitive nileids and cyclopygids both lack genal spines, the loss of which is normally accepted as an advanced character in trilobites. The family Taihungshaniidae also has the same glabellar form and thoracic structure, but genal spines are present, hence the Taihungshaniidae is the sister-group of Nileidae + Cyclopygidae, while it has paired pygidial spines as its own defining derived character. The most rational arrangement of these families results in the cladogram shown in Fig. 17. None of these early relationships are known in sufficient detail for a species-to-species tree, so the construction of a cladogram is a reasonable approach here, and stratigraphy is used only in the general sense of using early representatives of families to help the identification of characters useful in out-group comparisons (in this case the primitive form of the glabella which serves to link the three families). However, *within* the family Nileidae several species-to-species transitions are known, and one of these is of particular interest in the present context. There is a lineage leading from *Symphysurus arcticus* to *Peraspis erugata* which has been documented by a continuous series of samples from the early Ordovician of Spitsbergen (Fortey 1975, p. 38). In this lineage genal spines are progressively regenerated. As we have noted, genal spines are usually primitive, and their loss derived, and so here we have an example of secondary "pseudo-primitive" derivation, which is a minor parallelism with the truly primitive character state for the group Nileidae + Cyclopygidae + Taihungshaniidae. Genal spines are a character, being exoskeletal and peripheral, of low positional burden, and such a reversal is perhaps not surprising. But the point is that, without the stratigraphic species-lineage, stratigraphically well above the known origins of the family, it *would not have been possible* to decide whether the spinose nileid was primitive or derived. If a cladogram of the Nileidae had been constructed, using morphology and *a priori* notions of primitive and derived characters *alone*, there is no doubt that *Peraspis* would have found its place as the most primitive nileid, which the stratigraphic species-lineage demonstrates it is not. The stratigraphic evidence was invaluable in eliminating such a false trail. Small-scale "reversals" of this kind, involving characters of low burden, often occur in species-to-species lineages. The rejection by extreme cladists of such evidence is misguided.

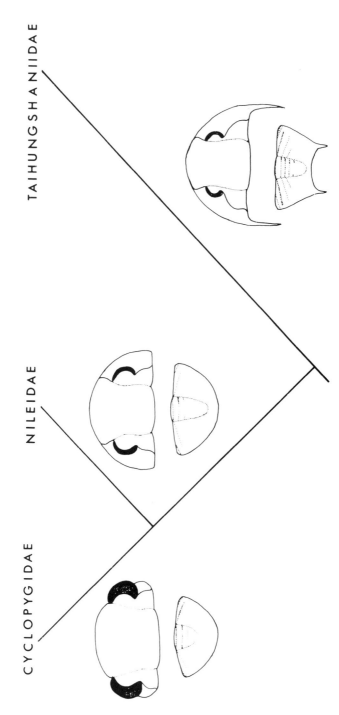

CYCLOPYGIDAE    NILEIDAE    TAIHUNGSHANIIDAE

Fig. 17. Cladogram to show relationships between families of trilobites in the superfamily Cyclopygacea, with cephalons and pygidia of a primitive representative of each family.

THE MORAL

To sum up, we would plead for tolerance. Sometimes stratigraphy is all-important in reconstructing phylogeny; sometimes it is misleading and ought to be disregarded; sometimes homologies can be recognized with certainty; sometimes they are highly problematical; sometimes fossils can be used in phylogenetic reconstruction without reference to stratigraphy; sometimes they can be used in combination with stratigraphy; sometimes they cannot be used at all; and sometimes phylogeny is in principle not reconstructable. The two important factors to be considered when deciding methodology will be the burden of the features compared, on the one hand, and the completeness of the record on the other. Phylogeneticists, whether of the stratigraphical or cladistic camp, should hesitate before, on the basis of their own experience, they legislate for other people. But each camp should be ready to learn from the other.

ACKNOWLEDGEMENTS

We gratefully acknowledge the help of Mr D. N. Lewis who drafted Figs 10 to 16 and reconstructed the phylogeny as shown in Figs 15 and 16.

REFERENCES

Bretsky, S. S. (1979). Recognition of ancestor-descendant relationships in invertebrate paleontology. *In* "Phylogenetic Analysis and Paleontology" (J. Cracraft and N. Eldredge, eds), pp. 113–163. Columbia University Press, New York.

Campbell, K. S. W. (1975). Cladism and phacopid trilobites. *Alcheringa* 1, 87–96.

Cracraft, J. and Eldredge, N. (eds) (1979). "Phylogenetic Analysis and Paleontology". Columbia University Press, New York.

Eldredge, N. (1973). Systematics of Lower and Middle Devonian species of the trilobite *Phacops* Emmrich in North America. *Bull. Am. Mus. nat. Hist.* **151**, 285–338.

Eldredge, N. (1979). Cladism and common sense. *In* "Phylogenetic Analysis and Paleontology" (J. Cracraft and N. Eldredge, eds). Columbia University Press, New York.

Eldredge, N. and Gould, S. J. (1972). Punctuated equilibria: an alternative to phyletic gradualism. *In* "Models in Paleobiology" (T. J. M. Schopf, ed.), pp. 82–115. Freeman, Cooper & Co., San Francisco.

Fortey, R. A. (1975). The Ordovician trilobites of Spitsbergen. II. Asaphidae, Nileidae, Raphiophoridae and Telephinidae of the Valhallfonna Formation. *Skr. Norsk. Polarinst.* **162**, 1–207.

Gingerich, P. D. (1976). Paleontology and phylogeny: Patterns of evolution at the species level in early Tertiary mammals. *Am. J. Sci.* **276**, 1–28.

Gould, S. J. and Eldredge, N. (1977). Punctuated equilibria: the tempo and mode of evolution reconsidered. *Paleobiology* **3**, 115–151.

Hennig, W. (1966). "Phylogenetic Systematics." University of Illinois Press, Urbana.

Jefferies, R. P. S. (1973). The Ordovician fossil *Lagynocystis pyramidalis* (Barrande) and the ancestry of amphioxus. *Phil. Trans. R. Soc. Lond. B.* **265**, 409–469.

Jefferies, R. P. S. (1979). The origin of chordates – a methodological essay. *In* "The Origin of Major Invertebrate Groups" (M. R. House, ed.), pp. 443–477. Systematics Association Special Volume No. 12. Academic Press, London and New York.

Jefferies, R. P. S. and Lewis, D. N. (1978). The English Silurian fossil *Placocystites forbesianus* and the ancestry of the vertebrates. *Phil. Trans. R. Soc. Lond.* **B282**, 205–323.

Løvtrup, S. (1977). "The Phylogeny of Vertebrata." John Wiley, London and New York.

Riedl, R. (1975). "Die Ordnung des Lebendigen". Paul Parey, Hamburg and Berlin.

Riedl, R. (1979). "Order in Living Organisms". John Wiley, London and New York.

Schaeffer, B., Hecht, M. K. and Eldredge, N. (1972). Phylogeny and paleontology. *Evol. Biol.* **6**, 30–46.

# 7 | Directions of Evolution in the Mammalian Dentition

P. M. BUTLER

*Department of Zoology, Royal Holloway College,
Egham, Surrey, England*

Mammalian teeth occupy a unique place in evolutionary studies. Owing to their durability they are abundantly represented in the fossil record, so that the time dimension can be taken into account when assessing character states. There are of course gaps, particularly during the Mesozoic, and even during the Early Tertiary their history is inadequately known in much of the world outside Europe and North America. Nevertheless, we have a reasonably good idea of the general course of mammalian dental evolution. This evolution can be interpreted in functional terms, for by studying the worn surfaces of teeth inferences can be made about their mode of action during life, thus relating structure to function.

The richness of the fossil record allows the use of a stratigraphical criterion for determining evolutionary polarity: primitive characters are older than derived characters. This means that character states that become more frequent as a group is traced back in time are likely to be more primitive than those that become less frequent. Thus Cope (1883a) found that in the oldest Tertiary mammalian fauna, from the Lower Palaeocene (Puerco) Beds of New Mexico, 38 out of 41 species had upper molars with three principal cusps, and 35 of these were triangular, whereas many Eocene and later mammals had quadrate molars with four cusps. In this way he was able to recognize the tritubercular type as the ancestral form of upper molar.

Systematics Association Special Volume No. 21, "Problems of Phylogenetic Reconstruction", edited by K. A. Joysey and A. E. Friday, 1982, pp. 235–244, Academic Press, London and New York.

There are many mammals from the Palaeocene and Eocene that show intermediate stages in the development of a fourth cusp, the hypocone (Osborn 1888), from a cingulum ledge at the posterior–internal margin of the tooth (Fig. 1). These mammals belong to many different families in several orders – Insectivora, Primates, Condylarthra, Artiodactyla, Rodentia and so on. We have here an example of widespread parallel evolution, constituting an evolutionary trend. It is because so many groups were evolving in the same direction that the polarity of change can be determined statistically, without the necessity of tracing individual phyletic lines.

Teeth are particularly prone to parallel evolution, and many characters can be tested against time in this way. Cope (1874) was able to show that in ungulates bunodont teeth, with cusps, went back farther in time than lophodont teeth, in which the crown is crossed by ridges, as in the tapir. Earlier workers, e.g. Rütimeyer (1863), had held that the lophodont condition was the more primitive and that cusps had been formed by the breaking up of ridges. Lophodont teeth evolved in tapirs, hyracoids, rhinoceroses, elephants and several other ungulate groups, as well as in rodents and kangaroos.

Another widespread trend is the evolution of hypsodonty, in which the height of the crown is increased by prolongation of its growth, and root formation is postponed or even never reached, so that the crown continues to grow throughout life. Hypsodonty evolved independently in many families, mainly ungulates and rodents, in North America during the Oligocene and Miocene. It is clearly an adaptation to compensate for the abrasion that results from eating grass or grit-contaminated food (Stirton 1947). Its evolution in North America was associated with the spread of savanna vegetation (Webb 1977), and a similar development took place independently during the Oligocene in South America (Webb 1978).

Hypsodonty illustrates the adaptiveness of an evolutionary trend. To find a functional explanation for the development of the hypocone it is necessary to consider some detailed aspects of molar occlusion. Cope (1883b) showed that tritubercular upper molars were associated with lower molars of a type that he called tuberculo-sectorial; today we speak of both types together as tribosphenic molars (Simpson 1936). Tribosphenic lower molars consist of two parts: an anterior triangle of three cusps, the trigonid, fitted into

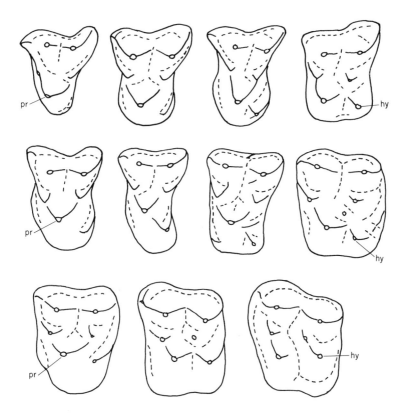

Fig. 1. Left upper molars to illustrate the evolution of the hypocone. (Top) Erinaceoid insectivores. From left to right, *Leptacodon ladae* (Palaeocene), *Litolestes ignotus* (Palaeocene), *Scenopagus mcgrewi* (Eocene), *Proterix loomisi* (Oligocene). (Middle) Primates. From left to right, *Purgatorius unio* (Palaeocene), *Palenochtha minor* (Palaeocene), *Hemiacodon aracius* (Eocene), *Microchoerus erinaceus* (Oligocene). (Bottom) Ungulates. From left to right, *Litoletes disjunctus* (Palaeocene), *Hyopsodus paulus* (Eocene), *Hyracotherium tapirinum* (Eocene). Anterior end to the left, external (buccal) above. hy, hypocone; pr, protocone.

the embrasure between two upper molars, and a posterior heel, or talonid, bit against the inner cusp (protocone) of an upper molar (Fig. 2). The functional operation of this system has been analysed in detail by the study of wear facets, aided by cineradiographic studies of living mammals, such as the opossum and tree-shrew, whose teeth retain primitive features (Crompton and Hiiemae 1970; Hiiemae and Kay 1973). It is now known that tribosphenic molars

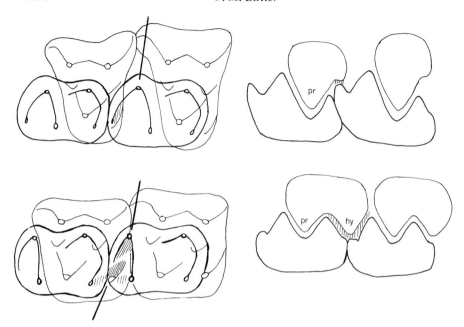

Fig. 2. (Left) Superimposed drawings of upper and lower molars to show the effect of hypocone development. The heavy straight lines show the direction of medial movement during the chewing stroke. Areas that come into contact with the hypocone are shaded. (Right) The same teeth in medial (lingual) aspect. hy, hypocone; pr, protocone. Anterior end to the left.

appeared early in the Cretaceous, and a few genera developed small hypocones before the end of that period (Crompton and Kielan-Jaworowska 1978; Butler 1977).

Most Cretaceous tribosphenic molars have high cusps with strongly developed crests, adapted for puncturing and cutting. The presence of such features in living insectivores, together with the small body size of Mesozoic mammals, indicates a mainly insectivorous diet, although it must be remembered that all the living Insectivora have departed from the primitive molar pattern in one way or another. The puncturing–cutting adaptations can be traced back to *Kuehneotherium* in the Late Triassic or earliest Jurassic, and throughout the Jurassic the molars of therian mammals functioned entirely in this way (Butler 1972b). Only in the Cretaceous was a crushing function added, by the development of the protocone on the upper molar and enlargement of the talonid on the lower molar. The trigonid retained its primitive structure and function.

The hypocone originated from a cingulum ledge behind the protocone. As the teeth closed together this cingulum met the edge of the trigonid of the more posterior lower molar, perhaps preventing the trigonid from penetrating too deeply between the upper teeth and damaging the gum. Widening of the cingulum and its elevation to form a cusp increased the area of contact. Eventually the hypocone filled the embrasure between the teeth, and it opposed the trigonid in much the same way as the protocone opposed the talonid. Ungulates, primates and rodents developed a more horizontal, grinding mode of chewing in which the trigonid was reduced in height to the level of the talonid, the anterior trigonid cusp (paraconid) degenerated, and the hypocone came to resemble the protocone in size and function (Fig. 2). In these mammals, hypocone development was an aspect of an adaptive shift from an insectivorous to an omnivorous or herbivorous diet.

In the carnassial teeth of carnivorous mammals the trigonid was modified in a different way. Its shear against the upper molar anterior to it was exaggerated, while the talonid–protocone crushing function was reduced (Fig. 3). This trend affected a number of lineages of placental creodonts and carnivorous marsupials. In modern Carnivora only two teeth in each jaw were modified, forming the deciduous and permanent carnassials; the more posterior part of the dentition was adapted for crushing (Butler 1946).

The length of cutting edges was increased in a different way in molars described as dilambdodont (Fig. 3). The lateral cusps of the upper molar (paracone and metacone) are V-shaped, together forming a system of crests in the form of a W; an inverted W is formed on the lower molar. Such teeth occur in some ungulates, including Perissodactyla (*Palaeotherium*, brontotheres, chalicotheres), Artiodactyla (*Anoplotherium*) and Pantodonta (*Pantolambda*), where they are believed to be adaptive for leaf-eating, but they also occur in insectivorous mammals, such as *Tupaia* and shrews, where the cutting up of insect cuticle is a more likely function. In dilambdodont molars the groove between the paracone and the metacone becomes deeper and extended laterally. A new cusp, the mesostyle, develops at the lateral end of the groove. The effect

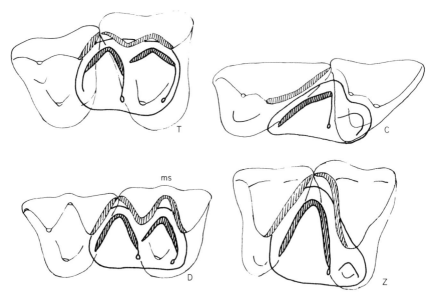

Fig. 3. Superimposed drawings of upper and lower molars to illustrate types of specialization of shearing crests (shaded). T, tribosphenic (primitive); C, carnassial; D, dilambdodont; Z, zalambdodont. Anterior end to the left. ms, mesostyle.

is to lengthen the crests that cut against the lateral talonid cusp (hypoconid) as it passes up the groove during chewing.

Throughout the modifications of molar pattern, the basic inter-relations of upper to lower cusps are retained: the hypoconid always occludes between the paracone and the metacone, the protocone always passes between metaconid and entoconid, and so forth. Homologous cusps have homologous wear facets, and the history of each facet can be traced back in time, some of them to the Triassic (Crompton 1971). This constancy of occlusal relations is presumably due to the necessity of functional continuity: trans-formation is possible only if all the intermediate steps are viable. A cusp that occludes in front of another cusp cannot come to occlude behind it, for at an intermediate stage the two cusps would meet tip to tip, resulting in malocclusion or the elimination of one of the cusps. The complexity of the occlusal system limits the number of ways in which it can change, and increases the probability of parallel evolution. Indeed, parallel evolution in the dentition is the rule, rather than the exception: comparatively few characters are unique to a single taxon. When thinking of occlusal relations we have to remember that they are only part of a larger functional system that involves the masticatory muscles and the neural mechanisms that control them. A basic pattern of chewing movements can be recognized in mammals (Hiiemae 1978).

The major trends of dental evolution can be reasonably well understood in terms of function. The processes of puncturing, cutting, crushing and grinding can be analysed, and deductions can be made about the physical properties of the food eaten (Lucas 1979). This can be done even for fossils but only in living mammals can the adaptations of teeth to diet be tested directly. There are some striking cases in which unusual diets are associated with unusual dentitions, such as the vampire bat and the ant-eating hyaena (*Proteles*). In phyllostomatid bats, modification of the cheek teeth, as fruit-eating has been substituted for insect-eating, can be traced in series of living genera. The more subtle differences that distinguish closely related genera and species have been investigated mainly in primates. For example, Seligsohn and Szalay (1978) made a functional analysis of the dentitions of two lemurs which feed respectively on tough leaves and on bamboo shoots, and Kay (1978), by a multivariate analysis of tooth measurements, showed that molar structure in Old World monkeys is related to the proportion of leaves to fruit in the diet. Freeman (1979) found differences in the teeth between beetle-eating and moth-eating molossid bats. More work needs to be done, but it seems reasonable to believe that even minor differences in the teeth are subject to natural selection.

What is known about the ontogeny and genetics of the dentition indicates that the relation between genes and morphological characters is very indirect (Butler 1978). Underlying the morphological pattern is a complicated sequence of mesenchyme–epithelial interactions, inductions and cell migrations whose genetical control is far from being understood. It is very difficult to imagine how a cusp, for example, which develops as a fold in a layer of epithelium, could arise as the specific consequence of a genetic mutation. The characters that we use to describe the dentition are unlikely to be represented individually in the genome.

In an analogous way, if we think of the dentition from a functional standpoint, as an organ concerned with the intake and comminution of food, the characters that we pick out on individual teeth are seen as evolving not independently but as parts of a larger system. It is the dentition that becomes adapted; the teeth and their cusps change in accordance with the part they play in the total functional complex.

It is in this light that we should think of evolutionary reversals. There are many cases where characters have been lost on some teeth, while other teeth have become more specialized. Dedifferentiation of the carnassial teeth has occurred in several families of Carnivora: bears, procyonids, mustelids and viverrids. In each case it is accompanied by an enlargement of the crushing area of the molars, indicative of an adaptive trend from a carnivorous to an omnivorous or herbivorous diet (Butler 1946). The widespread tendency for the posterior premolar to resemble the molars was reversed in the hedgehogs (Erinaceidae) when the first lower molar became specialized (Krishtalka 1976). Reduction or loss of canines was associated with enlargement and specialization of incisors in the rodents and several other orders. Within single teeth, changes of the total pattern may result in the loss of some cusps while others are enlarged; thus in the zalambdodont molars of some insectivores and marsupials (Butler 1972) secondary loss of the protocone and metacone accompanied an emphasis on the cutting crests of the paracone (Fig. 3). Even the common trend to reduce the number of teeth can be reversed. *Otocyon* is a dog with four molars (Van Valen 1964), and the manatee produces molars throughout life. The large number of molars that some Jurassic mammals such as *Spalacotherium* and the Dryolestidae have is quite possibly a derived character; it was accompanied by small tooth size, and increased the number of points and cutting blades in a given length of jaw.

Thus, when considered in isolation, the evolution of dental characters may be reversed, but such reversals are secondary consequences of progressive evolution of a wider functional system to which the characters are subordinate. Evolution of the dentition, like that of other complex structures, seems to be irreversible. Readaptation to more primitive conditions may result in some degree of convergence towards an ancestral state, but the effect of previous adaptations is never completely lost. The rodent *Deomys*, which has become readapted as an insectivore, has developed high molar cusps, slender incisors and a more vertical chewing action, but its dentition remains unmistakably that of a rodent (Lemire 1966).

The conclusions that I think may be drawn from a consideration of dental evolution in mammals may be summarized as follows.

(1) Primitive character states are older than derived states, and they should occur more frequently in earlier members of a phyletic group than in later members.

(2) Derived states tend to be produced independently in more than one phyletic line, by parallel evolution. This is due to limitations on the mode of change of complex systems, imposed by the necessity for functional continuity.

(3) Evolution is adaptive, and characters should be considered in relation to the functional system in which they take part, and to which they are subordinate.

(4) The evolution of complex functional systems is progressive and irreversible, even though reversed evolution of single characters frequently takes place.

REFERENCES

Butler, P. M. (1946). The evolution of carnassial dentitions in the Mammalia. *Proc. zool. Soc. Lond.* 116, 196–220.

Butler, P. M. (1972a). The problem of insectivore classification. *In* "Studies in Vertebrate Evolution" (K. A. Joysey and T. S. Kemp, eds.), pp. 253–265. Oliver and Boyd, Edinburgh.

Butler, P. M. (1972b). Some functional aspects of molar evolution. *Evolution* 26, 474–483.

Butler, P. M. (1977). Evolutionary radiation of the cheek teeth of Cretaceous placentals. *Acta palaeont. pol.* 23, 241–271.

Butler, P. M. (1978). The ontogeny of mammalian heterodonty. *J. Biol. buccale* 6, 217–227.

Cope, E. D. (1874). On the homologies and origin of the types of molar teeth of Mammalia Educabilia. *J. Acad. nat. Sci. Philad.*, ser. 2, 8, 71–89.

Cope, E. D. (1883a). Note on the trituberculate type of superior molar and the origin of the quadrituberculate. *Am. Nat.* 17, 407–408.

Cope, E. D. (1883b). On the trituberculate type of molar tooth in the Mammalia. Palaeontological Bulletin No. 37. *Proc. Am. phil. Soc.* 21, 324–326.

Crompton, A. W. (1971). The origin of the tribosphenic molar. *In* "Early Mammals" (D. M. Kermack and K. A. Kermack, eds). *Zool. J. Linn. Soc.* 50, Suppl. 1, 65–87.

Crompton, A. W. and Hiiemae, K. M. (1970). Molar occlusion and mandibular movements during occlusion in the American opossum *Didelphis marsupialis* (L.). *Zool. J. Linn. Soc.* 49, 21–47.

Crompton, A. W. and Kielan-Jaworowska, Z. (1978). Molar structure and occlusion in Cretaceous therian mammals. *In* "Development, Function and Evolution of Teeth" (P. M. Butler and K. A. Joysey, eds), pp. 249–287. Academic Press, London and New York.

Freeman, P. W. (1979). Specialized insectivory: beetle-eating and moth-eating molossid bats. *J. Mammal.* 60, 467–479.

Hiiemae, K. M. (1978). Mammalian mastication. A review of the activity of the jaw muscles and the movements they produce in chewing. *In* "Development, Function and Evolution of Teeth" (P. M. Butler and K. A. Joysey, eds), pp. 359–398. Academic Press, London and New York.

Hiiemae, K. M. and Kay, R. F. (1973). Evolutionary trends in the dynamics of primate mastication. *Symp. 4th int. Congr. Primatol.* 3, 28–64.

Kay, R. F. (1978). Molar structure and diet in extant Cercopithecidae. *In* "Development, Function and Evolution of Teeth" (P. M. Butler and K. A. Joysey, eds), pp. 309–339. Academic Press, London and New York.

Krishtalka, L. (1976). Early Tertiary Adapisoricidae and Erinaceidae (Mammalia, Insectivora) of North America. *Bull. Carnegie Mus. nat. Hist.* 1, 1–40.

Lemire, M. (1966). Particularités de l'appareil masticateur d'un rongeur insectivore, *Deomys ferrugineus* (Cricetidae, Dendromurinae). *Mammalia* 30, 454–494.

Lucas, P. W. (1979). The dental-dietary adaptations of mammals. *Neues Jb. Geol. Paläont. Mh.* 1979, 486–512.

Osborn, H. F. (1888). The nomenclature of the mammalian molar cusps. *Am. Nat.* 22, 926–928.

Rütimeyer, L. (1863). Beitrage zur Kenntniss der fossilen Pferde und zur vergleichenden Odontographie der Hufthiere überhaupt. *Verh. naturf. Ges. Basel* 3, 358–696.

Seligsohn, D. and Szalay, F. S. (1978). Relationship between natural selection and dental morphology: tooth function and diet in *Lepilemur* and *Hapalemur*. *In* "Development, Function and Evolution of Teeth" (P. M. Butler and K. A. Joysey, eds), pp. 284–307. Academic Press, London and New York.

Simpson, G. G. (1936). Studies of the earliest mammalian dentitions. *Dent. Cosmos* 78, 791–800.

Stirton, R. A. (1947). Observations on evolutionary rates in hypsodonty. *Evolution* 1, 32–41.

Van Valen, L. (1964). Nature of the supernumerary molars of *Otocyon*. *J. Mammal.* 45, 284–286.

Webb, S. D. (1977). A history of savanna vertebrates in the New World. Part I: North America. *A. Rev. Ecol. Syst.* 8, 355–380.

Webb, S. D. (1978). A history of savanna vertebrates in the New World. Part II: South America and the Great Interchange. *A. Rev. Ecol. Syst.* 9, 393–426.

# 8 | Computers Versus Imagination in the Reconstruction of Phylogeny

R. A. CROWSON

*Department of Zoology, University of Glasgow, Glasgow, Scotland*

I will take as my starting point Hennig's (1979) reprinted book "Phylogenetic Systematics", and the review of it in *Nature* by Martin (1980). I too have written a book advocating strictly phylogenetic classification (Crowson 1970) – but as an aim rather than a technique. The reviewer Martin insisted that the determining of phylogenetic relations and the formalization of a classification are logically separate processes. It is not altogether clear from his review whether he would advocate the retention of traditional taxa of modern organisms in the face of clear evidence for polyphyletic origins, but he evidently favours the retention of fossil taxa which would be paraphyletic in Hennig's sense. Martin praises without reservation Hennig's prescriptions for establishing phylogenetic relations, but objects to the use of the resulting dendrograms as the sole basis for classifications. My attitude is almost the reverse of his, in that I fully accept Hennig's ultimate aim but I am sceptical of his methods.

Hennig, as a German professor, strove to give phylogenetic classification academic respectability in his time by providing it with a formalized theory and more or less algorithmic procedural rules. It is these aspects of his book which have been most enthusiastically taken up by many young systematists, particularly in

Systematics Association Special Volume No. 21, "Problems of Phylogenetic Reconstruction", edited by K. A. Joysey and A. E. Friday, 1982, pp. 245–255, Academic Press, London and New York.

America, and it is the diverse dendrograms so produced which have tended to give substance to Martin's objection that Hennig's methods "inevitably lead to a proliferation of alternative classifications and instability of higher taxonomic categories". Martin himself describes such dendrograms as "phylogenetic hypotheses" and suggests that all we can do with them is to attach varying degrees of probability. I do not consider that the proper thing to do with a scientific hypothesis is to attach a probability figure; a hypothesis should rather be used as a basis for verifiable (or falsifiable, if you are a Popperian) predictions. We should use phylogenetic hypotheses to make predictions about as yet unobserved but observable characters of living or fossil representatives of the taxa concerned. Only dendrograms which have survived such tests should be used as bases for classifications.

According to Hennig's procedures, once you have correctly assessed the plesiomorph and apomorph (I prefer the good old terms primitive and derivative) states of all the "characters" in your taxon, and distinguished the polyphyletic from the monophyletic among the derivative characters, you are in the position to construct a correct dendrogram and a formal phylogenetic classification. All the difficulties lie in the preliminary determinations about character-states, and for these Hennig is unable to formulate any simple and objective algorithms. Four different criteria have been suggested and applied in determining the primitive state of a character in a taxon. The first (and, where applicable, in my view the best) is the condition in the oldest fossils of the group; the second, the condition in the most primitive living members of the taxon; the third, the condition in the most nearly related species outside the taxon; and the fourth, the condition in the majority of the Recent forms. The first requires the availability of suitable fossils, the second and third depend on prior phylogenetic ideas about the taxon concerned, and the fourth is clearly very unreliable. To distinguish monophyletic from polyphyletic derived characters, there is really only one general criterion, that the true phylogeny will be that involving polyphyly in the fewest and simplest characters, a form of the parsimony principle discussed later.

In establishing the primitive state of each character in a taxon we are in effect building up a specification for its ancestral species, and that specification is really the fundamental hypothesis emerging from our study. To my mind, the most important test of this hypothesis is the question: *Is the postulated ancestor imaginatively credible as a living species, with a definite mode of life to which its characters are well adapted?* A common way of evading this criterion is to postulate that an ancestral species must be what the Americans call "generalized", i.e. without specific adaptations to any particular mode of life. I see no reason to suppose that ancestral species were any more generalized in this respect than most of their descendants. They must have been capable of survival as members of more or less complex and competitive ecosystems.

The requirements for a true picture of phylogenetic history in any taxon are, not merely that the ancestral species should have an imaginatively credible set of characters and mode of life, but also that the lines from it to its various descendants must be through series of adaptations, each of which is ecologically and morphologically credible in relation to the circumstances of the time. I cannot see how it could be possible to develop algorithms for these requirements, so I can see no prospect of computers replacing the functions of the controlled human imagination in developing such hypotheses.

Imagination has recently been defined by Mary Warnock as "the faculty by which we go beyond what is given". Literally, it means image formation; in phylogenetic studies its major function is in the actualization of ancestral species as living organisms, while in the "hypothetico-inductive" philosophy of Popper, Medawar and others, it is essential for the formulation of hypotheses (note that these philosophers never suggest that hypothesis-formation is a function for computers). There is a rather different but scientifically important mental function which may also be called imaginative, one for which Koestler (1967) coined the word "bisociation". This is the sudden apprehension of a possible link between items of information from previously unconnected fields of study. In phylogenetic studies, one important type of bisociation is the apprehension of possible interactions or parallels between one's own taxon and other probably unrelated ones. This function too can hardly be put into algorithmic form, and must depend on the extent of the bisociator's knowledge outside his own taxon.

There appears to be a psychological antagonism between imagination, which tends to particularize, and abstraction, which tends to generalize. Imaginative minds tend to be what some psychologists have called "divergers", while lovers of abstraction and algorithms are mainly "convergers". Each type of mind has its characteristic virtues and vices, and its particular part to play in the development of science. In historical studies like phylogeny, I think the imagination, properly disciplined by respect for evidence, has a very important part to play.

A domain of phylogeny in which the algorithm-loving convergers have hitherto been dominant is protein systematics, and particularly the study of amino acid sequences of homologous polypeptide chains, e.g. by Dayhoff (1969), Goodman and Moore (1974), Romero-Herrera et al. (1978), Lyddiatt et al. (1978). There are obvious reasons why comparative data of this type are less amenable to traditional or imaginative treatment than are most others. For one thing, the data are obtained by difficult, complex and expensive techniques, requiring well-equipped biochemical laboratories and technicians, and information available is usually from very small and poorly representative samples of taxa. Secondly, we do not have and are unlikely ever to get data of this kind for fossil species. Thirdly, there is already good evidence that unit sequence changes (replacements of particular residues, terminal additions or deletions) are liable to be reversible and polyphyletic. Finally, we can very rarely attribute adaptive functions to such changes.

On the other hand, protein sequences lend themselves better than most types of systematic data to algorithmic treatment. The "alignments" of homologous proteins can be treated as matrices, and analysed in the manner of the "numerical taxonomists". Using algorithmic methods on cytochrome c sequences, Dayhoff (1969) and others have produced reasonable approximations to a "true" phylogenetic dendrogram of the Vertebrata and even of Eukaryota as a whole. It was soon realized that there were several different algorithms which could be used with such data, and they commonly produced different dendrograms. A "Principle of Parsimony" was introduced, according to which that dendrogram should be selected which involved the fewest total sequence changes in the evolution of the taxon. There is little empirical evidence to support this principle, except for the fact that its introduction did tend to produce dendrograms in slightly better accord with other lines of evidence.

In Crowson (1972), I suggested that the most systematically useful and probably the most biochemically suggestive way of treating protein sequence data was to attempt to define taxa by sequence particularities. Such definitions would be true hypotheses, in the sense that they could be tested by studying further species of the taxa concerned. They would also facilitate the detection of apparent correlations of sequence particularities with other classificatory characters of more evident adaptive significance, which might in turn suggest functional relations that biochemists might find worthy of investigation.

My suggestions were, however, ignored by the cytochrome c workers; all that came of their costly labours were computer-generated dendrograms in which the biochemists themselves could see little value and in which no one else appeared to be seriously interested. No effective links were established with systematists, and after about 1975 interest in cytochrome c sequences largely ceased outside Professor Boulter's laboratory at Durham. In my view, the greatest potential value of cytochrome c sequence data in zoology is in providing evidence for phylogenetic relations in major groups of Invertebrata when fossil evidence is scanty or lacking, as has been indicated by recent results from Professor Boulter's school (Lyddiatt *et al.* 1978). By analogy, one might expect its greatest systematic value in botany to be in providing evidence on interrelations of major groups of Gymnospermae, Pteridophyta, Bryophyta and so on.

In another group of respiratory proteins, myoglobin and its derivatives the haemoglobins, the development of sequence studies has taken rather different lines. In the first place, there has been a much greater tendency to expect functional differences in sequence particularities, because of evidence that the presence of unusual residues at particular points in human haemoglobin chains is apt to have visible phenotypic and pathological effects. Furthermore, because the globins are much more effective antigens than cytochrome c, there has been a good deal of serological work on them, which has served as a link between protein sequence studies and traditional systematics. The more recent workers in the field, notably Romero-Herrera *et al.* (1978) on myoglobin of Vertebrata, have paid much greater attention to other lines of evidence in selecting between alternative possible dendrograms, and have seriously

considered possible adaptive significances and parallels in the postulated substitutions. This seems to be a thoroughly healthy development, boding well for the future.

Looking at the 1978 data from the point of view of a systematist,
the first thing to be noted is the inadequately representative nature
of the recorded sequences. The Primates are heavily over-represented
in comparison with mammalian groups like Rodentia or Chiroptera,
and important groups like Xenarthra and Subungulata are entirely
unrepresented; non-mammalian vertebrates are represented only
by two birds, a shark and a lamprey. The second weakness is the
unreliability of some of the published sequences (errors in several
of those shown in the alignment are admitted in a footnote). However, it appears that some of the major taxa have distinctive residues.
Thus the Mammalia as a whole appear to differ from the few other
recorded vertebrates by having Gly at position 5, Ile at 30, Glu at 41,
Lys at 42, Asp at 53, Glu at 54, Ser at 117 and Ala at 144; the
two marsupials, one American and one Australian, have Gln at 102
and Phe at 103, not found in the placentals, while a single monotreme has Gly at 19, Gly at 48, Ala at 59, Ser at 100, Phe at 103,
Tyr at 106 — the Gly at 48 common to the birds, the Phe at 103
to marsupials.

At least some of the peculiarities of mammalian myoglobin (and
haemoglobins) might well be related to the origins of homothermy
in the group, and if so, one might expect some parallels in the Aves,
but until we have reliable sequences for a representative series of
Reptilia and Amphibia, such parallels will be difficult to detect.
I hope that in the not too distant future we may see myoglobin
sequences for at least a crocodile, a snake or lizard, a turtle, one or
two amphibians, a teleost and a lungfish.

To sum up, I am prepared to admit that algorithmic procedures
played a useful part in the early development of protein sequence
studies. Unless such studies succeed in breaking out of the limits
of such procedures, however, they are likely to prove unsatisfying
and to be virtually abandoned by biochemists. The direction of
progress, if there is to be any, must be towards integration of protein
sequence data with everything else that is known about the
organisms concerned. In this progress, imagination is likely to play
an increasing part.

Akin to algorithms are those abstract concepts which we all use to a greater or less extent when thinking about evolutionary history. In geology I would cite the Permanence of the Continents and Oceans and the Uniformitarian Principle, in biology such things as the Unity of the Organism, Dollo's Law, Homology, Recapitulation, the Principle of Parsimony, Metameric Segmentation, the Primitiveness of the Generalized, and the Competitive Exclusion Principle in Ecology. All of these embody at least some elements of truth, but all have led to false conclusions when unimaginatively applied. Some we have already discussed in this meeting. All of them merit extended consideration, but I have time to consider only two here.

The idea that all the parts and functions of an individual, at least in a higher animal, are directly or indirectly connected and mutually influencing has some force; one conclusion often drawn from it is the "*ex pede Herculem*" principle, the idea that the whole organism is somehow implicit in and could be inferred from any one of its parts. This idea, rarely openly avowed nowadays, seems to be present in the minds of those who set out to deduce the phylogeny of birds from comparative studies of the syrinx, of mammals from myoglobin sequences, of insects from the wing venation, or of Angiospermae from the gynaecium alone. In phylogenetic studies we need to take account of a converse generalization, which I have called the Non-congruence Principle and others refer to as Mosaic Evolution – stating that evolutionary change tends to operate on one character at a time. It is very characteristic of phylogeny, and of historical studies generally, that they constantly call for the use of contrary concepts, in the spirit of Hegelian dialectics and of some of the later writings of Darlington.

The second abstract concept I wish to consider here is Metameric Segmentation. Its root idea is a basic analogy between the organization of the chordate–vertebrate line on one hand and the annelid–arthropod one on the other. For these groups, the concept postulated ancestors with a body composed of a series of somites, with the mouth at one end and the anus at or near the other, rather like an annelid worm. From such archetypes, it was thought, higher groups developed by differentiation of originally similar somites and by aggregations of groups of them into more or less consolidated tagmata. This picture fits fairly well with acutal evidence in the annelid–arthropod line, and there was inculcated in the young of my generation a strong presumption that the evolution of the chordate–vertebrata line would follow a similar pattern. There was a major attempt, by Goodrich and others, to interpret the vertebrate head as a tagma built of fused somites. Interest was concentrated on those chordate types most suggestive of a worm-like ancestry – the Enteropneusta rather than Pterobranchia, Cephalochordata rather than Tunicata, and the ammocoete larva rather than the tadpole. A recent sign of this attitude was Attenborough's *Life on Earth* television series, referring to the oldest fossil Vertebrata as "like armour-plated lampreys". To a mind free of such preconceptions, I think it would be much more natural to describe them as "armour-plated tadpoles".

A tadpole, whether of a tunicate or an amphibian, is organized in two main parts, a large superficially unsegmented head containing not merely the brain and sense organs but also the entire alimentary canal and general viscera, and a segmented propulsive tail. We have very clear fossil evidence that the earliest Vertebrata had this type of organization, as far back as the Ordovician period. Compared with this pattern of organization the adult tunicate (except for Larvacea) has lost the tail, while *Amphioxus* and the ammocoete show extreme reduction of the head and its partial fusion with the anterior tail somites. If zoologists had taken the tadpole larva instead of *Amphioxus* as their prototype for vertebrate structure, they would have been far less reluctant to accept the ideas of Jefferies (1968).

I do not know just how Jefferies reached his ideas about the nature and affinities of Calcichordata, but I myself first came to suspect these fossils (then known as Carpoidea) as possible ancestors of Vertebrata around 1950, and I even suggested this idea to students at that time, though not venturing to put my ideas into print. I had already then come to accept the tadpole theory of vertebrate origins, and to question the doctrines of vertebrate head segmentation. The Carpoidea then struck me as the only known fossils which offered a prototype for the tadpole, and which occurred far enough back to be ancestors of Vertebrata. Another thing which struck me at that time was that Carpoidea were always more or less asymmetrical externally, and that deep-rooted traces of head asymmetry were to be seen in most of the lower chordate types, e.g. in the coelom formation of the Tornaria larva, the formation of the gill-slits in Cephalochordata, the sense-organ of the head of tunicate tadpoles, the diencephalon and anterior cardinal veins of Cyclostomata. I did not then know that among the oldest fossil Vertebrata, at least some (Astraspidae) showed only an imperfect bilateral symmetry of the head-shield plates.

A major difference of course between Jefferies' Calcichordata and the early Vertebrata lies in the material of the skeleton — calcite in the former, phosphatic in the latter. Can we find any parallels to such a replacement in the animal kingdom? Possible ones exist I think in Brachiopoda and among the aquatic Arthropoda. Among the supposedly more primitive Brachiopoda–Inarticulata, some have a calcitic shell, some a phosphatic one, while all Articulata are reported to have calcitic shells. The very interesting thing is that the Palaeozoic Inarticulata with calcitic shells, tend to occur in association with typically marine forms like Echinodermata and Trilobita, as do the Calcichordata, whereas phosphatic-shelled types like the linguloids tend not to occur in close associations with echinoderms or Trilobita, but do sometimes occur with early Vertebrata. In the Arthropoda, some aquatic forms show varying degrees of mineralization of the exoskeleton, involving calcium carbonate, calcium phosphate, or mixtures of the two. Modern marine Crustacea are said to have carbonate in the skeleton, as probably did the fossil Trilobita, but among Chelicerata, there have been reports of phosphatic mineralization in some of the Palaeozoic fossils — and most of the Palaeozoic fossils of Chelicereta (Merostomata) occur in associations with early Vertebrata and linguloid Brachipoda rather than with typical marine faunas.

Could it be that the origins of the vertebrate, lingulid and merostomatan lines were all conditioned by shifts from the sea proper into brackish or fresh water? We encounter here the difficulty that there are no known pre-Devonian fossils which are generally accepted as of freshwater organisms. It seems to me highly improbable that there were no fossilizable Metazoa in fresh water even in the Silurian period, at a period just before the first colonization of the land by vascular plants and by Arthropoda, both of which are much more likely to have made this step from fresh water rather than straight from the sea. The existence today of obviously ancient high-level groups of freshwater Crustacea like the Notostraca and Conchostraca makes it likely that their ancestors were established in fresh water already in the Lower Palaeozoic.

Those who argue for a marine origin of Vertebrata assert that the earliest fossils attributed to the group, consisting of isolated plates of *Anatolepis* occurring in late Cambrian and early Ordovician deposits, are associated with marine fossils. However, it has been pointed out that isolated plates of this type are very similar to those of Calcichordata, and I doubt whether any of them have been chemically analysed.

The hypothesis that the precursors of Vertebrata moved from the sea proper into the brackish or fresh water of estuaries and rivers might explain some of their differences from Calcichordata. Estuarine and river environments tend to be less stable than marine ones, and would place a greater premium on effective motility for long-term survival. The first Vertebrata seem to have differed from their calcichordate ancestors in having larger and stronger tails, and more perfect bilateral symmetry, both suggestive of higher motility (which would be needed anyway to make way upstream against currents). Possibly the first Vertebrata could even swim a little, whereas calcichordates could only wriggle slowly and painfully for short distances on the bottom, perhaps tail first as Jefferies has suggested.

On the predictive side, it might be suggested that a move into estuarine or river life should have had considerable effects on development. It seems probable that the sluggish bottom-living Calcichordata would have relied on ciliated, possibly tornaria-like, larvae for their dispersal, developing from small yolk-poor eggs. Such larvae however are ill-adapted to life in rivers, where they are liable to be swept down stream by currents; freshwater animals generally eliminate them from their life-cycles, producing larger and yolkier eggs and relying on adults for dispersal. Thus using my hypothesis the Calcichordata might be expected to have produced large numbers of small eggs, whereas the early Vertebrata should have produced much fewer and larger ones. Some indications of these circumstances might well be detectable in fossils. Obviously, the hypothesis, if correct, will have implications for other groups of Lower Palaeozoic fossils, notably in the Arthropoda. Its consequences, in any case, will need to be considered imaginatively rather than algorithmically.

REFERENCES

Crowson, R. A. (1970). "Classification and Biology". Heinemann, London.

Crowson, R. A. (1972). A systematist looks at cytochrome c. *J. Mol. Evol.* 2, 28–37.

Dayhoff, M. (1969). "Atlas of Protein Sequence and Structure", vol. 4. National Biomedical Research Foundation, Silver Springs, Maryland.

Goodman, M. and Moore, G. W. (1974). Phylogeny of Hemoglobin. *Syst. Zool.* 32, 508–532.

Hennig, W. (1979). "Phylogenetic Systematics". University of Illinois Press, Urbana.

Jefferies, R. P. S. (1968). The subphylum Calcichordata: primitive fossil chordates with echinoderm affinities. *Bull. Brit. Mus. (Nat. Hist.) Geol.* 16, 243–341.

Koestler, A. (1967). "The Act of Creation". Dell, New York.

Lyddiatt, A., Peacock, D. and Boulter, D. (1978). Evolutionary change in invertebrate cytochrome c. *J. Mol. Evol.* 11, 35–45.

Martin, R. D. (1980). Phylogenetic hypotheses. *Nature, Lond.* 284, 285–286.

Romero-Herrera, A. E., Lehmann, H., Joysey, K. A. and Friday, A. E. (1978). On the evolution of myoglobin. *Phil. Trans. R. Soc.* (B) 283, 61–163.

# 9 | Ecological Aspects of Phylogenetic Approaches to Taxonomic Classification

A. D. J. MEEUSE

*Hugo de Vries Laboratorium, University of Amsterdam, Amsterdam, The Netherlands*

Abstract: Of the various cases in which ecological aspects play an important part in phylogenetic approaches to taxonomic classification a few are briefly mentioned: *Ficus* and the Agaonidae, and several examples of host–parasite relations. The interactions between reproductive structures and animal vectors of pollen or diaspores are more extensively treated, in particular the flower–insect mutualism and its implications in the evolution of the flowering plants. The ecological pointers may just turn the scales when it comes to the acceptance or the rejection of certain "established" views concerning the phylogeny of the most advanced cycadophytinous seed plants. Recent palaeobotanic evidence seems to support certain ecological indications and suggest a predominance of dicliny and anemophily in early Magnoliophyta rather than the opposite. The taxonomic implications are by no means negligible because the current ideas regarding the "basic" (primitive) type of angiosperm may have to be rejected, which — owing to a different starting point — upsets all systems of classification based on a "ranalean" group of progenitors.

## INTRODUCTION

A classification is usually given the qualification "phylogenetic" when the resulting arrangement or "system" is supposed to reflect, one way or the other, the evolutionary history of the group of organisms concerned. It is a moot point how far the often scanty and incomplete fossil evidence can be incorporated in a classificatory

Systematics Association Special Volume No. 21, "Problems of Phylogenetic Reconstruction", edited by K. A. Joysey and A. E. Friday, 1982, pp. 257–268, Academic Press, London and New York.

system of recent taxa (sometimes there is even none at all) and whether it is possible to arrange fossil and recent taxa together in one system other than a cladogram.

Another debatable item is the postulation of certain general rules of phylogenetic advances or "trends", such as progressive reductions or oligomerizations and irreversible modifications of features. The question arises whether such evolutionary tendencies are indeed a sound basis of deduction, and one must also ponder over the possibility that there may be exceptions to the "rules" of evolutionary advancement. A secondary increase in the number of certain morphological entities or morphomes is more than likely in the case of the antennary segments of certain families of Coleoptera and the number of legs in sea-spiders (Pycnogonida) (Crowson 1970, p. 103), and the large number of ovules in the ovaries of Orchidaceae and Asclepiadaceae, which are advanced angiospermous families, can only be explained by a secondary increase associated with an adaptive anthecological syndrome (namely, the transfer of pollinia, each representing the whole pollen output of a theca, by insects).

One must also bear in mind that not all current interpretations of certain conditions are necessarily acceptable: the present author has repeatedly pointed out that not all oligomerous and unisexual angiospermous flowers (such as those of, e.g. "Amentiferae", *Euphorbia*, and Cyperaceae) need have arisen by a process of progressive oligomerization of various floral whorls and the ultimate complete elimination by reduction of perianth members and of one of the kinds of sexual organs. The indiscriminating application of this tenet of progressive disappearance has led to absurdities, such as the interpretation of the, in several respects still clearly "gymnospermous", Gnetatae as having much reduced "flowers". The users of that term in this context fail to realize that the so-called "established" postulates relating to the concept of the "flower" compel us to accept the erstwhile presence of an ambisexual (unreduced) floral structure in a taxon progenitorial to the recent chlamydosperms, i.e. of a gymnospermous form more ancient and more primitive than the latter but with "complete" flowers, which is incongruous to say the least. An evaluation of ecological and functional aspects pleads in favour of an alternative interpretation (compare Meeuse 1978b, 1979a, b, 1980a, b).

It is also known that there are numerous cases of correlative evolution (co-evolution), which means that an ecological interaction between two unrelated taxa persisted during their phylogenies, i.e. their interrelationship was constantly involved in their individual evolutionary advancement and diversification.

This suggests the significance of the eco-physiological, eco-biochemical, and eco-functional aspects with regard to phylogenetic processes and, consequently, their bearing upon problems of classification (compare Meeuse 1978b, 1979a, b, 1980a, b).

A few of these aspects will be discussed, but there are doubtless numerous other ones which ought to be taken into consideration whenever "phylogenetic" classifications are being attempted.

### CO-EVOLUTION OF PARASITIC ARTHROPODS AND THEIR HOST PLANTS

It is now an established fact that there is a chemical basis of a great many cases of parasite—host interactions (see summaries in Harborne 1972, 1977; Meeuse 1973, 1979b). Certain hereditary factors controlling the synthesis of the responsible bioconstituents are involved, and this explains why natural selection has played a part in the perpetuation of the link between a host group and its associated parasitic taxon. The ramification of a host taxon in space and time during its evolution is, accordingly, reflected in the phylogenetic diversification of the corresponding group of parasites. Although the mutualism between the large genus *Ficus* and the Agaonidae is nowadays of the nature of an obligatory symbiosis, the inter-relation must have begun as a host—parasite interaction which gradually changed until the progenitors of the recent agaonid wasps became the sole pollinators of figs. The classification of the representatives of the genus *Ficus* and that of the fig-wasps can be used to check the one against the other, and it is fairly generally known that the cooperation between the fig-specialist Corner and the student of agaonid systematics Wiebes has been most fruitful, because the one keeps benefiting from the experience and the views of the other. The agreement is upon the whole quite satisfactory but some minor discrepancies will still have to be ironed out (see Wiebes (1979) for details and references).

The host plants of the swallowtail butterflies (Papilionidae) predominantly belong to a limited number of dicotyledonous families: Magnoliaceae, Annonaceae, Lauraceae, Aristolochiaceae, and some other woody Polycarpicae; Rutaceae; Umbelliferae; and, rarely, Compositae. This host spectrum is, in combination with very cogent phytochemical pointers, clearly suggestive of phylogenetic links between these plant taxa in a certain sequential order:

Polycarpicae ⟶ Rutales-like forms ⟶ Araliales s.s. ⟶ Asterales
(for details see Meeuse 1979b)

The taxonomic implications are manifest: the sympetalous Asterales must be terminal members of a certain lineage and are not to be associated with other sympetalous groups such as Gentianales, Ericales, Lamiales, Scrophulariales, Rubiales (Loganiales), and Dipsacales any longer; in other words, the "Asteridae" of most of the recent systems of the angiosperms are heterogenous and the group must be dismembered (it is conceivably not simply biphyletic but at least triphyletic). As was repeatedly pointed out by the present author, the association of the apparently most primitive subfamily of the Papilionidae, the Troilidae, with the Aristolochiaceae, considered to be an advanced taxon among the ranalean assembly by most phanerogamists, seems to be incongruous, but he is inclined to accept the co-evolutionary indications as having a greater demonstrative force. A thorough inquiry into the phytochemistry and the taxonomic affinities of this plant family is clearly indicated.

The monotypic family of the Batidaceae has in the past usually been considered to be *insertae sedis* if it was not included (as a family or as a separate order) in the Caryophyllidae (= Centrospermae s.l.). *Batis* certainly does not belong to the latter assembly owing to the ultrastructure of the microplastids in its sieve-tubes (Behnke and Dahlgren 1976, p. 291). Biochemical pointers quite clearly suggest an affinity with the glucosinolate families united in the Capparidales, which is emphasized by the association of the American pierid butterfly *Ascia monuste philete* with Capparidaceae, Cruciferae, and *Batis* (Nielsen 1950, 1961). The present author would insert *Batis* at family rank among the Capparidales and reject the maintenance of a separate order Batidales (or "Batales") for this ecologically specialized (halophytic) taxon as still seems to be all the vogue. There are several other examples of such mutualisms with a taxonomic meaning of which only the position of the

Salicaceae is mentioned here (see Meeuse (1975b) for details), because both the host—parasite relations (involving certain Uredinales and the nymphalid genus *Atella* which are associated with both Salicaceae or Flacourtiaceae) and cogent phytochemical evidence confirm the taxonomic proximity of the two families (which seems, incidentally, now to be accepted by nearly all leading phanerogamists). Similar studies have been most fruitful and inquiries of this kind will undoubtly yield similar, and often far-reaching, taxonomically useful conclusions.

REPRODUCTIVE BIOLOGY OF SEED PLANTS

*1. Early Phases*

During the course of their evolution the spermatophytes gradually achieved an increasingly more efficient and reliable mode of sexual reproduction. Instead of being dependent on the successful germination of a spore in a favourable place to produce a prothallus, and, subsequently, on a rather vulnerable process of fertilization (the condition still obtaining among the recent mosses and most of the pteridophytes), the progressive retention of the megaspore and, later, the incubation of the megagametophyte inside the spore (= nucellus) wall, combined with the catching of microspores (later to become pollen grains) by the ovule or by one of its accessory organs, created a much more streamlined and far less hazardous method of progeny production ultimately culminating in the advent of the true seed, i.e. the derivative of a fertilized ovule already containing the young sporophyte of the next generation before it becomes severed from the mother plant. Ultimately a quantity of storage tissue in the primary or secondary endosperm, perisperm, or cotyledons enabled the young seedling to cater for itself during its early existence as an independent individual.

The eco-physiological and functional requirements involved include: the protection and the feeding (incubation) of the ovule and its derivatives, the increased efficacy of the pollen-catching contraptions (the pollination droplet, later the stigma), the furthering of pollen germination and, in later stages of advancement, the feeding of the pollen tubes (presence of sugars and boron). The more primitive phase was retained by the Coniferophytina whereas the Cycadophytina acquired a morphological adaptation to improve the reliability of the pollination process, viz., the salpinx or lagenostome forming a pollen chamber (for incubation and protection, and for providing a favourable "internal environment"), and bearing at its tip the pollination droplet. At about the same time a protective cupule originated. The evolutionary potential became greatly increased because after a while the cupule was shed with its contents instead of a megaspore or an ovule, and thus provided more protection (and nutrition, as the case may be) in addition to its functioning as a device furthering anemochory, hydrochory, or zoochory, and the pollination droplet started to attract insects owing to its sugar content. This paved the way for the success of the flowering plants (compare Meeuse 1978b, 1979b, 1980a, b, and appended bibliographies). Two plausible suppositions can be made: (1) as far as our fossil evidence goes, the sex distribution among the Cycadophytina remained of the kind known as dicliny for a considerable length of time (presumably because cross-pollination and, hence, genetic recombination was advantageous from an evolutionary point of view); and (2) the available data suggest a homology of the inner integument with the salpinx, and of the carpel with the cupule (this provides an evolutionary continuity, otherwise the Angiosperms "hang in the air" in a phylogenetic sense).

## 2. Further Developments

The Mesozoic cycadophytinous forms were still predominantly diclinous and only rarely did they begin to become dependent upon insect pollination for their reproduction, whilst some forms tended strongly towards angiospermy of the mature cupule. The most advanced surviving group of the Gnetatae (Meeuse 1978c) exhibits transitional features towards the angiosperms: the sex distribution, for instance, is not strictly of the diclinous kind but shows a marked tendency towards monocliny although remaining functionally unisexual in the sense that only the plants with well-developed ovules bear viable seeds. The acquisition of an incipient form of monocliny was an evolutionary advancement because it created an anthecological link between the functionally male and the effectively female plants: insects interested in pollination droplets began to visit both the female and the male individuals (the sterile ovules of the male plants are not "reduced" and anything but non-functional because they produce the sugary exudate necessary to lure insects) and the animals thus became effective pollen vectors. The efficacy of this zoophilous mode of pollen transfer is at best 50% (insects must visit both a functionally male and a functionally female plant to achieve a pollen transfer) but this is compensated by a greater economy (the amount of pollen need not by far be so high as in comparable anemophilous forms).

It is feasible that this morphological adaptation and the concomitant anthecological advancement as found in the recent Gnetatae took place in the immediate precursors of some magnoliophytic groups, but it is not a foregone conclusion that the progenitors of typically diclinous and anemophilous groups of Angiospermae also exhibited a partial sex reversal (compare also Meeuse 1978a). This used to be a moot point because it involves two tenets of old standing, viz., the supposed primacy of monocliny and entomophily in early flowering plants and their fairly generally accepted, allegedly monorheithric evolution. The first viewpoint is not confirmed by recently published palaeobotanical records (compare with Crépet 1979; Dilcher 1979; Meeuse 1979a); and a pleiophyletic advent of the Magnoliophyta agrees with an early diversification among pre- and protangiospermous groups, which is highly probable in the light of the fact that such characteristics as angiospermy and embryo sac formation (concomitant with the skipping over of the formation of differentiated archegonia) had developed in forms technically still gymnospermous (e.g. in Caytoniales, Gnetatae, and most probably some Cycadeoidales).

## 3. Ultimate Conditions in Angiosperms

Angioody (i.e. the inaccessibility of the micropyles to pollen grains at anthesis) is a common characteristic of all living Angiosperms as far as can be ascertained. Because angiospermy is not necessarily synonymous with angiody (the cupule or its derivative, the carpel, may close *after* pollen grains have gained access to the pollen chambers of the ovules, as in *Gnetum* and as was conceivably the case in the Caytoniales), a transition was necessary in the more or

less clearly monoclinous protangiosperms from an already entomophilous but still gymnoodic condition (with pollination droplets
as lures to attract insect pollinators) to the fully-fledged angiospermous condition. This was because the original attraction (the
micropylar exudate) ceased to exist. The advent of angioody,
originally explained as a protection of the ovules against especially
dystrophic insect visitors (an altogether inadequate interpretation),
must have had certain advantages overruling the anthecological
disadvantage, which are associated with the selective efficiency
and economy of the interactions between the stigmatic surface
and the pollen grains (compare also Mulcahy 1979). This also
explains the consistent occurrence of angioody in all primarily
anemophilous groups of flowering plants whose ovules did not
need any protection against such marauders.

The necessary compensation for the loss of the pollination droplet
resulted in several alternative, adaptive syndromes which apparently
originated during the transitional phase between gymnospermy
and angioody, such as nectarial secretion, floral scents, optical
lures (semaphylls), and/or the reliance on pollen as the only attract-
ant, or on deceit. These developments need now to be translated
in terms of classification.

### THE SIGNIFICANCE OF ANTHECOLOGICAL EVOLUTION IN THE
### CLASSIFICATION OF THE MAGNOLIOPHYTA

Primarily anemophilous and more or less strictly diclinous progenitors of angiospermous groups must have co-existed already with
entomophilous, and incipiently to manifestly monoclinous, ones.
This viewpoint, set forth in previous papers by the present author,
is not only compatible with recent finds of ancient angiosperms
with apparently unisexual blossoms but also follows from certain
plausible deductions indicating, among other things, the early and
more or less independent evolution of such anemophilous and
diclinous groups as Juglandaceae and Platanaceae (Meeuse 1978a,
1979a, 1980a). A conclusion reached on the basis of studies of
fossilized angiospermous reproductive structures of early Tertiary
age by Crépet (1979) is likewise affirmative. More sophisticated
zoophilous pollination syndromes had not yet developed among
these early flowering plants, which strongly suggests a relatively

late advent of more efficacious forms of entomophily rather than the opposite. This is undoubtedly associated with the relatively late, co-evolutionary establishment of the specifically anthophilous groups of insects such as the apids, hover flies and some other groups of Diptera, and advanced Lepidoptera (butterflies, hawk-moths).

The morphological pre-adaptation to entomophily, viz., an incipient monocliny and the presence of potential (intrafloral) semaphylls (the latter having become available owing to evolutionary changes of the male reproductive organs (Meeuse 1974)), already existed. However, the incidence of the beginning of ambisexuality must have been restricted to a few groups, because some strictly diclinous and anemophilous, recent taxa at the genus or the family level which appeared as early as the Cretaceous can hardly be derivatives of even incipiently monoclinous progenitors. The generic or familial characteristics include dicliny and, by inference, anemophily, features which were already present in the earliest forms referable to a recent taxon (e.g. a protoplatanaceous or protojuglandaceous form) and not exhibiting any clearer indications of a pre-existing entomophily or monocliny than their recent descendants. The prevalence of forms with unisexual blossoms among the earliest flowering plants (Dilcher 1979) is, likewise, significant. Such coincidences are only compatible with an early morphological and anthecological divergence, which almost certainly began at the still pre-angiospermous level of phylogenetic advancement. Most of the assemblies of flowering plants, almost universally accepted as natural major groups (Magnoliidae, Hamamelididae, Rosidae, Dilleniidae, Caryophyllidae, and presumably all or several major monocotyledonous superorders), may well have originated as independent lineages, each having started from a progenitorial group still at a pre-angiospermous level of evolutionary advancement. Classifications based on their common origin from a single (and traditionally phaneranthous, monoclinous, and cantharophilous) magnolioid group of ancestors are suspect (Meeuse 1977). Some special and distinguishing features were already present in each of the precursory groups of pre-angiosperms, or developed independently later (often as a convergence: sympetaly, unitegmy, tenuinucellate ovules).

Divergent evolutionary trends also developed in some or all

lineages leading to larger magnoliophytic assemblies. If one accepts, for instance, the incidence of alternative ecological adaptations in the floral region of the ranalean lineage culminating in holantho-cormoidal and anthoidal flowering units, respectively, the seemingly heterobathmic evolution of certain recent families of this assembly is simply and satisfactorily explained. Obviously closely allied families such as Magnoliaceae and Myristicaceae with an altogether different floral morphology (and almost always with a different sex distribution) need not be "explained" as the result of the derivation of the latter from the former by wholesale modifications, oligomerizations and reductions in the floral region of one of the two taxa (which "places" the one taxonomically as a much advanced derivative of the supposedly most primitive, other one). The two families simply evolved by parallel evolution and neither one is necessarily much more primitive (or more derived) than the other. One may also decide that the so-called flowers of such, and in many respects manifestly primitive, taxa as the homoxylous *Amborella* are not highly "derived", (i.e. oligomerized and/or reduced) but originated directly by the modification of unisexual gynoclads (see Meeuse (1978a) and appended references).

CONCLUSIONS

For the time being one should treat the various major groups of Angiosperms as more or less independent units, i.e. as of separate phylogenetic origin (see also Meeuse 1978a). A closer relationship between, for example, the Dilleniidae and (certain) Magnolidae need not be ruled out, but on the other hand there is evidence that plants of a nymphaealean type may be very old, so that they may represent a very early offshoot with perhaps some affinity with monocotyledons. The relationships between the latter group and the magnolioid—ranunculoid assembly are so manifest that a common origin is almost certain, but the actual connections (also in an ecological sense) are obscure (compare with Meeuse 1975a), and the postulation of an origin of all Liliatae from a single, nymphaeoid, ranunculoid, or piperoid group of ancestors is open to doubt. The primary task of the phanerogamist is to try and elucidate the position of each family within a major group. In this way the Caryophyllidae have become a natural assembly by the removal of certain families previously included in the assembly, and by the incorporation of other ones (in this case, on the basis of phytochemical, embryological, palynological, and certain nannomorphological distinguishing criteria). More cogent evidence concerning the delimitation and the mutual affinities of the major groups of Angiosperms will almost certainly be forthcoming only from additional records of fossil flowering plants of Cretaceous and Early Tertiary age. The recent reports based on finds of early Angiosperms (summarized in Meeuse 1979a) have heralded in a final turning point in that it is quite clear already what will be the ultimate guiding principles in phylogenetic Angiosperm taxonomy. The ecological approach has contributed towards this viewpoint and it is to be expected that the ecological aspects of the phylogenetic history of the Magnoliophyta will play a part in relevant deductions for some time to come.

### REFERENCES

Behnke, H-D. and Dahlgren, R. (1976). The distribution of characters within an angiosperm system. 2. Sieve-element plastids. *Bot. Notiser* **129**, 287–295.

Crépet, W. L. (1979). Some aspects of the pollination biology of Middle Eocene Angiosperms *Rev. Palaeobot. Palynol.* **27**, 213–238.

Crowson, R. A. (1970). "Classification and Biology". Heinemann, London.

Dilcher, D. L. (1979). Early angiosperm reproduction: An introductory report. *Rev. Palaeobot. Palynol.* **27**, 291–328.

Harborne, J. B. (ed.) (1972). "Phytochemical Ecology". Academic Press, London and New York.

Meeuse, A. D. J. (1973). Co-evolution of plant hosts and their parasites as a taxonomic tool. *In* "Taxonomy and Ecology" (V. H. Heywood, ed.), pp. 289–316. Academic Press, London and New York.

Meeuse, A. D. J. (1974). The different origins of petaloid semaphylls. *Phytomorphology* **23**, 88–99.

Meeuse, A. D. J. (1975a). Aspects of the evolution of the Monocotyledons. *Acta Bot. Neerl.* **24**, 421–436.

Meeuse, A. D. J. (1975b). Taxonomic relationships of Salicaceae and Flacourtiaceae: Their bearing on interpretative floral morphology and Dilleniid phylogeny. *Acta Bot. Neerl.* **24**, 437–457.

Meeuse, A. D. J. (1976). Origin of the Angiosperms — Problem or inaptitude? *Phytomorphology* **25**, 373–379.

Meeuse, A. D. J. (1977). Delimitation of the major taxa of the Higher Cycadophytina: Theoretical criteria against taxonomic practice. *Syst. Evol. Bot.*, Suppl. **1**, 13–19.

Meeuse, A. D. J. (1978a). Coincidence of characters and Angiosperm phylogeny. *Phytomorphology* **27**, 314–322.

Meeuse, A. D. J. (1978b). Nectarial secretion, floral evolution, and the pollination syndrome in early Angiosperms. I–II. *Proc. K. Ned Akad. Wet.,* C, **81**, 300–312, 313–326.

Meeuse, A. D. J. (1979a). Why were the early angiosperms so successful? – A morphological, ecological, and phylogenetic approach. *Proc. K. Ned. Akad. Wet.,* C, **82**, 343–354, 355–369.

Meeuse, A. D. J. (1979b). Mutualisms, ecology and taxonomy. *Symb. bot. upsal.* **22** (4), 32–38.

Meeuse, A. D. J. (1980a). Fundamental paleoecological considerations regarding floral evolution. *Glimpses Pl. Research* **5/6** (in press).

Meeuse, A. D. J. (1980b). Paleoecology, paleobotany, and angiosperm evolution – New developments and the possible extra-tropical origin of the flowering plants. *In* "Current Trends in Botanical Research" (M. Nagaraj and C. P. Malik, eds), pp. 255–266. Kalyani Publ., New Delhi.

Mulcahy, D. L. (1979). The rise of the Angiosperms: A genecological factor. *Science, N.Y.* **206**, 20–23.

Nielsen, E. T. and Nielsen, A. T. (1950). Contributions towards the knowledge of the migration of butterflies. *Am. Mus. Novit.* **1471**, 1–28.

Nielsen, E. T. and Nielsen, A. T. (1961). On the habits of the migratory butterfly *Ascia monuste* L. *Biol. Meddel. Kong. Danske Vidensk. Selsk.* **23**, 5–81.

Wiebes, J. T. (1979). Co-evolution of figs and their insect pollinators. *Ann. Rev. Ecol. Syst.* **10**, 1–12.

# 10 | Evolutionary Cladistics and the Origin of Angiosperms

CHRISTOPHER R. HILL

*Department of Palaeontology, British Museum
(Natural History), London, England*

and

PETER R. CRANE

*Department of Botany, University of Reading,
Reading, England*

**Abstract:** Established views on angiosperm phylogeny favour a monophyletic origin in the early Cretaceous, though some continue to argue for polyphyly and a few have argued for a significantly earlier time of origin. In part I of this chapter, general concepts of biological classification and phylogenetic reconstruction are reviewed. Particular attention is given to cladistics, its basis, strengths and limitations, since this approach has not previously been discussed in relation to angiosperm origin.

In part II, cladistic method is applied to the living seed plants. Six or seven uniquely shared characters (synapomorphies) appear to unite the angiosperms as monophyletic in evolutionary cladistic terms and three in methodological cladistic terms. The marginally most parsimonious result supports the view that coniferophytes, excluding *Ginkgo*, also form a monophyletic group and are the nearest living sister-group to angiosperms. However, a cladogram with gnetopsids alone as the nearest sister-group is not much less parsimonious. The results therefore suggest that a likely primitive morphotype could correspond to a combination of characters found in angiosperms and in almost any of the other living seed plant groups. This further defines a range of likely

Systematics Association Special Volume No. 21, "Problems of Phylogenetic Reconstruction", edited by K. A. Joysey and A. E. Friday, 1982, pp. 269–361, Academic Press, London and New York.

time of origin from Carboniferous to Cretaceous. Since most theories of origin previously proposed have accepted one or other particular gymnosperm group as ancestral, the theories have conflicted. Likewise the postulated times of origin have been polarized between Carboniferous *versus* Lower Cretaceous. The views reached from cladistic analysis therefore appear to reconcile these theories and suggest that a broader theory is appropriate, which may or may not be predictively useful. Like other speculations it is potentially open to examination from new fossil discoveries.

## PART I: REVIEW OF PHYLOGENETIC THEORY

### INTRODUCTION

### 1. The Theoretical Problem

When Darwin wrote a century ago that the origin of flowering plants was an "abominable mystery", he referred to a major unsolved problem of Systematic biology (Sporne 1971). Attempted solutions have centred mainly on comparison of extant angiosperms, aimed chiefly at elucidating the most primitive living group; and this and other ideas have then been used by palaeobotanists as a guide to recognizing likely gymnospermous ancestors. In the last 20 years many fossils once thought to be ancestral to angiosperms have been discounted as such, or, like *Sanmiguelia* (Tidwell *et al.* 1977) they have proved doubtfully attributable owing to limited preservation of diagnostic characters (Ash 1976; Daghlian 1978; Doyle 1978). Cretaceous and Tertiary records of detached angiosperm-like organs have been re-examined in detail: leaves, pollen grains (Hughes 1976; Hickey and Doyle, 1977) and more recently flowers (Dilcher 1979); and progress has been made in determining the extent to which these organs may properly be compared with those of extant taxa (Dilcher 1971; Doyle 1978; Crepet 1979). Perhaps the main practical advance has been Dilcher's recent (1979) recognition that magnoliid flowers apparently do not pre-date amentiferous ones in the fossil record. Established ideas place the time of origin in the early Cretaceous, with diversification having occurred mainly during the succeeding Cretaceous and early Tertiary (Beck 1976; Doyle 1977, 1978; Crepet 1979).

Although there has thus been much recent progress in knowledge of early angiosperms, the main theoretical outlook on the problem of their origin has altered surprisingly little in the past twenty years (Takhtajan 1959). Like the question of the relationships of angiosperms to one another, this problem is phylogenetic, and it concerns the broad question of their relationships to other seed plants. Exactly which group or groups may have given rise to angiosperms, and how to define an angiosperm, are distinctly relevant questions that have propagated a variety of conjectures (e.g. Melville 1963; Meeuse 1965, 1977; Krassilov 1977; Nair 1979). Faced with such a diversity of speculations many botanists and palaeobotanists in recent years have tended to avoid proposing wider phylogenetic judgements; and phylogenetics in general is no longer held in the great esteem that it once was (Harris 1954, p. 290, 1961, pp. 179–180; Davis and Heywood 1963, p. 32; Gould and Delevoryas 1977, p. 396). Where phylogenetic diagrams have occasionally been offered (e.g. Chaloner and Boureau 1967; Doyle 1977, fig. 1) they have sometimes been developed too intuitively or tentatively to facilitate full understanding of how they were arrived at. As Harris *et al.* (1974, p. 85) succinctly point out, it has appeared distinctly possible in many such cases to reach other equally plausible conclusions from the same published evidence. Aware of limitations both of preservation and of loose speculation, most palaeobotanists have therefore concentrated on discovering facts of permanent value. Despite this, certain broader phylogenetic speculations have been widely accepted; for example, the definition of the ovule as an integumented megasporangium containing a single functional megaspore (Smith 1964), and the awareness that psilophytes do not form a natural group (Banks 1968).

A reason for the difficulties mentioned by Harris *et al.* (1974) may be that few explicitly formulated rules of procedure have appeared to be available. In addition it has perhaps been overlooked that science can be considered to advance through active interplay between observations and theories, and that rarely can genuine growth of knowledge be adequately explained as a process of dispassionate accretion of new observations alone, divorced from theoretical preconceptions and assessment (Popper 1959, 1972; Lakatos 1968, 1970, 1974; Magee 1972). Thus the diversity of speculations available on angiosperm origin appears to us encouraging. However the important and perhaps most difficult task is how such speculations can be sifted, in order to discriminate those ideas that seem nearest to the truth. Such a theoretically based choice, in turn, may lead to new and valuable observations, suggesting new practical possibilities to be explored, and replacement theories. That there has been apparently rather little debate of this nature concerning angiosperm origin was the initial stimulus for this chapter. However, both the inherent difficulties of this task and our own considerable disabilities will be only too evident in what follows. The foregoing appraisal would itself seem to be a matter of negative fault-finding had there not apparently been relevant progress in phylogenetic theorizing in some other areas of Systematics. Over the past thirty years, the methodology of both old and new approaches has been rendered more explicit and there is some indication that phylogenetic theory in general is moving towards a firmer basis, regardless of whether one adheres to the controversial cladistics school or not.

We begin our review with the question of what is the evolutionary (phylogenetic) tree, which leads to the practical question of how organisms are classified, and then returns with hindsight to reconsider evolutionary tree reconstruction and the ancestor problem. Three approaches to classification have usually been distinguished on their goals, e.g. by Bock (1973, 1979), normally termed "evolutionary systematics", "phenetics", and "cladistics", though on grounds of the essentials of their method of classifying similarities there are basically two approaches: evolutionary systematics + phenetics, based on clustering of overall similarity, and cladistics, based on derived similarity. Numerical taxonomy is not treated separately since it covers developments integral to all schools, including evolutionary systematics.

## 2. The Evolutionary Tree

The ultimate ideal in phylogenetics would be to reconstruct the single real tree of evolution in detail, as it actually happened. In the interim, the form of the tree (Fig. 3) can be postulated to have ultimate twigs representing terminal species (e.g. B, E, D), with the stem base representing the species (A) at the point of life's origin and its branches the species (e.g. C) between. A continuous (progressive, anagenetic) component connects each terminal species back in a single ancestor—descendant line of descent to the base; and a discontinuous (diversity, cladogenetic) component is represented by the successive splits of the stem into branches and twigs. Neither component can be imagined separately as fully explaining the tree. If evolution was entirely anagenetic there would have been only a single stem leading to a single extant species (Fig. 1), whilst if entirely cladogenetic there would be a diverse array of species but with only a single point connecting them (Fig. 2).

Given a fully sampled and completely preserved fossil record in time and space, the true form of the real evolutionary tree could be directly revealed and, as it were, proven in detail. As palaeontologists widely recognize, however, such complete examination is for practical purposes unattainable. Whether described fossil species exactly match "biological" species is more doubtful than it is even for most described extant species (Bremer and Wanntorp 1979). Correspondence can at best, therefore, be regarded as approximate, fragmentary and pragmatic, and thus systematists are inevitably forced to rely on theoretical interpretations of what the tree may have been like, given the limited available, rather than absolute desirable, information. Numerous reasonable assumptions are involved, for example the uniformitarian view that present day biological species concepts hold good uniformly for the past. The practical problem of phylogenetics, then, is twofold: (a) to determine the evolutionary theory that seems best to fit the available observations, and (b) in view of inherent difficulties to be as explicit as possible about the assumptions involved. Such hypotheses are inevitably fragile, though since explicit they are open to criticism and therefore to improvement.

Thus the form of tree shown in Fig. 3 is a model, reflecting current Darwinian theory (Patterson 1978), in contrast to the model in Fig. 2 which would be favoured by special creationists. Conceivably the form of the real tree was unlike either model and lies outside present limits of human understanding.

## 3. Classification

(a) *Similarity and difference.* Phylogenetic theories have traditionally been based on two sources of evidence: similarity of taxa and their time of occurrence. Stratigraphy (time) alone would, however, be

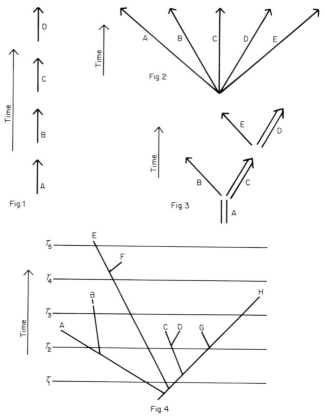

Fig. 1. Anagenesis, the continuous component of evolution.

Fig. 2. Cladogenesis, the discontinuous (splitting) component.

Fig. 3. An evolutionary tree; double arrows represent the anagenetic component of species A to D and the divergence of species B, C, D and E represent the cladogenetic component.

Fig. 4. Hypothetical evolutionary tree of eight taxa A–H showing their stratigraphic relations. If fossils are sampled at time planes $T_2$ and $T_3$ and their relative times of occurence alone are used to establish phylogeny it can be clear only that the collective entity $A + B + (E + F) + (C + D) + (G + H)$ gave rise in some way to $B + (E + F) + H$ and finally at $T_5$ to E. Reference to character similarities and differences is needed, in addition to time, to resolve the taxa and their nested relationships.

almost valueless, since there could be little point in discovering that fossil A occurred earlier or later than B if there was not also an idea that A and B were more similar to one another than to their contemporaries C, D, G or H (Fig. 4; Davis and Heywood 1963, pp. 33–34, fig. 2; Colless 1967; Schaeffer *et al.* 1972; Cracraft 1974; Hecht 1976). Biological classification seen in this light is thus of basic importance to theories of phylogenetics.

If phylogenetic theories therefore depend on comparative similarities, it may well be asked why ideas have varied so widely. A reason may be that although perception of similarities and differences is one of Man's most basic intellectual tools, it is highly problematic. Lombard and Bolt (1979) aptly quote from Plato's theory of knowledge, "a cautious man should above all be on his guard against resemblances; they are a very slippery sort of thing". Figures 5 and 6 for example show individuals having markedly different levels of complexity in their similarities. The individuals of Fig. 5 can easily be grouped, but in Fig. 6, which corresponds to the complexity found in living organisms, there are many ways of forming groups according to which point of view is adopted.

(*b*) *The sub-class relation.* How, then, are biological classifications actually formed? Representing similarities in observed characters such as of morphology *sensu stricto*, anatomy, biochemistry, chromosomes and DNA sequences, "objective", straightforward comparison results in a complex network as in Fig. 6. To obtain greater resolution of the similarities they are normally grouped and ordered into a hierarchical scheme of classes and sub-classes (here using the terms "class" and "sub-class" in the logical sense, which is equivalent to the non-commital biological concepts of "taxon" and "sub-taxon"). In this way various classification hierarchies, for example Fig. 7, can be achieved from the multivariate situation in Fig. 6, in which the sub-classes of grouped similarities are now more or less nested within one another like Russian dolls, at progressively more generalized levels of universality in the hierarchy. (The phrase "level of universality" used here is equivalent to the biological concept of "rank".) Hierarchical classification can thus be summarized in logicians' terms as a hierarchy of nested sub-classes of similarities and differences at progressively greater levels of universality, or, less clumsily, as representing in some measure the so-called "sub-class relation" of syllogistic logic (e.g. Popper 1959, pp. 64–68, 115). Translated into biological terms it is a hierarchy of nested sub-taxa based

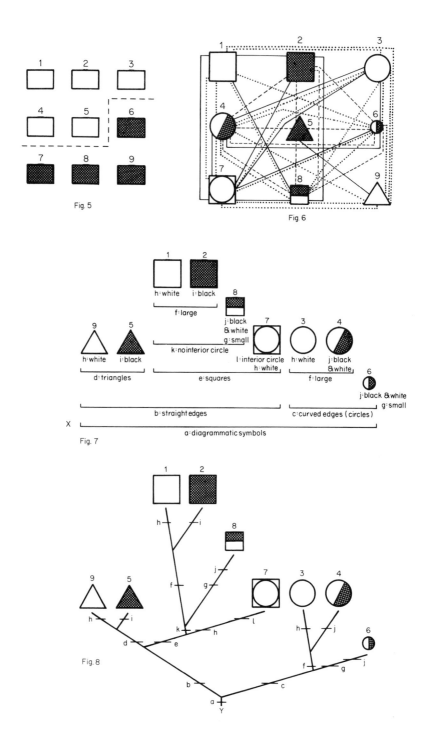

Fig. 5

Fig. 6

Fig. 7

Fig. 8

on discriminating the similarities shared by individuals from differences that separate them, at progressively higher ranks.

Two additional points are basic and require some elaboration. First, independent of any evolutionary interpretation, classification hierarchies can be expressed diagrammatically in the form of tree diagrams. Figure 8 is such a representation of Fig. 7, in which the lines may be imagined as representing the way in which the shared similarities have been abstractly generalized, or simply as pictorial boundaries of the sub-classes as in Fig. 7, but connected up to display their nested arrangement more graphically. Secondly, the term "classification" can apply equally to such diagrams as to written classifications, i.e. their more familiar form. Figures 9–11 attempt to illustrate these points and to show the requirements to form the simplest hierarchy: three taxa arranged in a single class and sub-class (Popper 1959, p. 115; Nelson 1972). Figure 10 is a written classification of the diagrammatic classification in Fig. 11, which in turn is based on the individuals and their grouped similarities shown in Fig. 9. Each figure represents identical raw information.

By way of example, the taxon Tracheophyta is united as a broad class by the universal or "generalized" similarity that all its member individuals have lignified conducting cells of a certain kind of

---

Fig. 5. Nine symbolic individuals of the same shape and size but differing in colour. Individuals 1–5 can readily be separated as similar to one another and different from 6–9, as indicated by the dashed line. Both sets are subsets of the universal set of rectangular symbols.

Fig. 6. Nine more complex individuals connected up to display their mutual similarities and differences, resulting in a highly complicated network. For simplicity, only three characters are compared: shape (□, △ or ○), indicated by unbroken lines; presence of black coloration, indicated by dashed lines; and presence of white coloration indicated by dotted lines.

Fig. 7. Similarities of the individuals in Fig. 6 expressed as a hierarchy of sub-classes. Class X is the most universal, encompassing all the characters a–l used to achieve sub-classes in the hierarchy. The figure shown is one of many hierarchies that could accommodate these individuals into some kind of sub-class relation. In this version shape has been weighted at the expense of colour.

Fig. 8. Tree diagram rooted at Y, representing the hierarchy shown in Fig. 7 with characters a–l indicated on the branches.

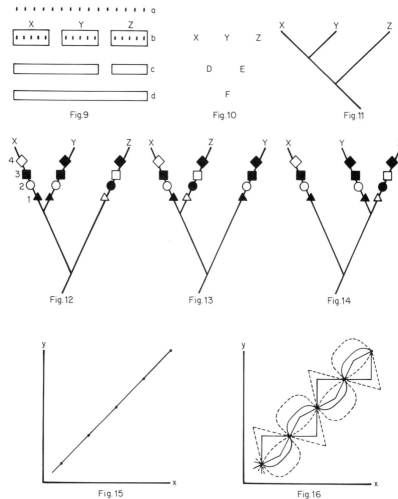

Fig.9

Fig.10

Fig.11

Fig.12

Fig.13

Fig.14

Fig.15

Fig.16

Figs 9—11.  Contrast between Fig. 9, sub-class groupings as related mentally
to individuals (a), Fig. 10 the equivalent written classification, and
Fig. 11 the equivalent classification diagram, in which the lines
can be considered to indicate directions of mental abstraction.
Each diagram is based on identical information and is merely
symbolized differently. a = individuals; b = individuals grouped
on their similarities into first abstract level (equivalent to named
species X, Y and Z); c = grouped into next more universal level
(equivalent to named genera D and E); d = grouped into next
more universal level (equivalent to named family F). Figure 11
illustrates graphically how X + Y represent a sub-class nested into
class X + Y + Z which represents a greater level of universality.

construction unique to them. Nested within this class the sub-taxon Gnetopsida, for example, is a narrower sub-class at a more restricted level of universality, for its members share some of these cells in a special form as vessels, and thus in a less generalized condition. Perception of what is a more generalized as opposed to a less generalized condition of a particular character therefore depends intimately on relative levels of universality (ranks), and *vice versa.* Conducting cells in the universal sense of tracheids + vessels occur more generally in plants than vessels. Thus, given any taxon, a way initially to "test" whether a character state found in it may be more or less generalized in its distribution is to raise the level of universality for the purpose of comparison, sometimes termed "out-group comparison". If found only in that taxon it is less generalized (relatively speaking), whilst if also found in others it is relatively more generalized and cannot strictly be used to define the taxon since it would represent logically redundant information.

Thus, amongst living cycads it may be considered whether the taeniopterid leaf of *Stangeria* is more or less generalized relative to the much more dissected leaf form of *Cycas* (Chamberlain 1935). By raising the level of universality these genera can be compared with fossil cycads, Bennettitales, Pentoxylales and taeniopterid pteridosperms. The taeniopterid leaf form occurs in more of these groups than the *Cycas* leaf form, suggesting conjecturally that the *Stangeria* leaf form is generalized relative to that of *Cycas*, though the position is by no means clear cut. A simpler example is perhaps the monolete spore relative to the trilete spore in pteridosperms and pteridophytes: Tschudy and Scott (1969, p. 15, table 2-2) indicating the monolete condition to be apparently the less generalized state.

---

Figs 12–14. The three possible classifications for taxa X, Y, Z based on shared similarities of the characters in Table I. Figure 12 is the most parsimonious, since similarities in characters 1, 2 and 3 place taxon X closer to Y than to Z, whilst only the similarity in character 4 places Y closer to Z.

Fig. 15. Observed points interpreted parsimoniously as representing a straight line.

Fig. 16. Some of the infinite ways in which the same points could be connected, possibly correctly but less parsimoniously than in Fig. 15.

Out-group comparison is a complex concept, intimately bound up with the fundamentals of hierarchical classification. It involves a regress (Lakatos 1962; Colless 1967), by which is meant that a thing can be related to other things only by reference to a more universal thing or essence, and in turn to a still more universal thing or essence, and so on. The regress of out-groups used for Systematic comparisons can terminate logically only in a comparison with characters or assumed characters as expressed in the most universal out-group, i.e. by all living organisms. In practice the regress is therefore used in reverse, representing the postulate that the simplest, least complicated forms of all life, i.e. bacteria and viruses, provide the concept of the most generalized condition. These organisms provide the point for fixing those universal generalized features such as presence of nitrogen-containing nucleic acid that (perceived as relatively less generalized features) serve to discriminate life from non-life such as free nitrogen. Thus a basic postulate used for determining generalized *versus* less generalized characters is to regard a simple character condition as generalized in the first instance, and its more complex state as relatively less generalized. A leaf with open dichotomous venation is regarded as more generalized in venation than one with net venation, and a megasporangium as more generalized than an ovule. Essentially the same assumption, but stated differently, is that additional characters are also to be regarded as relatively less generalized. When working with a particular group in practice, the usual procedure is to refer to only the nearest out-group or a few near ones in the regress. This is for one good reason alone, that we trust 2000 years of taxonomy to have arrived at a usable outline of classification, rendering the regress "harmless" for practical purposes. (To seriously attempt the alternative would be insane.)

Another major assumption for determining the extent to which a character may be generalized is ontogeny, in the sense of von Bauer (Nelson 1978; Eldredge and Cracraft 1980). Thus the needle-like juvenile leaves of conifers may be regarded as representing the generalized leaf form for the group; similarly for the deeply dissected young leaves of *Ginkgo*. The phyllotaxis of leaf primordia of *Lycopodium* accords with the Fibonacci series (relatively generalized), whilst the adult leaves do not (less generalized). Stelar ontogeny of pteridophytes suggests that the protostele is the generalized condition, and so on. Frequently the result will be consistent with the notion of simple to complex, discussed above.

(c) *The principle of parsimony (Occam's razor)*. When comparing a set of observed characters for our simplest out-group comparison: three taxa X, Y, Z (Table I), most systematists would intuitively

Table I. Distribution of character states for three hypothetical taxa X, Y, Z, displaying four characters (1—4). Pairs of the same shade (black or white) represent shared similarities. For example, for character 2 taxon X and Y have similar conditions while taxon Z is different. See Figs 12—14

| Taxa | Characters | | | |
|:---:|:---:|:---:|:---:|:---:|
| | 1 | 2 | 3 | 4 |
| X | ▲ | ○ | ■ | ◇ |
| Y | ▲ | ○ | ■ | ◆ |
| Z | △ | ● | □ | ◆ |

classify them as in Fig. 12, which represents the most economic arrangement of their similarities. Similarities in three of the four characters place taxon Y closer to taxon X than to Z, whilst similarity in only one character (4) places Y closer to Z than to X. Figures 13 and 14 represent the same similarities but are less economic, group Y + Z in Fig. 14 being supported only by the similarity of a single character (4), whilst in Fig. 13 the group X + Z is not united by any similarities. This illustrates the principle of economy, or "parsimony" in classification.[1] * It is used because similarities in different characters generally have a mozaic (heterobathmic) distribution amongst related organisms, corresponding to the complex situation shown symbolically in Fig. 6. A classification based on only one kind of character, for example comparative biochemistry of an enzyme, would not necessarily be exactly coincident with one based solely on another, for example comparative anatomy of the leaf trace xylem. Weighting of one character, e.g. biochemistry, in preference to others presents a

---

*Superscript numbers in square brackets refer to numbered notes at the end of the chapter.

way of overcoming the need for resorting to parsimony in such a
situation. Views however differ widely on whether characters should
be weighted, and if so, by how much and why. Similarly there are
varying measures of parsimony, and Figs 15 and 16 merely introduce
its function in a very intuitive way. Many, including ourselves, feel
that it is preferable in the first instance to avoid character weighting,
i.e. by giving each character equal weight. This too is complicated
by the problem of arbitrariness and of what can properly be regarded
as a unit character (Sneath and Sokal 1973, pp. 72–75).

These limitations are mentioned briefly to emphasize that the
very way in which classifications are formed involves procedural
assumptions. This inevitably means that any classification will
be an approximation to, a measurement of, the unknown "real"
classification. There is no fully objective way of proceeding,
regardless of the school of thought to which one might adhere.

(d) *Homology and analogy.* Though often thought of in evolutionary terms both homology and analogy were formulated in
terms of their purely classificatory meaning before Darwin's "Origin"
(Owen 1847). Homology indicates some kind of perceived essential
similarity whilst analogy represents similarity having no essential
significance. (Naturally it is of prime importance to distinguish
the two.) Although the terms are widely used, however, the meaning
attached to them varies widely.

In an earlier draft we defined the terms in a purely methodological
sense. Thus, for the character states in Fig. 12, the similarities
which unite and define taxon X + Y are homologous similarities,
whilst the similarity of character (4) between Y and Z, which conflicts with these three on grounds of parsimony, is an analogous
one. At the generalized level of Tracheophyta the vessels of species
of *Equisetum* and *Selaginella* are homologous with those of
gnetopsids and angiosperms, but at the narrower levels of universality
of these latter groups, few, if anyone, would propose that *Equisetum*
and *Selaginella* should be classified intimately with them, and their
vessels are widely regarded as analogously similar. This is because
xylem vessels are only one of the characters used to classify these
taxa, and it would not therefore be parsimonious of the whole
range of characters to classify *Equisetum* and *Selaginella* intimately
with angiosperms and gnetopsids. Homology may therefore be

defined as a similarity at a given level of universality which is consistent with parsimonious arrangement of similarities in a range of other characters. Analogy is the converse, in which similarity is inconsistent with parsimony.

While this gives a guide to how homology can be recognized it is a limited explanation since it fails to explain what homology actually is. It also suggests that homology can be very objectively defined in terms of method, whereas frequently it is bound up with initial, largely intuitive choices about which character of species A is really comparable with which in species B. Relative position in the organism provides one criterion, as does perceived closeness of similarity (see Hull 1967, 1968).

Definition of homology is for methodological purposes perhaps best examined and understood under those conditions where it becomes most problematical. This is normally when a considerable transformation occurs from a generalized to a less generalized condition. Most botanists, for example, accept homology between the angiosperm and gymnosperm embryo sac, but several regard the angiosperm embryo sac as homologous with only an archegonium of the gymnosperm embryo sac. The criteria involved in making a choice between these alternative hypotheses are by no means straightforward, and notably the choice will tend to be made before having considered the parsimony of a classification in any explicit manner. An equally difficult example is the widely accepted homology of the integument of the ovule with the telomes of a cupule as seen in the fossil *Genomosperma* (Smith 1964). Ovules of intermediate form appear to support this homology and an argument could be developed in terms of recognition through the parsimony of a classification. However, this does not necessarily confirm the homology as correct. At least in records of extant radiospermic seeds known to us there is no evidence from ontogeny to support the homology of integument with telomes, since the integument primordium is an undivided ring. In platyspermic seeds there is some support, since there sometimes are two separate primordia.

A still more difficult example is the male cone of *Austrotaxus*, in which the microsporophylls resemble other Taxaceae but are each subtended by a bract. According to the rule adopted generally for seed plants the microsporophyll should be homologous with an

axillary shoot, but there appears to be no independent evidence
for this in *Austrotaxus*.

(e) *Differing outlooks on homology*. Classification on the lines so far
discussed has been practised in botany at least from the times of
Theophrastus (Davis and Heywood 1963). Historical developments
reduce mainly to two interrelated factors: a gradual refinement
of what properly constitutes "similarity" for purposes of comparison
demanded by developing viewpoints, and an increasingly explicit
understanding of how biological classifications are arrived at.
Linnaeus (1753) in his sexual system used mainly similarities in
the number and arrangement of sexual parts, leading to a classification that was practically useful and for a time highly popular
despite its gross weighting. The now much adopted use of a range
of characters — which heterobathmy in our view dictates as desirable — was attributed to Adanson (1763–4) by de Candolle, whilst
as already discussed, Owen perceived a narrower basis for recognizing
similarities by his discrimination of homology from analogy. Reviews
of homology, mainly in methodological terms, are given by Hennig
(1966, pp. 93–94), Jardine and Jardine (1969), Sneath and Sokal
(1973, pp. 75–90) and Patterson (this volume); see also Hull (1967,
1968).

A relatively recent development which can be traced back to
Woodger's work in the 1930s and 1940s (references in Jardine 1969),
is the description of classification procedures in explicit mathematical or logical terms, and this has provided a key for unravelling the
methodological differences between cladistics and other approaches.
Implicit in our recognition in this chapter of the relevance of the
sub-class relation of syllogistic logic is the already widely held
notion that the sub-class hierarchy can represent what logicians
refer to as "proper" sub-classes, in which a sub-class (taxon) is
only to be distinguished by attributes unique to it; since any
relatively more generalized character states will represent logically
redundant information or "noise" and thus cannot cognitively
provide a defining statement for that sub-class. Properly such
attributes belong to a more universal class, representing the back-
ground necessary for the less generalized ones to be perceived.
Cladistics for example attempts to define only proper sub-classes,
using only such relatively less generalized features as are necessary

for cognitive class definition, whilst evolutionary systematics and phenetics also use the redundant, relatively generalized attributes. Thus their methods can equally result in a classification hierarchy but in terms of the sub-class relation it is less "proper", and consequently it is sometimes somewhat less resolved. The truth of this is correspondent whether viewed in the more usual terms of set theory (e.g. Sharrock and Felsenstein 1975), information theory (Riedl 1978; Farris 1979) or graph theory (Estabrook 1978), and in this chapter we use class logic because of its simplicity and therefore utility for ordinary discussion.

To summarize the position: in terms of homology and analogy all modern systematists agree that analogous similarity is to be disregarded as a criterion for defining taxa, though cladistics uniformly uses only the less generalized (*derived*) components of homology (termed synapomorphy) for defining taxa (Nelson, unpublished). Phenetics and evolutionary systematics also use synapomorphies but in addition they use relatively more generalized (*generalized*) homology (termed symplesiomorphy), even when redundant for forming proper sub-classes; that is, they emphasize similarities at inappropriately narrow levels of universality in terms of the sub-class relation. Basically, cladistic homology equals synapomorphy. The effects of these differences, based on Fig. 17,

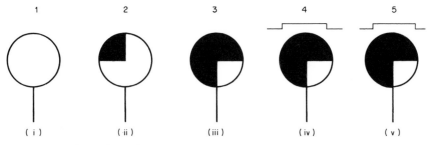

Fig. 17. Five hypothetical taxa displaying the following characters: a = blackness wholly or partly absent (1—5), b = uniform size (1—5), c = circular shape (1—5), d = possession of projections (1—5), e = some blackness present (i.e. a new character acquired) (2—5), f = one quarter black (2), g = three quarters black (3—5), h = hats present (4, 5). Each taxon has a unique character (i)—(v), and taxon 1 is at the most plesiomorphous end of the morphocline, the polarity of which is fixed on grounds of common absence of a character equals generalized: i.e. it is the least "advanced" taxon.

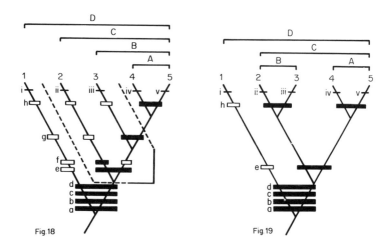

Fig. 18. A cladistic diagram (cladogram) based on Fig. 17. With the exception of f, which is an autapomorphy, the black rectangles indicate synapomorphies, i.e. shared similarities at their proper, "derived", levels of universality. Other autapomorphies are indicated symbolically by the numbers (i)–(v) on the projections, as shown in Fig. 17. Open rectangles indicate plesiomorphies, i.e. "generalized" character states relative to the corresponding synapomorphies.

Fig. 19. A possible phenetic or evolutionary systematics diagram based on partly weighted assessment of the characters in Fig. 17. Black rectangles here indicate intuitive similarities. Of these a–d and h are used as synapomorphies, as in Fig. 18 but the symplesiomorphy "absence of hats" is also emphasized for hat an improperly restricted level of universality, to unite group B. The similarity relationships are consequently less resolved (less neatly nested) than in Fig. 18. The supra-specific taxa can be accommodated in fewer ranks. Note that in Fig. 19 taxa A, C and D are equivalent to taxa A, C and D of Fig. 18 but taxon B (a grade group) is represented in Fig. 18 only by arbitrarily dividing off the portions marked with the dotted line. Each group A, B, C and D in Fig. 19 is monophyletic in the evolutionary systematics sense whilst cladistically only groups A, C and D are monophyletic, and group B is paraphyletic. Definitions of monophyly thus vary depending on the point of view adopted.

are illustrated in Figs 18 and 19. It can be imagined from these that by emphasis on symplesiomorphy, evolutionary systematics and phenetics (Fig. 19) would tend to result in fewer taxa and ranks in the classification hierarchy. Differences of approach thus reduce at base to the manner in which homology is used in classification.

An interesting historical example of the use of synapomorphous homology concerns the attribution of *Stangeria* to the cycads after its initial attribution to the ferns based on symplesiomorphy (Arber and Parkin 1907; Chamberlain 1935). It was attributed to the Ferns since its leaves resemble fronds of *Lomaria*. They are also coiled early in development (like those of ferns and of some other cycads). Since these are objective similarities to ferns this was a tenable view on the evidence at first available, though *Stangeria* was attributed firmly to the cycads when its seeds and other derived characters were discovered. *Stangeria* can, of course, only be attributed to cycads on this kind of character, since plesiomorphous characters such as its multiflagellate spermatozoids and to a lesser extent the venation and coiling of young leaves cannot and do not define cycads as distinct from ferns. They represent phenetically secure but symplesiomorphous homology which is redundant at that level using proper sub-classes. Similar considerations applied to the recognition of Carboniferous pteridosperms as distinct from ferns towards the close of the nineteenth century. Like Arber and Parkin's historic diagram (1907), these changes in classification were essentially cladistic, as is much of existing botanical classification.

## 4. Evolutionary Interpretation of Classification

Up to this point we have mainly been concerned with the patterns of morphological similarities and differences that provide a basis of classification. We have attempted (on purpose) to introduce classification in terms of its procedures, speaking for example of "generalized" *vs* "less generalized", rather than in the more familiar form of "primitive" *vs* "advanced", and we continued to speak of homology in these terms. This is tedious and to some extent provides a false air of objectivity, though it has some value in clarifying the concepts of phylogeny reconstruction. As Eldredge and Cracraft argue (1980), there is a pragmatic case for such separate consideration of "pattern" as distinct from "process" (see also Hull (1965, 1967, 1968)). It is now appropriate to introduce this notion of process into our discussion, i.e. to consider, in outline, the evolutionary role and interpretation of classification.

Biologists widely accept that the evolutionary process has

occurred as "descent with modification", marked by successive adaptations to environmental conditions, governed by competition and natural selection acting on variation (e.g. Patterson 1978). Phylogenetics may be defined in one (rather abstract) sense as the formulation of hypotheses about the historical sequences of evolutionary adaptations. From a procedural viewpoint we may therefore ask how classification, based solely on morphology, can contribute to hypotheses (a) of adaptations and (b) of historical sequence. From Jefferies (1979, pp. 444–445) – for example – it is clear that "relative recency of common ancestry" (the historical sequence) is to be equated with "relative degree of similarity" in a classification. The sequence of sub-classes in a classification thus represents a presumed historical sequence and the characters defining the sub-classes represent hypotheses of successive adaptations. In a sense, *classifications* become *phylogenetic classifications* when *interpreted* in terms of evolutionary theory, by the simple device of incorporating an assumed time axis for the hierarchy of similarities, and the assumption therefore that the similarities represent adaptations through time. Taxa at a given level of universality, in which narrower taxa are nested, are then understood to be taxa of relatively ancient origin which thus encompass ancestral species that gave rise with time to descendant taxa. Generalized and relatively less generalized characters can correspondingly be regarded as "primitive" and "advanced" (derived). Homologous similarity indicates common ancestry while analogous similarity indicates evolutionary convergence or parallelism. Groups defined on the basis of homologies are therefore interpreted to be "monophyletic", having apparently descended from a single common ancestor, and those united by analogies are interpreted to be "polyphyletic" – for example a group *Equisetum* + *Selaginella* + gnetopsids + angiosperms (e.g. based on presence of vessels).

Since cladistics has the potential to achieve more resolved classifications than other approaches it results potentially in a narrower, more resolved estimate of phylogenetic relationships and monophyly, in the abstract sense of phylogenetics referred to above. In this sense it is more consonant with Darwin's view that evolution has occurred by descent with modification.

The foregoing continues to emphasize recognition of process as dependent on perception of pattern, i.e. a purely methodological view. In many ways this reduces the notion of phylogenetic reconstruction to simple terms which are readily understood by the practitioner. The ensuing discussion (particularly pp. 301–315) should however be sufficient to indicate that the pattern/process interface is neither so simply subliminal nor as coolly objective as we have so far indicated. It is a very slippery interface and has yet to be fully explored (Hull 1965, 1967, 1968, 1970, 1979, 1980). Does, for example, synapomorphy equate exactly with Darwin's "modification"?

## 5. Comparison

The three schools of classification have been compared by Hull (1970), Cracraft (1974), Mayr (1974) and in Cracraft and Eldredge (1979). The following is a brief survey of salient points.

(*a*) *Evolutionary systematics.* Useful references to the very wide range of individual methods and concepts are Simpson (1953, 1961); Mayr (1969); Bock (1973, 1979); Harper (1976); Gingerich (1977, 1979) and, in botany, Sporne (1948, 1974, 1977, 1980). A uniting feature of this school is its emphasis on process as well as pattern and thus on ancestry. A major aim is to identify primitive ancestral organisms or groups and thus to arrive at classifications that can be interpreted as representing segments of the evolutionary tree, displaying the possible course of ancestor–descendant evolution. For this reason evolutionary systematics has sometimes been termed the "palaeontological approach", though in general it also places considerable emphasis on knowledge of evolutionary process at the present day and interprets fossil evidence in this light.

Like phenetics but unlike cladistics, symplesiomorphy as well as synapomorphy is used as a criterion of monophyly. Thus some groups, like group B in Fig. 19, would tend to be considered by evolutionary systematists as monophyletic, whilst when cladistically resolved as in Fig. 18 they would be "paraphyletic". The relative effect on classification (Figs 18 *vs* 19; see also the figures in Jefferies 1979, p. 447) is a bunching up of intersections of proper sub-classes (cladistically monophyletic groups) into paraphyletic "grade groups", incorporating the idea of a primitive "common stock", for example progymnosperms, trimerophytes, pteridosperms, pteridophytes, gymnosperms and proangiosperms. Some evolutionary systematists now accept that few ancestral species as opposed to such abstractly ancestral grade groups have been recognized in the fossil record (Harper 1976; see also Engelmann and Wiley 1977, p. 2). Although abstract grade groupings are useful, for example in ecological and stratigraphic studies, their potentially inhibiting effect on phylogenetic research progress can however be considerable. Harris' (1960) resumé of angiosperm origin provides an example, based partly on his outstanding research achievements (1933, 1940) demonstrating the occurrence of pollen within the micropyle of the seed of *Caytonia*. For whilst Thomas (1925) and indeed Harris himself (1954) had discovered several fascinating derived similarities between Caytoniales and angiosperms, having securely demonstrated one plesiomorphous similarity in common with gymnosperms, Harris apparently concluded that Caytoniales should be treated in all respects as dominantly gymnospermous (see also Harris (1937) for other emphasized symplesiomorphies). Despite his own reservations (Harris 1951) there is little doubt that, thus "relegated" to the gymnosperm grade, phylogenetic interest in *Caytonia* has declined. Thomas' later discussions became increasingly obscure and perhaps overlooked the greatest difficulty, which is the incomplete preservation of this plant; and this did little to reawaken positive interest such as renewed fieldwork. However the truth remains that so long as such taxa which are not fully like living angiosperms are relegated to the gymnosperm grade there will plainly be no possibility of finding a link between gymnosperms and angiosperms on the lines of Arber and Parkin's "hemiangiosperms", for the blockage is caused by a point of view, regardless of factual observations (cf. Harris 1951).

This conceptual difficulty arises partly from the tendency to use symplesiomorphous characters inconsistently, to define some taxa but not others. If applied consistently there would only be one monophyletic group of organisms — life itself — based on the ultimate symplesiomorphy of nucleic acid (which is also of course a synapomorphy in relation to non-life). To achieve resolution, evolutionary systematics is constrained to emphasize symplesiomorphy to a limited extent only, and it is therefore inevitably forced to rely selectively and often heavily on some symplesiomorphies as criteria of homology but not on others (for example the supposedly fern-like foliage of all pteridosperms). Despite this argument against seeking the lowest common denominator of similarity as a basis for classification, an implication of the preceding paragraph is that differences of evolutionary classifications from cladistic ones in practice may frequently be rather slight, representing mainly a difference in emphasis. Nor are they rigid, since, methodologically, cladistics does not adhere absolutely to strict application of synapomorphy. The genera *Cuscuta* and *Orobanche* lack chlorophyll and are therefore an exception to the synapomorphy that all higher plants possess chlorophyll, yet no one disputes that they are nested well within the angiosperms on grounds of parsimonious comparison.

(*b*) *Phenetics.* Phenetics attempts to achieve an objective estimate of similarity, sometimes quantitatively, sometimes qualitatively (Davis and Heywood 1963; Sneath and Sokal 1973; McNeill 1979). Prior weighting of characters is in general avoided, on grounds that weight is indicated with least circularity in hindsight from the classification itself rather than from preconceived notions. The majority of phenetic studies are aimed at practical rather than phylogenetic classifications, though since phylogenetic interpretation can in principle be applied to any hierarchical classification based on similarity, this means that hierarchical phenetic classifications are nevertheless open to phylogenetic interpretation. Phenetic similarity, being broader than that of cladistics, encompasses both synapomorphy and symplesiomorphy (Fig. 19) and in this it resembles evolutionary systematics, though the choice of which symplesiomorphies to adopt in phenetics is often differently based. A wide range of methods is available, which aim to be internally consistent.

Figures 20–23 attempt to illustrate the contrast of cladistics with phenetics (and to that extent also with evolutionary systematics). Figure 20 shows a cladistic diagram that can be assembled from standard information (Table II) about the early vascular plants *Cooksonia*, *Rhynia* and *Asteroxylon* shown as restorations in Figs 21–23. Figure 20 shows *Rhynia* cladistically more closely related to *Asteroxylon* than to *Cooksonia*, whilst reference to Table II and Figs 21–23 indicates that *Cooksonia* is more objectively (phenetically) similar to *Rhynia* than to *Asteroxylon* (see also Hennig 1966, p. 92).

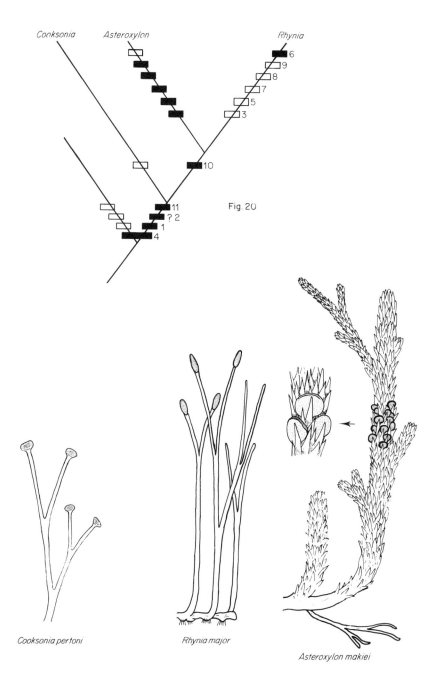

Cooksonia  Asteroxylon  Rhynia

Fig. 20

Cooksonia pertoni

Rhynia major

Asteroxylon makiei

Table II. Character-state matrix for the three genera *Cooksonia*, *Rhynia* and *Asteroxylon*, based on the references cited for Figs 20–23. Unique features of *Rhynia* and details of the spores for each genus have been omitted. + = apomorphous, − = plesiomorphous

| Character | Sister group, e.g. *Nematothallus* | Cooksonia | Rhynia | Asteroxylon |
|---|---|---|---|---|
| | | Taxa | | |
| 1. Conducting elements continuous tubes (−) or discrete cells (tracheids) (+) | — | + | + | + |
| 2. +/− phloem | — | ? + | ? + | ? + |
| 3. Protoxylem endarch (−) or exarch (+) | — | — | — | + |
| 4. Spore wall without (−) or with (+) Y mark | + | + | + | + |
| 5. Sporangia terminal on main axes (−) or lateral (+) | | — | — | + |
| 6. Sporangia equally as broad or broader than long (−) or longer than broad (+) | — | — | + | — |
| 7. Sporangial dehiscence unspecialized (−) or by transverse slit (+) | — | — | — | + |
| 8. +/− enations | — | — | — | + |
| 9. +/− monopodial branching | — | ? − | — | + |
| 10. +/− stomata with guard cells | — | — | + | + |
| 11. Erect axes unbranched (−) or branched (+) | — | + | + | + |

Figs 20–23. Cladogram showing relationships of *Cooksonia*, *Rhynia* and *Asteroxylon*, based on Table II and the restorations shown in Figs 21, 22 and 23 based on Edwards (1979), Kidston and Lang (1921), Lang (1937) and Lyon (1964). Magnification: Fig. 21 × 2.5; Fig. 22 × 0.25; Fig. 23 × 0.5.

(These figures are used solely to illustrate a difference of approach and are far from what is to be expected from the cladistic study of a wider range of taxa, and from revisions of Cooksonia and Rhynia now in progress by Drs D. E. Edwards and D. S. Edwards).

Phenetics is by far the most widely used approach in botanical classification today. Like some cladistic and evolutionary systematic studies (e.g. Sporne 1980) it aims at its best to provide explicit, tabulated information which is of value for subsequent studies of whichever school. Numerical phenetics (included in Sneath and Sokal (1973)) also recognizes the need for computers if large numbers of characters are to be adequately sorted, and such techniques have been adopted by evolutionary systematists (e.g. Sporne 1980) and by cladists (see below). It needs to be stressed that phenetic similarity is quite as logically consistent a viewpoint as is derived similarity used by cladists, when expressed in the logic of set theory (Buck and Hull 1966). Thus there is nothing inherently illogical about the methods of phenetics, and therefore in principle at least, about those of evolutionary systematics. The methods are simply differently logical.

A difficulty of explicitness can however occur with phenetics where intersecting sub-classes are involved. Since the basic objective comparison is in practice a multivariate network (Fig. 6) rather than a classification, this problem arises on converting such a network into hierarchical form, when it is found that phenetic class boundaries can often be defined only by use of percentages of similarity. As Nelson (unpublished) and Humphries (1979, pp. 102–103) state, this means that the defining synapomorphies and parallelisms can not explicitly be illustrated and totalled. Therefore taxon boundaries can be difficult to standardize and this means that many phenetic hierarchies could perhaps be erected, more or less arbitrarily, to represent one phenetic network. Inter-retrievability of written and diagrammatic classification may therefore suffer, though this is a purely methodological point. The main disadvantage of phenetics is its atheoretical view of Systematics as a mere accretion of pattern, divorced from the theory of evolution, and consequently its exclusive emphasis on the method of Systematics rather than its aims (Hull 1965, 1967, 1968, 1970).

(c) *Cladistics.* Cladistics, known also as phylogenetic systematics, has been discussed by a number of zoologists since the "classic" accounts of W. Hennig appeared in 1950 and 1966, principally by G. J. Nelson, N. I. Platnick, J. Cracraft, N. Eldredge and E. O. Wiley. Selected references are Nelson (1970), Cracraft (1974), Bonde (1975, 1977), Eldredge and Tattersall (1975), Tattersall and Eldredge (1977), Andersen (1978), Cracraft and Eldredge (1979), Gaffney (1979), and Eldredge and Cracraft (1980), with a comprehensive bibliography by Dupuis (1979). Books are also in preparation by Nelson and Platnick and by Wiley. Botanical reviews have been given by Bremer and Wanntorp (1978), Estabrook (1978) and by Humphries (1979).

Historically the position of most cladists has been that synapomorphies could essentially be perceived as adaptions, as outlined above. In tune with the methodological concept of proper sub-classes, most cladists also hold the notion that species, as the terminal units, should be defined by adaptations unique to them (so-called autapomorphies). This is because a species falling between two sub-classes can only separate the two classes if it itself has such a uniquely derived character, i.e. a unique adaptation in evolutionary terms. Since this criterion of relative uniqueness is thus built into cladistic procedure, most cladists recognize that ancestors cannot be differentiated using cladistic method. Therefore in terms of uniquely derived characters, anagenesis cannot be distinguished from cladogenesis (Figs 24–26), and cladistic diagrams represent diversity relationships alone. Thus a cladistic diagram need not correspond exactly to the ancestor–descendant pattern of the unknown segment of the evolutionary tree that it may represent; as in Fig. 26 it may represent only the "end-points" of change. As Eldredge and Tattersall (1975), Tattersall and Eldredge (1977), Platnick (1977) and Wiley (1979a) point out, a cladistic diagram is to be regarded as a rather abstract hypothesis of relationships and therefore not a tree, even though it looks like one. Ancestors are conceived by most cladists as solely hypothetical constructs.

## FURTHER ASPECTS OF CLADISTICS

Having attempted to introduce and briefly compare the elements of approaches to phylogenetics, it may now be useful to consider

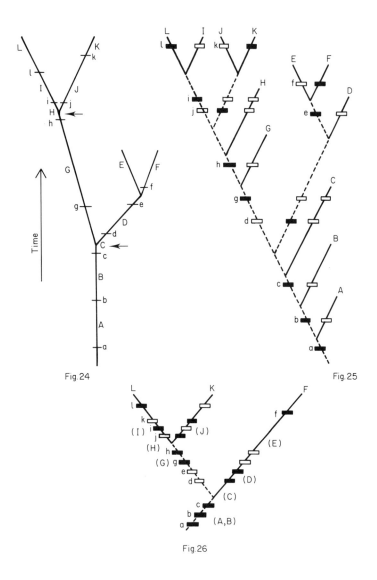

Fig. 24

Fig. 25

Fig. 26

Figs 24–26. Figure 24 is an "actual" segment of the evolutionary tree for twelve hypothetical species A–L. Twelve changes in the state of a single character occur. Most of the tree represents anagenetic change but there are two points of true cladogenesis (arrowed). Given each of the species from this tree they could be represented in cladistic format as in Fig. 25. However by more rigorous study based on unique as well as shared derived characters, species A, certain aspects in more detail, with particular emphasis on cladistics.

## 1. Heterobathmy

Hennig (1965, p. 107) emphasized that the empirical basis for cladistic resolution of diversity lies in the position symbolically illustrated in Fig. 6; see also Hennig (1966, fig. 20). Even when closely related species are compared, their similarities characteristically appear as a mozaic of generalized and derived characters that can only be resolved into a hierarchy of sub-classes by means of a highly branched diagram. This mozaic was termed "heterobathmy" by Takhtajan (1959; see also Arber and Parkin 1907, p. 35; Davis and Heywood 1963, pp. 34–35; Hennig 1966, pp. 86–87). Heterobathmy is intuitively familiar to taxonomists and we merely include two examples by way of illustration.

First, Florin's summary of coniferous cone evolution illustrates heterobathmy amongst fossils. Figures 27–30 show female cone scales of *Cordaites* and of selected members of the Lebachiaceae, illustrating reduction, modification of shape and planation of their parts (Florin 1951). The mozaic distribution of apomorphous and plesiomorphous conditions of their characters is shown in Table III. In Fig. 28 the number of megasporophylls is reduced to one, regarded by Florin as an apomorphous condition, whilst the ovuliferous shoot remains more or less unmodified. Figure 30 represents the converse, so much so that the ovuliferous shoot has been almost completely lost, a condition more apomorphous than in most living conifers. (In living conifers the presumed remains of the lebachiacean ovuliferous shoot are represented by the ovuliferous scale, or, in araucarians, by the apparently homologous "ligule".) Figure 31 shows a cladistic diagram based on Table III. It would clearly be impossible to arrange Figs 27–30 in an

---

B, C, D, E, G, H, I and J could not be discriminated from F, K and L (but see text) since they have no unique characters. Figure 26 is the fully resolved cladistic diagram equivalent to Fig. 24 and shows the same relative degree of cladistic relationship. Black rectangles = apomorphy, open rectangles = plesiomorphy, dashed lines = entirely hypothetical constructs.

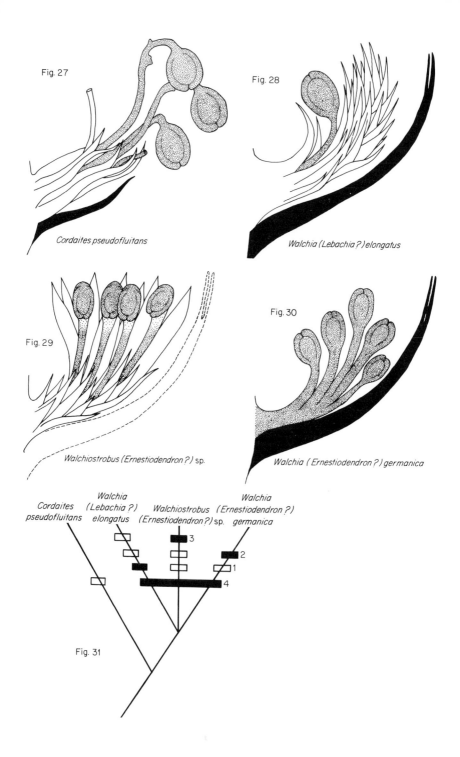

Fig. 27

*Cordaites pseudofluitans*

Fig. 28

*Walchia (Lebachia ?) elongatus*

Fig. 29

*Walchiostrobus (Ernestiodendron ?) sp.*

Fig. 30

*Walchia ( Ernestiodendron ?) germanica*

*Cordaites pseudofluitans*     *Walchia (Lebachia ?) elongatus*     *Walchiostrobus (Ernestiodendron ?) sp.*     *Walchia (Ernestiodendron ?) germanica*

Fig. 31

Table III. Character-states for female cone scales of *Cordaites* and selected Lebachiaceae, based on restoration shown in Figs 27–30. Apomorphous character state (+), plesiomorphous (−), following Florin's views. If leaf form and other characters of the whole plants were included heterobathmy would probably appear still more marked

| Taxa | Character | | | |
|---|---|---|---|---|
| | 1, megasporophylls reduced in number (+) | 2, sterile units of ovuliferous shoot absent (+) | 3, ovuliferous shoot flattened (+) | 4, bract scale forked (+) |
| *Cordaites pseudofluitans* | − | − | − | − |
| *Walchia* (*Lebachia?*) *elongatus* | + | − | − | + |
| *Walchiostrobus* (*Ernestiodendron?*) sp. | − | − | + | + |
| *Walchia* (*Ernestiodendron?*) *germanica* | − | + | − | + |

unbranched lineage, as if they were "homobathmic", without using reversals of character states.

Secondly, heterobathmy amongst living angiosperms is well illustrated amongst those taxa often considered generally primitive (Table IV; Stebbins 1974, p. 240).

## 2. Written Classification

Mayr (1974) acknowledged that cladistics may achieve greater resolution of taxa than evolutionary systematics, though pointing

---

Figs 27–30. Female cone scales of Lebachiaceae and *Cordaites*, redrawn from Florin (1951). Megasporophylls stippled, bract scales black or dashed where conjectural, and sterile units of the ovuliferous shoot white.

Fig. 27.    *Cordaites pseudofluitans.*
Fig. 28.    *Walchia* (*Lebachia?*) *elongatus.*
Fig. 29.    *Walchiostrobus* (*Ernestiodendron?*) sp.
Fig. 30.    *Walchia* (*Ernestiodendron?*) *germanica.*
Fig. 31.    Cladogram based on Table III.

Table IV. Matrix showing heterobathmy amongst some primitive Angiosperms (Magnoliales). Identification numbers refer to the following characters, for which the plesiomorphous condition is given first, followed by the apomorphous condition. 1: Wood lacking vessels/wood with vessels. 2: Perianth of undifferentiated tepals/perianth absent or of clearly differentiated sepals and petals. 3: Laminar stamens/stamens with distinct anther and filament. 4: Pollen anasulcate/pollen not anasulcate. 5: Carpels incompletely closed at pollination/carpels closed at pollination. 6: Gynoecium apocarpous/gynoecium of a single carpel or syncarpous. In many cases there is more than one kind of apomorphous condition and the morphoclines are presented in simplified linear form. They reflect established views. Data from Bailey (1949); Bailey et al. (1943); Bailey and Swamy (1948); Bean (1973); Hutchinson (1959); Takhtajan (1969); Walker (1974b)

| Family | Species | Characters 1 | 2 | 3 | 4 | 5 | 6 |
|---|---|---|---|---|---|---|---|
| Amborellaceae | Amborella trichopoda | − | − | + | ± | + | − |
| Annonaceae | Anaxagorea costaricensis | + | + | ? | − | ? | − |
| Austrobaileyaceae | Austrobaileya maculata | + | − | − | − | + | − |
| Calycanthaceae | Calycanthus floridus | + | − | + | + | + | − |
| Canellaceae | Canella winterana | + | + | + | − | + | + |
| Degeneriaceae | Degeneria vitiensis | + | + | − | − | − | + |
| Eupomatiaceae | Eupomatia laurina | + | + | + | + | − | + |
| Himantandraceae | Himantandra belgraveana | + | + | + | − | − | + |
| Magnoliaceae | Magnolia campbellii | + | − | + | − | + | − |
| Magnoliaceae | Michelia fuscata | + | − | + | − | − | − |
| Trochodendraceae | Trochodendron aralioides | − | + | + | + | + | + |
| Winteraceae | Drimys piperita | − | + | + | + | − | − |

out that this could result in a cumbersome increase in number of ranks and named taxa when expressed in words, relative to accepted evolutionary classifications (see also Harris 1962–1963). Nelson (1972), Griffiths (1974) and Patterson and Rosen (1977) have suggested straightforward methods of presenting cladistic classifications in writing while preserving the information resolved by their equivalent diagrams, and Bonde (1977) gives a thorough review (see also Wiley 1979b). Given these methods it appears that cladistic diagrams can in fact be retrieved consistently from their written representations and need not be especially cumbersome.

## 3. Numerical Cladistics

Wiley (1975) points out that in formulating a classification every character condition should be checked independently against every other one. Polarity of character states (whether primitive or derived) should be checked in reverse, and if a series of transformations (morphocline) is envisaged it should also be checked in the middle. However, the tediousness of doing this juggling of characters by hand, for any but the simplest of situations, needs scarcely to be emphasized, and points clearly to the need for a much wider adoption of computer-based sorting methods in cladistics than is now practised. Farris *et al.* (1970), Estabrook (1972, 1978), Mickevich and Johnson (1976), Estabrook and Anderson (1978), Funk and Stuessy (1978) and Estabrook and Meacham (1979) illustrate the diversity of progress being made in this area. They exemplify how the principle of parsimony is itself open to different points of view. Unfortunately for the ordinary worker some articles have been obscurely or even acrimoniously written (Farris and Kluge 1979) which does little to clarify numerical cladistics.

## 4. Methodological or Evolutionary Cladistics?

In the past few years cladistics has become split into two approaches: (a) evolutionary cladistics (Hennigian cladistics), and (b) method-ological cladistics, also known as "modern cladistics" and "trans-formed cladistics". Like evolutionary systematics the former con-ceives of phylogenetics more in terms of evolutionary process and the scientific aims of Systematics (Ashlock 1974; Whetstone 1978), while the latter conceives of cladistics as a relatively abstract programme of research with the emphasis on method and its explicitness rather than on its relation to process. Both approaches attempt methodologically to obtain proper sub-classes in a nested hierarchy on the basis of parsimony of derived characters (syn-apomorphies). According to methodological cladistics, however, synapomorphies are regarded as relatively idealized indications of organizational status, which have no necessary assumed connection with environmental adaptations. Cladistic taxa are therefore seen as "natural" rather than monophyletic. The notions of "simple to complex" and of ontogenetic sequence (as discussed on p. 280) are the main bases for determining whether a character is generalized or derived. Use of ontogeny is particularly interesting since, as it represents a sequence in time, the resulting cladogram already incorporates an assumed time axis (Nelson 1978; C. Patterson, personal communication). (If not already incorporating an onto-genetic concept of time, the cladogram may still be interpreted phylogenetically by the assumption that simple to complex synapomorphies became progressively more derived with time, as discussed above.) The result of such analysis is a classification based on idealized comparative morphology alone, with as little input as possible from evolutionary ideas of adaptation. The interest of such a method is apparent not so much in itself, but when it is appreciated that as far as possible all reference to extrinsic factors of time and space (stratigraphy and biogeography) has thus far been avoided. Therefore it is possible to proceed to check such factors semi-independently, as a kind of "test". From use of stratigraphy, for example, cladograms can be used to assess whether the phylogeny really does exhibit "advancement" with time, according to general evolutionary theory (Patterson 1978, p. 129). Evolution has often been considered a fact proven by fossils, though what this often tends to mean palaeontologically is that it is an inference, based on twofold congruence between successive derivativeness of characters (phylogenetic classification) and relative time (based on stratigraphy). This has now become a more highly corroborated threefold congruence with "absolute" time, based on rates of decay of radioactive elements. Initially, since the time axis of the phylogeny is assumed (or inherent on ontogenetic grounds), observed first stratigraphic occurrences of taxa can be used to calibrate it. If the sequence of the phylogenetic diagram is inconsistent with the sequence of first recorded fossil occurrences this might be taken to indicate that the theory of evolution by "advancement" is incorrect. A similar approach to historical biogeography has been proposed and discussed by Platnick and Nelson (1978) and Rosen (1978), and termed "vicariance biogeography".

Where there is inconsistency, as just indicated, this may suggest that received inferences such as advancement with time in terms of increasing degree of synamorphy are incorrect, e.g. that the cladogram does not represent the true historical sequence of acquisition of adaptations. This in turn may suggest there is something wrong, in evolutionary terms, about the cladistic notion of derivativeness. Equally, however, the result may reflect imperfections in the fossil record. This would tend to be argued out in connection with particular circumstances, and situations might frequently be such that it would be difficult to make a choice, i.e. they would be *ceteris paribus* (Lakatos 1968, 1970). It is a reasonable expectation that, as more information is attained, the more evident will be the choices.

Similar comparisons can potentially be made in relation to ontogenetic shifts with phylogeny, e.g. neoteny. If initial comparison is based on what are judged to be comparable stages of the life cycle (Hennig's "semaphoronts") it may be possible to achieve a still more parsimonious phylogeny by invoking such shifts (Jardine and Jardine 1969; Gould 1977; cf. Nelson 1978). Ontogenetic shifts might be interpreted as representing changes in "regulator genes".

In general, methodological cladistics has set out to achieve, and has actually achieved, an explicit system of abstract phylogeny based on relativities, which can apparently be conducted without reference to adaptive process in any particularly specific way. In this it represents a genuine attempt to overcome the lack of explicitness that has sometimes been evident in evolutionary systematic discussions, together with sometimes baseless and all-embracing opinions about adaptive value of characters and other preconceptions. It attempts to minimize such postulates and to separate them from interpretative steps as far as possible, thus making it clear when interpretative steps are involved. This explicit attention to method is an important contribution to phylogenetic theory in general. However, there are indications that methodology may be becoming the major aim of methodological cladistics, and that in most respects the approach is therefore becoming as atheoretical as phenetics. It is worth examining the current spectrum of study in this regard, and reflecting why so much attention has been placed, say, on biogeography, while so little has been given to the equivalent programme of stratigraphic "testing", as outlined above. When attention is sometimes given to stratigraphy, a convention is apparently adopted that the fossil record is uniformly imperfect, such that any inconsistencies would not reflect on the phylogeny (C. Patterson, personal communication). Similarly, scant attention has been given to the question of ontogenetic shifts, perhaps because this has little interest if not further related to aspects of process such as regulator genes, which in theory provide independent tests. (Little is in fact known about regulator genes, though this need not and should not diminish interest in them. The ability to experimentally "neotenize" organisms would be fascinating.) It is in fact clear that methodological cladistics is excluding more and more of that which is of evolutionary interest from its research programmes, apparently with the goal of becoming more "scientific". We find this view impossible to accept in context of science being based on theories (Hull 1965, 1967, 1968, 1970, 1979, 1980). In many respects it is fair to suggest that what is known about evolutionary processes at the present day is difficult. It provides systematists with few hard and fast guidelines with which to interpret the vast majority of taxa and their relationships, particularly fossil, or by which to independently test them, but to divorce process from pattern to the extent that seems current in methodological cladistics simply withdraws from the philosophical problem involved, which has been discussed by Hull (1965).[2] The research programmes of methodological cladistics appear in some respects rather opposed to scientific development of evolutionary biology, and in particular, palaeontology. Such lack of interest in evolutionary processes and fossils is consistent only with essentialist aims (Hull 1965). Since synapomorphies are conceived of foremost as idealized defining statements of "phylogenetic" status, rather than as historical adaptations, there is no reason to expect them to have advanced with time. Hence any interest in checking semi-independently against stratigraphy, for example, can, in a methodological straightjacket, be seen as inherently peripheral. It is apparently sometimes accepted that a cladogram is mainly of interest as a system of classification based on certain rules, which hence is to be accepted for what it is, a catalogue, rather than to be assessed as it relates to exploring scientific aims. Any concept of testing such a phylogeny in genealogical terms would be seen as trivial or impossibly difficult. It is very difficult to see how methodological cladists thus conceive of historical biogeography fitting into the scheme, for certainly it involves some theorizing. Perhaps the main apparent *consistency* of methodological cladism currently is its very avoidance of stratigraphic evidence in evaluating history! We can understand this only in terms of convention, however, for historical biogeography uses geological evidence of history based on the same sorts of uniformitarian assumptions as are used by palaeontologists to evaluate fossils. Possibly some methodological cladists are retreating into a phenetics-like position, unaware of the criticisms that have been very effectively raised by Hull in his classic 1965 and later papers. This is in no way a criticism of the explicitness of cladistic method, but we see it as a means towards an aim, in our case the evolutionary tree, and if it turns out to be limited in some respects for this purpose we will be glad to modify aspects or even, if necessary, reject the whole of it.

The contrast between methodological and evolutionary cladistics can further be elaborated by way of an example and the attitudes that might likely be adopted to it. It follows from the method adhered to by both approaches that in the first instance an attempt will be made to minimize cases where the polarity of a character appears (in evolutionary terms) to have become reversed. Such a character change occurs in the apparent reduction of thickness and architectural complexity of sporopollenin in the embryo-sac wall of some conifers, with the apparent subsequent loss of all sporopollenin from the embryo sac of angiosperms. Most conifers, together with heterosporous spore plants, possess a considerable thickness of sporopollenin in the embryo sac or megaspore wall, and typically the wall is of considerable architectural complexity. The reduction of sporopollenin in the embryo-sac wall of some conifers, and its absence from angiosperms, would therefore be regarded methodologically as a generalized feature, since absence or relative thinness resembles the condition found in out-groups (such as most algae and homosporous spore plants). Such a symplesiomorphous character would therefore simply be omitted from the analysis at the level of seed plants (B. G. Gardiner, personal communication). In other words the classification would be based solely on those characters that do display increase in complexity. Since there is no input from ideas of evolutionary processes it is in fact difficult to see how such an apparent loss character could be discriminated, on morphological grounds, in any other way than this, and in the absence of the full knowledge of how angiosperms actually arose (hence the rationale of the cladistic method). Similarly it is the complex features of the Podostemonaceae and of *Orobanche* which indicate on morphological grounds that they are angiosperms, rather than their simple features – and the fact that there happen to be more features of high burden than of low burden (Riedl 1978).

Beyond this point, views will diverge. A methodological cladist would see no further reason to examine such a character reversal, once it is recognized. The evolutionary cladist, on the other hand, would conceive of the symplesiomorphy as a "synapomorphy", i.e. representing a genuine adaptation in the Darwinian sense. (Thus, as a matter of incidental interest, he would perceive strictly proper sub-classes throughout the cladogram, in contrast to the methodologist.) The importance of this is not that he or she would believe such a "synapomorphy" is thus confirmed as an adaptation, but that this would be accepted as an interesting interim hypothesis to be pursued critically. The aim would be to test it independently in relation to evolutionary process, to consider whether such a loss character really could be reconciled as an adaptation historically to environment. For the example in hand, loss of sporopollenin appears to occur in general in more substantially enclosed ovules, perhaps reflecting a changing osmotic regime that is advantageous to the angiosperm; but it is not quite general (cf. *Araucaria* and *Hirmerella*). That the physiological details have scarcely been examined renders this whole area of botanical evolution a most interesting open field for fundamental research today. Similarly the evolutionist will wish to test the rather obvious adaptations of Podostemonaceae, *Orobanche* and *Cuscuta*. The sole point of this discussion is not to suggest the classification might necessarily become reinforced by this, or be upset by the discovery of further simplicities, nor that adaptive value need be easy or difficult to perceive for every character, but rather that the investigator would actually be stimulated to pursue such unexplored areas. In our view avenues of knowledge about process in the world should be explored, not abandoned, the more especially so if they are difficult to test (Hull 1965, 1980).[2]

## 5. Hybridization and Reticulate Evolution

When genetically isolated species of living plants hybridize they normally produce sterile progeny, though as is well known these may sometimes become fertile as a result of chromosome doubling (allopolyploidy). Such progeny thus become genetically isolated from their parent species and constitute new species (White 1978). Polyploidy is estimated to have occurred in 35–95% of living pteridophytes and up to 47% of angiosperms, with only 1.5% in conifers, none in cycads (or *Ginkgo*) and few substantial records amongst animals. White implies that most of these estimates represent some kind of allopolyploidy.

Expressed in an evolutionary diagram the result of allopolyploid speciation, like ordinary hybridization, would be an inverted dichotomy (Fig. 32), resulting, if often repeated, in a reticulum. In a cladistic diagram the character distribution shown in Fig. 33 would correspond with a morphologically exact intermediate hybrid or allopolyploid, though more likely the characters would be

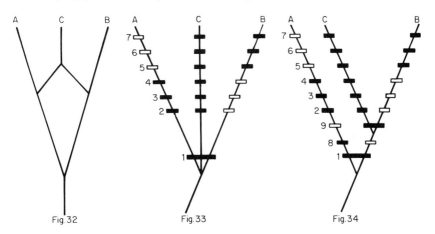

Fig. 32. An inverted dichotomy in an evolutionary tree, indicating that taxon C is a hybrid or an allopolyploid product of A + B. Taxon C forms a genetically but not a cladistically distinct monophyletic group.

Fig. 33. Polychotomy in a cladistic diagram, indicating possible hybridization or allopolyploidy. Taxa A, B and C share derived character 1 but the derived conditions of characters 2 to 7 are shared with equal parsimony between A + C and B + C. C is therefore exactly intermediate.

Fig. 34. Resolution into two dichotomies when taxon C is not exactly intermediate in morphology between taxa A and B.

resolved by the method into two dichotomies showing an approximation to this, as in Fig. 34. The extent to which cladograms of living and fossil ferns and angiosperms compared to those of gymnosperms and animals may actually accord to these patterns remains to be seen. Certainly, as Bremer and Wanntorp (1979) point out, the likelihood of distinguishing former genetic speciation events from geographical isolating events at the sub-species level is remote, since from cladistic character analysis there is no currently apparent way to distinguish intra-specific hybridization from allopolyploidy, and, for fossils, to test in general for barriers of genetic as opposed to geographical isolation. This is of particular interest to the evolutionary cladist, since there is plainly a likely limitation here in regard to the capture of process by cladistic pattern. Study might reveal that cladistic pattern is less adequate, or no more adequate, for revealing hybrids than pattern revealed by objective similarity. Such an argument of course becomes futile without reference to actual examples, and readers may wish to consult Bremer and Wanntorp (1979) and Humphries (1979) for initial studies.

## 6. Cladistic Species Concept

Hybridization and genetic isolation are pertinent to the cladistic concept of what constitutes a species. Hennig's recommendation (1966) that cladistic species are to be defined by unique characters (autapomorphies) has long been accepted by cladists, since without autapomorphy it would be impossible to separate the sub-classes in the sub-class hierarchy. If a cladistic hierarchy is thought of as a set of inverted Russian dolls, each separated by matchsticks (species), then it can be imagined that they would collapse into one another on removal of the matches. This is because autapomorphies provide the so-called "difference class" or "complement class", whilst synapomorphy in contrast defines the limits of any one sub-class as a whole (Popper 1959, p. 115). Theoretically, any one sub-class could contain an infinite number of species (or other included sub-classes) based on different autapomorphies, and one species is merely the minimum necessary number. To be quite explicit, a species is recognized and distinguished from others by autapomorphy, and its apparent relationship to others is established by synapomorphy. For convenience of discussion such a cladistic species can be termed a "cladistically resolved unit" (CU). Genetics provides independent tests of CU's for living organisms, i.e. it can contrast CU's with real (genetic) species. (Since genetic species are essentially time-dimensional "chronospecies" (Wiley's (1978) "evolutionary species") uniformitarian extrapolation of genetic results backwards in time is necessary for interpretation of real species.) The minute proportion of extant taxa that have been studied genetically reveal, as is well known (e.g. Sylvester-Bradley 1956) that there is no necessary congruence between such a CU and real species. Genetic species delimitation is not only highly complex but is such that any universal definition of the species will range from clones of phenotypically identical individuals, in asexually reproducing species, to sexually reproducing populations composed of highly variable individuals (Hull 1965; Patterson 1978; White 1978). As Wiley (1978) points out, a single CU might represent several possibilities, such as just a small part of the second kind of genetic species or it may include several of the first. Where CU's represent individual sub-specific populations (Bremer and Wanntorp 1979) it is even conceivable that they need not in all instances have advanced, phenotypically speaking, with time. During the actual duration of the species several CU's might become recognizable (perhaps by temporary isolation) and then (by inter-breeding) become unrecognizable, subsequently becoming recognizable again. Under these circumstances the notion of testing the hierarchy of synapomorphies with time as discussed above may be difficult.

In this regard, certain paradoxical aspects of the sub-class relation are interesting (Lakatos 1962). On the one hand a regress is involved, the choice of any one synapomorphy being justifiable only by invoking a higher-level synapomorphy (by out-group comparison) and so on. This is harmless (see above). In the opposite direction, however, the sub-class relation is by design a highly discriminatory tool that could lead if unchecked to a series of finer and finer splits until the terminal CU's would correspond to the individual organism (Jardine 1969). Thus by using human fingerprints as autapomorphies and synapomorphies it is possible to describe all humans as individual cladistic species, and certainly on general characters, men and women or babies and older people. The implications for the human species concept, which is known from observation of process, are empirically trivial, though less so for most living and all fossil taxa, where extrapolation of genetic results is needed to interpret species and which is inevitably a limitation of palaeontology and of herbarium taxonomy. The reverse implications, regarding a CU as a real species are however indefensible, for this would result in the assignment of different "semaphoronts" to different species, as well as different sexes, a problem familiar to palaeobotanists and one which much of palaeobotanical training is aimed at solving.

The point of this discussion is not to suggest that anyone would knowingly adopt such CU's in practice, but that everyone (including methodological cladists) adopts a species concept which corresponds as far as possible with what we know of present day genealogical process, since there is no very close correspondence with cladistic pattern. (A current methodological cladists' stance is that an autapomorphy can be regarded as much a plesiomorphy as an apomorphy (Engelmann and Wiley 1977), though what this means is obscure to us.) Methodological cladistics cannot therefore be entirely divorced from theoretical preconceptions.

Evolutionary cladists will be curious to compare the extent to which the pattern of genetic species is captured by cladograms relative to, say, phenetic diagrams. It may be that species based on phenetic comparison turn out to be more adequate, as is possibly the case for hybrids, and if so this would represent a limitation of cladistic method in these areas. Thus the cladistic species concept is not straightforward, either logically or empirically, and proper discussion is needed in relation to actual examples.

## 7. Ancestors

Ancestors in cladistics are perceived initially as hypothetical constructs (Fig. 35), which enable a hierarchical classification to be constructed from sister taxa. Ancestors are inevitably implied but not necessarily known, and Patterson and Rosen (1977) suggest they are tautologous. Vrba (1979, p. 225) even states that "the probability of . . . recognizing (ancestors) is zero", and Vrba (1979, p. 226) also assumes heterobathmy is so marked amongst organisms "in reality" that she might be taken to imply that every taxon will always prove eventually to have a derived character distinguishing it from all others. This is not surprising in view of the CU as a species concept as discussed above. Nevertheless, although ancestors may be tautologous (at any rate as lines connecting sub-classes), a cladistic ancestor definition can be proposed, and there is no reason why a logically tautologous entity should not be empirically feasible and thus recognizable. A cladistic ancestor can be described as a unit of description which shares one or more derived characters (synapomorphies) with its apparently nearest sister taxon but which otherwise differs in being entirely generalized (plesiomorphous) relative to it (Figs 36, 37). In other words its relationship is defined but its uniqueness is not, since it lacks autapomorphies. It is in our view preferable to accept such an entity as a cladistic ancestor than to reject it on logical grounds. However, the meaning of such an ancestor needs to be compared with that of a CU, for since an ancestor lacks a difference class it can only arbitrarily be separated as an "individual" or "species" within its class. Thus in logical as well as biological terms such a cladistic ancestor is the whole class

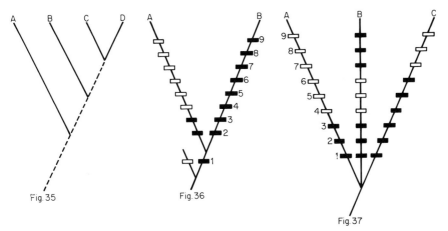

Fig. 35.

Fig. 35.   Cladogram showing conjectured ancestors (stem groups) as a dashed line.

Figs 36 and 37. Cladograms in which taxon A may be recognized as an ancestor. In Fig. 36 two taxa A and B are united by the synapomorphous similarity of characters 2 and 3, but taxon A is otherwise entirely plesiomorphous relative to taxon B. In Fig. 37 three taxa A–C are united by synapomorphies 1–3; taxa B and C each have three unique apomorphies but taxon A is entirely plesiomorphous in comparison.

of its descendants and is not a cladistically resolved "species". *Asteroxylon* might thus, for sake of discussion, be all lycopsids (Banks 1968). Once an ancestor is recognized empirically, the cladogram (or part of the cladogram) to which it pertains becomes essentially a hypothesized segment of tree, broadening the concept of monophyly to the truly genealogical sense of monophyly used by evolutionary systematists. Methods of representing such ancestors in written classifications have been proposed by Bonde (1977) and Wiley (1979b). The value of a cladistic ancestor in terms of evolutionary process needs to be assessed in the same light as hybrids and cladistic species, that is, unless an atheoretical approach is to be adopted. The cladistic concept of ancestor is as a morphologically generalized form, which need not necessarily correspond to what a real genealogical ancestor may have been like.

## 8. The Ancestor Problem

Ancestors defined cladistically appear to be recognized relatively rarely in cladograms. This may arise purely from limitations of the cladistic method, or from limitations of any method of recognizing fossil ancestral species on purely morphological grounds. The variety of ways of recognizing ancestors and species needs perhaps to be acknowledged in publications, such that it will be clear to readers which colour of spectacles is being adopted. Hence, above, we have referred to evolutionary cladistic ancestors, and these would be "tested" against time of occurrence, the expectation being that they would have originated earlier than their perceived descendants. An evolutionary systematic ancestor on the other hand might be initially recognized as such, on the basis of earlier occurrence in time as well as on morphological similarities (and thus cannot be "tested" in a comparable way). The test for an ancestor of whichever school of recognition would most powerfully be based on increasing knowledge of fossils and stratigraphy, and of present day process.

## 9. Evolutionary Trees

It is clear from this discussion that cladistic diagrams may be far removed from our initial, ideal goal of evolutionary tree reconstruction. Concerning hybrids, and the recognition of species as both descendants and ancestors, cladistic method has to be assessed in relation to what is known of present-day process. Perhaps the main function of cladistic method for the evolutionary taxonomist will lie in its greater resolution of relationships at higher levels of the taxonomic hierarchy, in just that idealized sense of plant morphology which inevitably applies to such abstractions, rather than to the actual genealogical tree that did not of course consist of higher taxa but of species (Wiley 1978). But we have merely introduced some possible limitations of cladistic method, mainly for further discussion rather than intending to provide definitive solutions.

The evolutionary cladist will wish to pursue the possibility of how far a cladogram can be converted into a hypothesis of the evolutionary tree. The only information that will be left over from a detailed cladistic analysis will concern the full stratigraphic ranges of species, as far as they are known, and it is worth considering whether this residual information might be used to progress further (Krassilov 1977). Once an ancestor is recognized, it can be identified as such in a cladistic diagram (Figs 38, 39) and that part of the diagram could then be regarded as a "minimal" segment of the evolutionary tree (in cladistic terms). Conjecturally it could represent any of the situations shown in Figs 40–43 (Eldredge and Tattersall 1975; Platnick 1977; Wiley 1979a). Calibrating the time axis by reference to stratigraphic ranges may indicate which alternative could be preferred on a parsimonious interpretation of the available stratigraphic information. It might suggest that branching of the tree segment with time was less parsimonious than indicated by the arrangement based on synapomorphies in the initial cladistic diagram (Figs 36 and 37) and, if circumstances are appropriate, that a persisting ancestral species could be recognized (Figs 40–42).

Such a tree model would also aim to include all available geographical information, and it would attempt to fully model individual speciation events in the past, in relation to particular circumstances. This, of course, is not easy and is beset with numerous limitations. For example, on a cladistic basis, ancestors might be recognized more rarely amongst megafossil plants and vertebrates than amongst other less morphologically complex organisms in which the stratigraphic record may be relatively

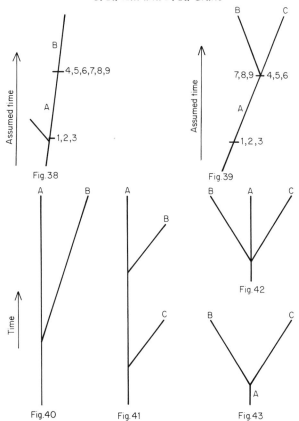

Figs 38 and 39. Minimal evolutionary cladistic trees, obtained for Figs 36 and 37 respectively when the ancestral taxon A is placed at the appropriate parts of the diagrams.

Figs 40—43.   Examples of "real" segments of evolutionary tree that could correspond with the same information in Figs 36 and 37 but that are less parsimonious except for Fig. 43. Figure 40 corresponds to Fig. 36. Figures 41—43 correspond to Fig. 37. In Figs 40—42 the ancestral taxon A gives rise to B, or to B and C, without itself changing, whilst in Fig. 43 it becomes extinct. In Fig. 41 taxon A gave rise first to C and then to B, but the converse is equally possible, and the situation could only be resolved stratigraphically.

complete. It is likely that there will be great difficulties in relating "herbarium" species to genetic species, and so on, though such difficulties do not in our view present a basis for therefore adopting a purely methodological approach. While some cladists have adopted a "naive falsificationist" view of tree modelling (Lakatos 1968, 1974), others (e.g. Tattersall and Eldredge 1977) have been less severe.

We will finish this rather long discussion by reference to the view of evolutionary tree speciation developed by Eldredge and Gould (1972) and Gould and Eldredge (1977). These writers pointed out that observed morphological changes within species, between their first and last recorded fossil occurrences, were too minor in scale to explain production of the observed overall diversity of species by slow, gradual, anagenetic change. They concluded that the observations in general are more consistent with a rapid, dominantly cladogenetic mode of speciation ("punctuated equilibria"). A species may therefore persist more or less unchanged for a long period after its inception, but the transition from its ancestral species would be relatively abrupt and of transient duration. This theory is difficult to test in relation to the alternative "gradualist" view, and would be examined in relation to knowledge of particular circumstances. For example, non-sequences and unconformities need to be taken into account, or with plants, non-deposition.

However one looks at "punctuated equilibria", it represents a genuine contribution to evolutionary theory in terms of process, based on observations of fossils. More exactly, it is based on observed stasis of complete stratigraphic ranges of species, i.e. precisely the evidence which methodological cladists perceive as the least interesting, and which is left unaccounted for by cladistic analysis.

PART II: ORIGIN OF ANGIOSPERMS

OUTLINE AND DISCUSSION OF EVOLUTIONARY SYSTEMATICS APPROACHES

## 1. The Primitive Ground-plan ("Morphotype")

Ideas about angiosperm origin have been based mainly on comparisons of extant taxa. Aimed at establishing the most primitive living group, two conflicting proposals have dominated the literature: Apetalae (Engler and Prantl 1897–1915; Wettstein 1901) *vs* Magnoliidae (Bessey 1897, 1915; Arber and Parkin 1970;

Cronquist 1968; Takhtajan 1969; Thorne 1976). Evidence has come from numerous studies, notably on vascular anatomy (Bailey 1944; Dickison 1975) and pollen (Wodehouse 1935, 1936; Walker 1974a, b, 1976; Walker and Doyle 1975). The majority of botanists now accept the magnoliid hypothesis and regard the unisexual-flowered Apetalae as too advanced to form a tenable hypothesis of the most primitive group (Sporne 1971, 1974).

Although this summarizes the general position, most botanists are aware of heterobathmy and would not consider any one living angiosperm or angiosperm family as entirely primitive, though hypothetical "morphotypes", which combine together all the supposedly primitive characters of angiosperms, have been proposed (Arber and Parkin 1907; Takhtajan 1969). Amongst other characters, such a (magnoliid) morphotype would have vesselless wood (Bailey 1944; Takhtajan 1969), monosulcate atectate pollen (Walker and Skvarla 1975) lacking stratification in the endexine (Doyle et al. 1975; Doyle 1978), an elongated floral axis bearing numerous leaf-like fertile and sterile parts, including conduplicate carpels open at anthesis, and laminar stamens (Sporne 1974). Its leaf would be simple, with pinnately camptodromous venation (Takhtajan 1969) and paracytic stomata (Baranova 1972).

While this magnoliid morphotype represents established views (Sporne 1971; Takhtajan 1969; cf. Banks 1961 and Dilcher 1979), its acceptability from a cladistic viewpoint clearly depends on the effectiveness of established classifications in resolving relationships within the angiosperms. A detailed cladistic reappraisal of intra-angiosperm relationships is undoubtedly needed to "test" it, since nearly every relevant classification and phylogenetic study has so far been based on the approach of evolutionary systematics, ranging from methodologically explicit studies like Sporne's (1948, 1949, 1974) and Lowe's (1961) to considerably less explicit ones like Hutchinson's (1964).

## 2. Heterobathmy Amongst Fossil Gymnosperms

The morphotype concept and other considerations of living angiosperms have been widely used as a guide for recognizing more or less directly ancestral groups amongst the gymnosperms; and since the majority of gymnosperms are extinct, attention has focused largely on the fossil record. Of special interest are a number of mainly Triassic to Cretaceous gymnosperms which possess certain characters apparently as advanced as the corresponding ones of living angiosperms. Many of these are reviewed by Dilcher (1979) and Doyle (1978): Caytoniales, Bennettitales, Czekanowskiales (Harris *et al.* 1974; Krassilov 1977), the cycad *Dirhopalostachys* (Krassilov 1975, 1977; see also Mamay 1976), and *Irania* Schweitzer (1977). In all but *Irania* the seeds and their manner of insertion are known, and were more or less enclosed, as they are in extant Araucariaceae. Enclosed seeds are also to be found amongst extinct conifers, in the Cheirolepidiaceae (Hirmerellaceae) (Harris 1979). Some of these plants had megaspore membranes apparently lacking sporopollenin, and some of the enclosing structures possessed papillose and perhaps "stigmatic" surfaces – for example, the pimply band of *Leptostrobus* capsules (Krassilov 1973, 1977). Bennettitalean male organs are inserted in a helix and some Bennettitales had seeds with a double integument (Harris 1954), while *Irania* and some Bennettitales in addition possessed bisexual strobili.

Whilst these characters may be homologous with angiosperm ones at a broad level of universality, most palaeobotanists – as did Arber and Parkin in 1907 – recognize that in each taxon they occur in combination (heterobathmically) with other uniquely advanced ones. Even when intuitively applied, the parsimony principle suggests that these taxa are to be regarded as in some way distinct both from angiosperms and from one another (Harris *et al.* 1974, p. 85). Bennettitalean female organs are uniquely specialized, as is the *Leptostrobus* female inflorescence and apparently also the male reproductive organs of Caytoniales. None of these groups seems to correspond straightforwardly with the accepted definition of an angiosperm (Harris 1960; Sporne 1971, 1974).

To bridge the gap between these gymnosperms and the angiosperms has proved highly problematical; and in the absence of clearly formulated phylogenetic principles it is to be expected that many mainly intuitive attempts have been made to bridge it (Harris *et al.* 1974, p. 85). Some of these interesting speculations have invoked apparent emphasis on symplesiomorphy, others have invoked polyphyly.

## 3. Search for Common Primitiveness

Searching further for common primitiveness (symplesiomorphy), the origin of the diverse Mesozoic gymnosperms would itself be sought by considering their own primitive morphotypes (Arber and Parkin, 1907). Certain Carboniferous pteridosperms appear on these grounds to be sufficiently generalized as to have given rise both to angiosperms and to Mesozoic gymnosperms. A "Carboniferous origin" for the angiosperms broadly along these lines has been proposed by Daber (1975), Harris (1937) and by Long (1966, 1977). In certain Lower Carboniferous seed ferns Long describes carpel-like cupular structures which are more generalized than corresponding structures of Mesozoic pteridosperms. He suggests that although these later seed ferns had more advanced leaf-form and venation, they possessed reproductive structures which had "overshot the possibility of evolving into [angiosperm] carpels" (Long 1977, p. 32), holding that "It is therefore quite feasible that the ancestors of the angiosperms are represented in some of the relatively more primitive pteridosperms of the Lower Carboniferous and Upper Devonian".

Although only implicated by these workers, Fig. 44 attempts to summarize their viewpoint in diagrammatic form. It illustrates

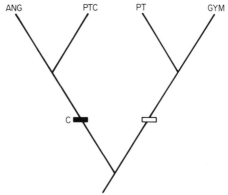

Fig. 44. Outline of a phylogeny which might accord with views on a "Carboniferous origin" of angiosperms, discussed in the text. ANG, Angiosperms; PTC, Pteridosperms with carpel-like cupules; PT, Pteridosperms without carpel-like cupules; GYM, all other Gymnosperms; C, presence of carpel-like cupules.

how by apparently emphasizing symplesiomorphy it is therefore possible to overcome the effects of heterobathmy. Long's work in particular is interesting, for it indicates unequivocally that advanced cupular characters did occur in Palaeozoic pteridosperms, as is demonstrated also for leaf form and venation by the Permian *Gigantopteris* (possibly pteridospermous). These cupules at least suggest a model of how later carpels may possibly have evolved from telomes. Furthermore, Fig. 44 does not necessarily imply that all uniquely angiospermous features need have arisen at an initial time of divergence in the Carboniferous; indeed Long refers to Carboniferous angiosperm *ancestors*, and not to Carboniferous angiosperms.

A difficulty, however, is that the scheme in Fig. 44 is potentially a relatively unparsimonious arrangement of similarities, and may seem to invoke a great deal more parallelism in seed-plant evolution than is necessary to correspond with the facts. The simple point can be readily understood that by invoking enough emphasis on symplesiomorphy and thus on parallelism it is possible to derive any plant group, however advanced, from any other, if this is what is required (Banks 1961).

## 4. Polyphyly Versus Monophyly; Are the Angiosperms Monophyletic?

A second way of bridging the gap is to accept heterobathmy but to invoke a polyphyletic origin of angiosperms (Just 1948; Melville 1962, 1963; Meeuse 1965, 1972, 1974, 1975a, b, 1977, 1979a, b; Hughes 1976; Krassilov 1977; Nair 1979); This raises the central question of whether a monophyletic or polyphyletic origin is to be preferred, and indeed of how to recognize an angiosperm; particularly since if the primitive morphotype is pressed to its conclusion it would become insufficiently distinct to be recognized as either a gymnosperm or an angiosperm. Even thought of as solely an advanced gymnosperm it would be more primitive in some respects than living gnetopsids or conifers. It is of course impossible to characterize angiosperms as a cognitively distinct group using shared primitive characters. In his reply to suggestions of polyphyletic origin, Takhtajan (1969, pp. 6–7) listed a number of implied synapomorphies:

(i) Uniform stamen structure, with a characteristic endothecial layer in the anther wall.

(ii) Presence of carpels with stigmas.

(iii) Constancy of the relative position of the androecium and gynoecium on the floral axis.

(iv) Characteristic gametophytes, the female with specialized nuclei and "double fertilization".

(v) Presence of sieve tubes.

He considers the chances of this combination having been derived independently to be infinitesimally small. Transposed into the terms used in this chapter, the conclusion would come from the most parsimonious comparison of a wider range of characters and taxa. Although statistically probable this conclusion does not however suggest that a monophyletic origin is "right" or "proven"; only that it appears to be the simplest hypothesis given presently available empirical evidence and theoretical predilections. Nor can such characters necessarily define the essential angiosperm (using the term essential in a non-philosophical sense to mean a truly genea-logical angiosperm). Since the concept of angiosperm is partly man-made it can be expected to alter in the light of new evidence as much as it could be expected to remain stable. Thus, if parsi-monious comparison of a still wider range of evidence suggested polyphyly, the present heuristic would properly be abandoned; and if the term angiosperm were to be retained subsequently it would change in meaning. Takhtajan's and the following proposals of implied angiosperm synapomorphies are discussed on p. 342. Meanwhile it is noted that Sporne (1974, pp. 23–24) accepts only two characters as implicitly synapomorphous:

(i) "Double fertilization";

(ii) companion cells in the phloem, formed from the same mother-cells as the sieve tube elements.

He points out that the 8-nucleate embryo sac (often considered to be uniquely shared) occurs only in the majority of angiosperms, and that several have as few as 4 nuclei or as many as 16. Niklas *et al.* (1980, p. 52) mention three "definitive features":

(i) "Double fertilization";

(ii) the reduced non-archegoniate mega-gametophyte;

(iii) tectate-columellate pollen with a laminated endexine.

Character (iii) is misprinted as "laminated", and should read "non-laminated". Character (ii) however, somewhat better expresses the condition of the angiosperm embryo sac, which as a good synapomorphy has 4–16 nuclei (*vide* Sporne) whereas all other extant plants have 125 or more.

## 5. Stratigraphy of Detached Organs

Several palaeobotanists have developed a chronological ("stratophenetic") approach to phylogeny construction, one of the most historically interesting studies being by Chandler (1923). The approach has sometimes been relatively cautious (Chaloner 1970a) but more recently it has become bolder (Doyle and Hickey 1976; Hughes 1976; Chaloner and Sheerin 1979). It attempts to gather phylogenetic evidence directly from the fossil record of detached organs such as pollen grains, secondary wood, leaves and seeds. These detached organs comprise the major proportion of fossil angiosperm material, partly because flowers and other relatively ephemeral and decay-susceptible tissues or organs are rarely preserved, and partly because terrestrial plants are normally broken up during their transport and deposition (Chaloner 1970a; Hill 1974, fig. 8). From their characters and relative times of occurrence in the fossil record, time of origin and relationships are worked out, as with evolutionary systematic approaches pertaining to relatively complete organisms (Banks 1968). Doyle (1978) implies that the approach has led to "unprecedented progress toward solution of Darwin's abominable mystery".

This stratigraphical–morphological approach has three apparent limitations from a phylogenetic viewpoint: (a) it does not uniformly recognize relationship on cognitive grounds; (b) it implicitly neglects heterobathmy; and (c) it has sometimes become detached from evolutionary aims and theorizing, in a comparable fashion to methodological cladistics.

(*a*) *Cognitive recognition.* Characters known not to occur uniquely in living angiosperms have occasionally been applied as if they were unique to them. For example the defining characters of angiosperm leaf-shape and venation discussed by Hickey (1977, pp. 157–163) do not exclude the leaves of *Gnetum*, which are indistinguishable from those of many angiosperms, though Hickey and

Doyle (1977, p. 68) and Hickey and Wolfe (1975, p. 546) recognize this. Leaves of *Gnetum* are simple and have pinnate camptodromous venation with anomocytic or paracytic stomata (Rodin 1966; Inamdar and Bhatt 1972). Some species uniquely possess a certain kind of sclereid, but others apparently lack it.

Doyle and Hickey (1976, fig. 16) and Hickey and Doyle (1977, fig. 26) figure an incomplete presumed angiosperm leaf that on the evidence presented looks remarkably similar to the Mesozoic fern *Hausmannia*. In their venation a number of angiosperm leaves (as incomplete fragments) resemble fronds of certain ferns, as is noted by Hickey and Doyle (1977, p. 68), while as Foster and Gifford (1974, pp. 563–566) point out, the Ranalean angiosperms *Kingdonia* and *Circaester* exhibit a venation pattern, shape and rank which is remarkably similar to some Triassic–Cretaceous fossil leaves. Until a more cognitive definition of angiosperm leaf characters is formulated it appears that some isolated fossil leaves that might be supposed to be angiosperms may equally be gnetalean, and in a few cases, may even be fern or other gymnosperm leaves (work in progress). It seems unreasonable to postulate that since the fossil record of *Gnetum* is mainly unknown (or unrecognized or unrecognizable) that there was none. There is a fairly impressive fossil record of *Ginkgo*, for example (which has only one extant species). The nub of this problem is that cognitively angiospermous leaf characters are limited to some extant taxa within the group, and do not in straightforward terms provide defining characters for the group as a whole. Beck (1976, p. 4) presents a comparable view.

(*b*) *Heterobathmy*. The full extent of heterobathmy is likely to be only partly determinable from most detached organs. Organ "phylogenies" (or trends of individual characters) are frequently equated with real phylogeny, which necessarily involved whole organisms (Davis and Heywood 1963, pp. 34–35, fig. 2; Schaeffer *et al.* 1972; Hecht 1976). Put simply, the tooth form of a sheep can at the present day be contrasted with that of a dog. This provides definitional criteria that are inevitably and reasonably extrapolated backwards in time to provide a frame of reference for detached fossil teeth. Such a procedure can, however, lead to difficulties of the excluded middle in coping with intermediates, and thus with the evolution of one type of whole organism from the other (Davis and Heywood 1963, pp. 13, 34–35). In view of the almost universal observation of heterobathmy at the present day it is unlikely that a detached tooth showing linking characters would provide exact evidence of the time of origin for sheep or dogs. Comparably, it seems unlikely that the synapomorphies which define living angiosperms were all acquired at the same time. The first "angiosperm" may have been "gymnosperm" which acquired no more than double fertilization and an embryo sac with, say, 32 cells, while conversely there may have been advanced "gymnosperms" that had tricolporate pollen but retained a laminated endexine, rather as some extinct "conifers" had apparently tectate pollen (Pettitt and Chaloner 1964; Chaloner 1976). Beck (1976) again argues along similar lines.

The morphological–stratigraphic approach to detached organs has been fairly widely accepted as providing convincing evidence for a Barremian–Aptian time of angiosperm origin (Beck 1976). Evidence has been based partly on apparent synapomorphies such as the essentially non-laminated pollen endexine, though mainly on preservable apomorphies that characterize many but not all living angiosperms, such as leaf characters discussed above, advanced tectate pollen wall structure and apertures, and certain wood and seed characters (Hughes 1976; Doyle 1978; Hughes *et al.* 1979). Claims based on pollen have sometimes tended to be exaggerated, e.g. Brenner's (1976) implication that the eastern Mediterranean region was a likely centre of angiosperm origin, or search for the same ends in equatorial Africa (Doyle *et al.* 1977; see also Martin 1978).

A difficulty over the Barremian–Aptian conception is that uncritical acceptance could give rise to circular argumentation (cf. Beck 1976, pp. 1–2, 4). Pre-Barremian problematic fossils such as the Valanginian *Montsechia* (work in progress) may have come to be regarded as definitely gymnosperms, whilst post-Barremian *problematica* have conversely been accepted as necessarily angiosperms. For example pre-Barremian *Sanmiguelia* (Tidwell *et al.* 1977; Beck 1976, p. 4), *Phyllites* sp. (Seward 1904), *Furcula* (Harris 1932), *Scoresbya* (Harris 1932; Kräusel and Schaarschmidt 1968; Krassilov 1973), and post-Barremian *Conospermites*, *Dicotylo-phyllum* sp. and *Halyserites* (Krassilov 1973; Knobloch 1978), amongst other examples (Daghlian 1978). Doyle (1978, pp. 367–368) perhaps rather eclectically invokes special emphasis on heterobathmy when it is useful, in support of the view that the relevance of pre-Cretaceous fossils has yet to be established. Certainly if a pre-Barremian fossil is to be regarded as gymnospermous until proved otherwise, and a post-Barremian fossil as an angiosperm until proved otherwise, this will tend to metaphysically circumscribe the possibility of finding intermediate forms. Reports by Cornet (1977 and unpublished) of Triassic–Jurassic pollen having apparently angiospermous characters therefore deserve most careful but open-minded scrutiny. There would seem to be an equal case for attempting to accept or reject both pre- and post-Barremian plants as angiosperms. Many fossil plants, both pre- and post-Barremian, were neither fully like living gymnosperms nor fully like living angiosperms, as is well documented by the genera cited above (p. 317), though not perhaps as fully acknowledged as might be.

(c) *Proven or fallible knowledge?* These considerations focus attention on the need, where conditions allow, for plant fossil organs to be linked together, on the grounds of field association and structural agreement. This can shed further empirical light on the extent of heterobathmy amongst early angiosperms. Further reconstructions of Barremian or earlier Cretaceous seed plants would undoubtedly be of the greatest interest. Reconstruction of fossil plants is widely practised in palaeobotany, and fructifications do occur in Cretaceous leaf floras (Doyle and Hickey 1976, p. 147), though few detailed restorations have been proposed from the Cretaceous apart from important pioneering studies by Dilcher *et al.* (1976) and Vachrameev and Krassilov (1979).

In common with other recent discussions (e.g. Knoll 1980) the foregoing concentrates on limitations of detached organs, and thus it is desirable to point also to the strengths of the approach. There can be little doubt that the fully documented studies of Wolfe, Doyle, Hickey and others have been most valuable, while those of Banks and Chaloner on early land plants in particular have achieved explicitness and economy in terms of synthesis and data presentation. Furthermore, not only is the pollen record relatively complete and readily sampled, but, as with that of seeds, it is composed of semaphoronts (Hennig 1966). Full description of detached organs is a long-term preliminary to reconstruction and in the interim every scrap of such evidence can, and clearly should, be used to formulate hypotheses as to "exact" time of origin of angiosperms — so long as these are recognized to be bold and inevitably fragile hypotheses. Even considering the potentially superior estimates obtainable from reconstructed "whole" plants, few fossil plants are really known whole and the majority may never be reconstructed, owing to limitations of preservation. The morphological–stratigraphic approach to detached organs will therefore undoubtedly continue to be a major contributor to hypotheses of angiosperm origin.

It is interesting that a stratigraphic-morphological approach to phylogeny which is dominantly methodological has been proposed by Hughes (e.g. 1976), despite his idealistic view that "genuine historical information derived from fossil evidence" will produce a detailed hypothesis of phylogeny (a spirit with which we fully concur). However, he suggests that we "set aside all theoretical phylogeny . . . until a genuine surplus of Cretaceous fossil information has been accumulated and has been stored under entirely neutral descriptors" (1976, p. 172). Hughes has carefully developed a computer-based approach that enables observations to be dispassionately stored. Apart from its obvious practical relevance to biostratigraphy, however, such storage seems phylogenetically limited, in a similar sense to phenetics and methodological cladistics. Indeed such an atheoretical approach seems untenable, for apart from an interim theoretical viewpoint it may be asked what leads Hughes to look specially at the Cretaceous, and how — without theory or convention — he can determine when a "genuine surplus" of observations has been acquired (Magee 1972). The whole basis of his research is, of course, theory-soaked: from the currently established assumption that fossils relevant to angiosperm origin occur in the Cretaceous, to his assumptions surrounding choice of computer program and uniformitarianism. As with the CU discussed in Part I, one may ask what is Hughes' conception of an evolutionary species, and whether such an approach is genuinely phylogenetic.

OUTLINE OF A CLADISTIC APPROACH

## 1. Provisional Cladistic Diagrams (Cladograms)

(a) Data and methodology. The diversity and processes of the living world provide a uniformitarian frame of reference for study of diversity and processes in the fossil world. One task of a cladistic approach is therefore to determine the "sister-group" relationships (most parsimonious proper sub-classes) of living seed plants (Figs 45–48), which can be extended to include fossil plants. The following characters were used for preparation of the data matrices in Tables V and VI. Identification numbers are given on the left and their plesiomorphous/apomorphous states are shown as far as possible in simple binary form.

1. Secondary thickening absent/present (in ferns apomorphous only in Botrychium (Sporne 1966, p. 22, 1974, p. 87; Bierhorst 1971, p. 132; Takhtajan 1969, p. 43).

2. Apical meristem of shoot without distinct tunica and corpus/ with distinct tunica and corpus (Johnson 1951; Sporne 1967, 1974).

3. Shoot apex with apical cell/group of cells (meristem) (Bierhorst 1958, 1971; Sporne 1966).

---

Fig. 45. Cladogram based on Tables V and VI, showing plesiomorphous character states ⬜; apomorphous character states ■. Character states marked ◣ are apomorphous in only some members of a taxon. Numbers at the left refer to the whole row of character states across the clado-gram. With the exception of character 25 no autapomorphies are shown (see text). * indicates apparent parallelism. Character group X is shown twice for practical graphical reasons. The same cladogram is shown in "skeletal" form in Fig. 48, and two slightly less parsimonious versions in Figs 46 and 47. The relative extent of parsimony was judged in all cases by comparing the total number of parallelisms (see legends to Figs 46–48. (Note that in any cladogram the taxa can be rotated on their nodes.) The total number of parallelisms is proportional to the number of taxa in any given cladistic comparison and also the depth of resolution. If the gnetopsid genera taken here separately were to be treated as one group the result may well be different. ANG, angio-sperms; GNE, Gnetum; WEL, Welwitschia; EPH, Ephedra; CON, Conifers, GIN, Ginkgo; FER, ferns; EQU, Equisetum; LYC, lycopsids.

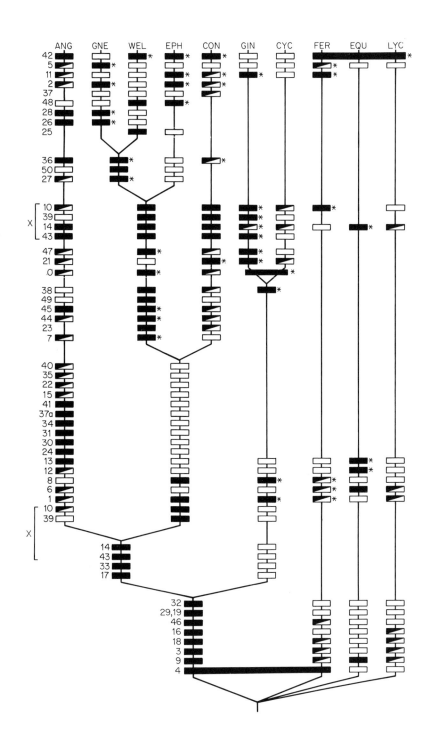

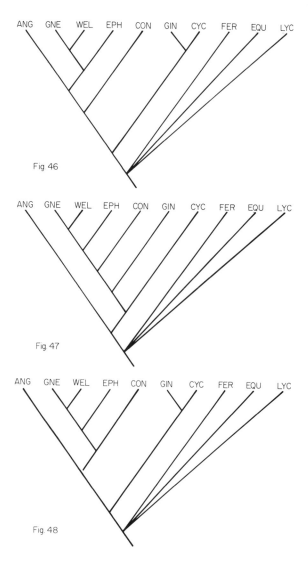

Figs 46–48. "Skeletal" cladograms, which are more or less equally parsimonious. Figure 46: 27 parallelisms excluding ▰character states, 47 parallelisms including ▰ character states. Figure 47: 24 parallelisms excluding ▰, 45 parallelisms including ▰. Figure 48: shown in more detail in Fig. 45, 25 parallelisms excluding ▰, 42 parallelisms including ▰. Abbreviations as in Fig. 45.

4. Shoot vascular system without leaf gaps/with leaf gaps (Sporne 1966; Bierhorst 1971).

5. Leaves having 1–2 orders of venation/2 or more orders.

6. Water conducting cells of the *primary metaxylem* "semaphoront" are tracheids alone/also with vessels. (Apomorphous in some pteridophytes, e.g. *Pteridium, Selaginella, Equisteum* – in which vessels are short in length; and in most angiosperms (Bierhorst 1958, 1960, 1971)).

7. Water conducting cells of the *secondary xylem* "semaphoront" are tracheids alone/also with vessels (Sporne 1967; Bierhorst 1971).

8. Thickening of the tracheid walls in the *protoxylem* "semaphoront" is without bordered pits/with bordered pits (Bierhorst 1958, 1960).

9. Thickening of the tracheid walls in the *centrifugal primary metaxylem* "semaphoront" is without bordered pits/with bordered pits (Bierhorst 1971).

10. Thickening of tracheid walls of *secondary xylem* "semaphoront" with multiseriate bordered pits/also with 1–2 seriate (Chamberlain 1935; Bierhorst 1971).

11. Thickening of tracheid walls of *secondary xylem* "semaphoront" with multiseriate bordered pits/with 1–2 seriate bordered pits (Chamberlain 1935; Bierhorst 1971).

12. Food conducting cells of the phloem are sieve cells/sieve tubes. (The sieve tubes of *Equisetum* occur in the roots only (Esau 1969.))

13. Axial parenchymatous component of vascular tissue includes specialized "albuminous" cells/"companion" cells, the latter developing from the same initial cells as the sieve elements. (The companion cells of *Equisetum* occur in the phloem of the roots only. Some conifers, e.g. *Pinus*, have a radially disposed system of albuminous cells either alone or in combination with an axial system (see Esau 1969). The sieve cells in gnetopsids are possibly more closely associated in position with the albuminous cells than in conifers, but in a different way from angiosperms.) (This study accepts as a postulate that absence of phloem in *Mourera aspecta* and some other angiosperms is secondary.)

14. Leaves bearing the sporangia (= sporophylls) not aggregated into strobili/more or less aggregated into strobili. (This study

Table V. Data matrix for the selected two state characters of living seed plants listed in the text. Ferns, *Equisetum* and lycopsids are included as selected out-groups. Plesiomorphous state shown (−), relatively apomorphous (+). Those marked +/− are apomorphous in only some members of a taxon. Doubt is indicated by ?, and no apparent applicability by a blank entry

| | Most Angiosperms | All Angiosperms | Gnetum | Welwitschia | Ephedra | Conifers | Ginkgo | Cycads | Ferns | Equisetum | Lycopsids |
|---|---|---|---|---|---|---|---|---|---|---|---|
| 1 | + | +/− | + | + | + | + | + | + | +/− | − | +/− |
| 2 | + | +/− | + | − | + | +/− | − | − | +/− | − | − |
| 3 | + | + | + | + | + | + | + | + | +/− | − | +/− |
| 4 | + | + | + | + | + | + | + | + | + | − | − |
| 5 | + | +/− | − | − | − | − | − | − | +/− | + | − |
| 6 | + | +/− | + | − | + | − | − | − | +/− | + | − |
| 7 | + | +/− | + | + | + | + | + | + | +/− | − | +/− |
| 8 | − | − | + | + | + | + | + | + | + | − | −/− ? |
| 9 | + | +/− ? | + | + | + | + | + | + | +/− | + | +/− |
| 10 | + | +/− | − | − | − | +/− | + | +/− | +/− | − | − |
| 11 | + | +/− | + | + | + | − | − | − | + | − | − |
| 12 | + | +/− | + | + | + | − | − | − | − | + | |
| 13 | + | + | + | + | + | + | +/− | + | − | + | |
| 14 | + | + | + | + | + | + | + | +/− | − | + | +/− |
| 15 | + | +/− | + | + | + | + | +/− | + | − | − | |
| 16 | + | + | + | + | + | + | + | + | − | − | +/− |
| 17 | + | + | + | + | + | + | + | + | +/− | − | − |
| 18 | + | + | + | + | + | + | + | + | +/− | − | +/− |
| 19 | + | + | + | + | + | + | + | + | − | − | − |
| 20 | +/− ? | +/− ? | + | + | + | +/− | + | +/− | | | |
| 21 | − | +/− | − | − | − | + | + | +/− | | | |
| 22 | + | +/− | −? | −? | −? | − | − | − | | | |

| No. | 1 | 2 | 3 | 4 | 5 | 6 | 7 | 8 | 9 | 10 | 11 |
|-----|---|---|---|---|---|---|---|---|---|----|----|
| 23 |  |  |  | − | − | +/− | + | + | + | + | + |
| 24 | − | − | − | − | − | − | − | − | − | +/− | + |
| 25 |  |  |  | − | − | − | − | + | + | + | − |
| 26 |  |  |  | − | − | − | − | − | + | + | + |
| 27 |  |  |  | + | + | − | − | + | + | + | + |
| 28 | − |  |  | − | − | + | + | − | + | + | + |
| 29 |  |  |  | − | − | − | − | + | − | + | + |
| 30 | − | − | − | + | + | − | − | − | − | + | + |
| 31 |  |  |  | − | − | + | + | − | + | +/− | + |
| 32 |  |  |  | − | − | + | + | + | + | + | + |
| 33 |  |  |  | − | − | − | − | + | − | + | + |
| 34 |  |  |  | − | − | − | − | − | − | − | − |
| 35 | − |  |  | − | − | +/− | − | − | + | − | − |
| 36 |  |  |  | − | − | +/− | − | + | − | +/− | + |
| 37 |  |  |  | − | − | − | − | − | − | + | + |
| 37a |  |  |  | + | + | −/−? | + | − | − | + | + |
| 38 | − | − | − | − | + | + | + | + | + | + | +? |
| 39 |  |  |  | − | − | + | − | + | − | + | +/− |
| 40 |  |  |  | − | − | − | − | − | + | +? | + |
| 41 |  |  |  | − | − | − | + | − | + | +/− | + |
| 42 | − | − | − | − | + | − | + | +? | − | + | + |
| 43 |  |  |  | − | − | + | + | + | − | + | − |
| 44 | + |  |  | − | − | + | + | + | − | + | − |
| 45 |  |  |  | − | − | +/− | + | + | + | +/− | − |
| 46 | + | + | + | + | + | +/− | + | + | + | − | − |
| 47 | − |  |  | − | + | + | + | + | + | ? | − |
| 48 | − | − | − | − | − | +/− | + | + | + | − | − |
| 49 | − | − | +/− | − | − | − | + | + | − | −? | − |
| 50 | ? | ? | ? | − | − | − | − | + | + | − | − |

Table VI. Data matrix of Table V more or less sorted for congruence of character states

| | Most Angiosperms | All Angiosperms | Gnetum | Welwitschia | Ephedra | Conifers | Ginkgo | Cycads | Ferns | Equisetum | Lycopsids |
|---|---|---|---|---|---|---|---|---|---|---|---|
| 42 | + | + | − | +? | + | + | − | − | + | + | + |
| 5 | + | +/− | + | − | − | − | − | − | +/− | − | − |
| 11 | + | +/− | − | − | + | +/− | + | − | + | | |
| 2 | + | +/− | + | − | + | +/− | − | − | | | − |
| 37 | | − | − | | | +/− | − | − | | | |
| 48 | − | − | − | + | + | − | − | − | − | − | − |
| 28 | + | + | + | −? | − | − | − | − | | | |
| 25 | | | | + | − | − | | − | | | |
| 26 | + | + | + | + | − | +/− | − | | − | − | − |
| 36 | + | + | + | + | − | − | − | − | | | |
| 50 | − | − | + | + | − | | − | − | ? | ? | ? |
| 27 | − | +/− | + | + | + | +/− | + | − | | | |
| 47 | + | +/− | + | + | − | + | + | +/− | | | |
| 21 | − | +/− | − | − | + | +/− | + | + | | | |
| 38 | − | − | +? | + | + | +/−? | + | − | − | − | − |
| 49 | − | −? | + | + | + | − | − | − | | | |
| 44 | +/− | +/− | + | + | + | +/− | − | − | − | − | − |
| 45 | + | + | + | + | + | +/− | − | − | − | − | − |
| 23 | | | + | + | + | +/− | − | − | − | − | − |
| 7 | + | +/− | | | | | | − | | | |
| 40 | + | +/− | − | − | − | − | − | − | − | | |
| 35 | + | +/− | − | − | − | − | − | − | | | |
| 22 | + | +/− | −? | −? | −? | − | − | − | | − | − |

| Taxon | 1 | 2 | 3 | 4 | 5 | 6 | 7 | 8 | 9 | 10 | 11 |
|---|---|---|---|---|---|---|---|---|---|---|---|
| 15 | + | +/− | − | − | − | − | − | − |  | + |  |
| 41 | + | + | − | − | − | − | − | − | − | + |  |
| 37a | + | + | − | − | − | − | − | − |  | − |  |
| 34 | + | + | − | − | − | − | − | − |  | + |  |
| 30 | + | + | − | − | − | − | − | − |  | − |  |
| 31 | + | + | − | − | − | − | − | − |  | + |  |
| 24 | +/− | + | − | − | − | − | − | − | − | − | +/−? |
| 12 | + | +/− | − | − | − | − | − | − | +/− | +/− | −/− |
| 13 | +/− | + | + | + | + | − | − | − | +/− | +/− | +/− |
| 8 | − | − | − | − | − | − | − | − | − | − | − |
| 6 | + | +/− | + | + | + | + | − | − | +/− | + | +/− |
| 1 | + | + | + | + | + | + | − | − | +/− | + | +/− |
| 20 | +/−? | +/− | + | + | + | + | + | − | +/−? | + | + |
| 10 | + | +/− | + | + | + | + | + | − | +/− | − | + |
| 39 | − | − | +/− | +/− | + | + | + | − | + | + | + |
| 14 | + | + | +/− | +/− | +/− | + | + | − | − | +/− | +/− |
| 43 | +? | + | +/−? | + | + | + | + | − | − | − | − |
| 33 | + | + | + | + | + | + | + | − | − | − | − |
| 17 | + | + | + | + | + | + | + | − | − | − | − |
| 29 |  |  | + | + | + | + | + | + | − | − | − |
| 32 | + | + | + | + | + | + | + | + | − | − | − |
| 19 | + | + | + | + | + | + | + | + | − | − | − |
| 46 | + | + | + | + | + | + | + | + | +/− | − | +/− |
| 16 | + | +/− | + | + | + | + | + | + | − | − | +/− |
| 18 | + | + | + | + | + | + | + | + | +/− | + | +/− |
| 3 |  |  | + | + | + | + | + | + | +/− | − | +/− |
| 9 | + | +? | + | + | + | + | + | + | + | + | +/− |
| 4 | + | + | + | + | + | + | + | + | + | + | − |

accepts as a postulate that *Lemna* female flowers and *Euphorbia* male flowers are secondarily reduced in angiosperms).

15. Strobili unisexual/bisexual. (Abnormal specimens of *Gnetum*, *Ephedra* and some conifers can be bisexual (Chamberlain 1935; Maheshwari and Vasil 1961).

16. Female (mega) gametophyte free-living/retained on the sporophyte. (In *Selaginella rupestris* retention occurs until young sporophytes germinate (Sporne 1966, p. 89.))

17. Male gametes motile/non-motile (Chamberlain 1935; Chaloner 1970b).

18. Spores homosporous (of essentially uniform size)/markedly heterosporous (of distinctly two sizes: large megaspores and small microspores). (Sussex 1966; Chaloner 1970b.)

19. Seeds absent/present.

20. In mature seed the integument (or innermost integument where there are 2 or 3) is free from the nucellus almost to the base/integument and nucellus adnate for more than half their length (Chamberlain 1935; Harris 1954; Foster and Gifford 1974, p. 388; also based on figures in Netolitzky 1926; Sporne 1967; Martens 1971; Foster and Gifford 1974).

21. Ovule and seed with two or three integuments/one integument. (The postulate that the seed envelopes of gnetopsids are cupular integuments is here accepted and thus the relative parsimony of alternative viewpoints has not been assessed (Sporne 1974)).

22. Nucellar cuticle relatively heavily cutinized/relatively thinly cutinized. (No detailed studies known to us; based on Harris 1954.)

23. Megaspore wall thickly impregnated with sporopollenin/relatively thin and with much reduced amount of sporopollenin (Thomson 1905; further references in text).

24. Megaspore wall with reduced amount of sporopollenin/apparently the cell wall only, without sporopollenin (Harris 1954; discussion in text).

25. Archegonia present/megagametophyte apparently with archegonial initials only (Chamberlain 1935; see discussion by Martens (1971) and Singh (1978)).

26. Embryo sac with archegonial initials/without archegonial initials (references as for 25). (The parsimony of the alternative

postulate that the angiosperm embryo sac is more similar to an archegonium than a reduced megaprothallus is not here compared.)

27. Ovule development monosporic/tetrasporic (Sporne 1967, p. 177, 1974, p. 162; Martens 1971).

28. Megaprothallus cellular at time of fertilization/micropylar end remains relatively free nuclear (Sporne 1974; Martens 1971).

29. Archegonia with neck canal cells/without neck canal cells (Chamberlain 1935, p. 328).

30. Megaprothallus composed of 125 or more nuclei at time of fertilization, all of which subsequently form an endosperm (which may include multiploid nuclei (see Singh 1978))/megaprothallus of 4–16 nuclei, only one of which (normally a fusion product of two) will form the diploid-multiploid endosperm nucleus (Sporne 1974; Foster and Gifford 1974).

31. Ovum fertilized only/ovum + one other nucleus fertilized to form endosperm nucleus ("double fertilization"). (Sporne 1974; cf. Moussel 1978.)

32. Microspore (pollen) without "pollen-tube"/germinating to form a pollen tube (Chamberlain 1935; Chaloner 1970b).

33. Function of "pollen-tube" haustorial/for transfer of male gametes (Chamberlain 1935; Chaloner 1970b).

34. Pollen endexine as seen in T.E.M. sections laminated throughout/laminated only under germination apertures (discussion in text).

35. Pollen ektexine granular or spongey/tectate-columellate (see text for discussion).

36. Pollen grain containing prothallial cells/in normal development without prothallials (Sporne 1974).

37. No papillose (stigmatic) surface for pollen reception/stigmatic surface present on integument (e.g. *Pseudotsuga*; see Doyle 1945).

37a. No papillose (stigmatic) surface for pollen reception/stigmatic surface present on a structure external to the integument. (In the conifers *Eutsuga*, *Saxegothaea* and in Araucariaceae pollen is received on a structure outside the integument but apparently not on a papillose surface (see Doyle 1945).)

38. Nuclear divisions occurring during the early stages of embryogenesis are separated by cell walls/free nuclear (Sporne 1974,

p. 39; plesiomorphous condition also apparently occurs in *Sequoia*; Chamberlain 1935; Sporne 1967, p. 143).

39. Ovule symmetry radial (radiospermic)/bilateral (platyspermic). (Sporne 1967; Bierhorst 1971.)

40. Seeds naked/enclosed at anthesis.

41. Pollen at time of fertilization has 4 or more nuclei (inclusive of nuclei formed earlier and by that time aborted)/3 nuclei (Sterling 1963; Sporne 1967, 1974).

42. Polyploidy absent/present (in gymnosperms certainly only in *Sequoia sempervirens*, *Juniperus* (2 species), *Fitzroya* and some species of *Ephedra* (see White 1978)).

43. Secondary xylem manoxylic/pycnoxylic (Sporne 1967).

44. Strobili simple/compound (Sporne 1967).

45. Mäule reaction absent/present (Nishio 1959; Maheshwari and Vasil 1961).

46. Primary metaxylem of stem with protoxylem in mesarch or exarch position/in endarch position (Bierhorst 1971).

47. Male sporophylls (stamens) laminar in appearance/microsporangia attached to a stalk-like structure.

48. Pollen not polyplicate/with 5–20 prominent ridges (Erdtman 1965).

49. Integumentary micropyle relatively unmodified/forming a long tube (see Maheshwari and Vasil 1961, pp. 130–131; Sporne 1974, p. 132).

50. "Feeder" in embryo absent/present (Sporne 1967).

Several of the characters shown in this list as simple two-state binary comparisons are much oversimplified at the level of resolution of the taxa utilized. Within the conifers, for example, the thickness of the megaspore membrane varies widely, and sporopollenin is almost lacking in Taxaceae (J. M. Pettitt, personal communication). Several entries in Tables V and VI are therefore made as +/−, indicating that only some species are apomorphous, the rest plesiomorphous. A column representing most, as opposed to all, angiosperms is included in the tables for interest's sake.

Certain characters initially considered for inclusion were rejected. They include features that would give dominantly +/− entries at our level of analysis: for example, (a) presence of short shoots (which may perhaps be historically linked with the origin of strobili);

(b) shape of nucellus and pollen chamber (Singh 1978); (c) details of relative enclosure of ovules, and seeds (*Araucaria*, podocarps, gnetopsids, *Reseda*); (d) most chemicals; (e) number of cotyledons (up to 4 in angiosperms). Some characters were excluded through difficulty of deciding how to treat them: (a) DNA per chromosome (Sparrow *et al.* 1972; Price *et al.* 1974); (b) other chromosomal characters. Several were excluded because our knowledge of them was too superficial, for example number, structure, and manner of insertion of pollen sacs. Thus the character list is by no means exhaustive and is open to critical improvement based on existing literature besides new information.

As far as possible, postulates about what constitutes a primitive angiosperm were avoided (i.e. magnoliid *vs* amentiferous). Postulates about the homologies are included as indicated in the character list, for example the angiosperm embryo sac is considered to be essentially comparable to a megaspore rather than a single archegonium (Sporne 1974). The level of resolution is very coarse and involves acceptance of numerous decisions in the literature, as does the use of out-groups and the particular out-groups selected. Thus, while attempting to be explicit, the whole exercise was "theory-soaked" and well laden with preconceived ideas.

Manual sorting of Table V led to Table VI and from this a number of possible cladograms were drawn up. The number of parallelisms was contrasted for (a) monothetic (+) states only, and (b) also for (+/−) entries. Bearing in mind that the aim is also to maximize the number of proper sub-classes, the least number of parallelisms gives the most parsimonious, and therefore preferred, result. The cladogram in Fig. 48 (= Fig. 45) is marginally more parsimonious than Fig. 47, with Fig. 46 still less so. For reasons of space only the marginally most parsimonious one is shown in detail, in Fig. 45.

The "autapomorphies" listed below include several that represent complexes of characters:

*LYCOPSIDS*

(a) Unique phyllotaxis in mature shoots (of most).

(b) In heterosporous species, megasporangia borne basipetally to microsporangia.

(c) Sporophylls are microphylls, sporangia adaxial or axillary.

*EQUISETUM*

(a) Carinal canal vascular system.

(b) Spores with elaters.

*GINKGO* and *CYCADS*

(a) No strict autapomorphies known to us.

*CONIFERS*

(a) Compound female cone-scale?

(b) Juvenile leaves uniformly needle-like.

*EPHEDRA*

(a) Strobilar and floral organization.

(b) Kind of "double fertilization" (Moussel 1978).

(c) Ephedrine.

*WELWITSCHIA*

(a) Unique growth habit; three pairs of decussate leaves, with anastomosing parallel veins, one pair of leaves dominating and lasts the life of the plant (Sporne 1967).

(b) Unique inflorescence and floral organization.

(c) Unique fertilization process.

*GNETUM*

(a) Unique strobilar form and floral details.

Note that a parallelism, or synapomorphy of a taxon as a whole, may function methodologically as an autapomorphy.

(*b*) *Interpretation.* In Fig. 45 the proportion of synapomorphies which also occur as parallelisms is high. The diagram indicates that some nested taxa appear to be defined by several synapomorphies, e.g. seed plants as a whole, while others have relatively few, although whether the occurrence of numerous congruent synapomorphies should increase confidence as to the "correctness" of the result in evolutionary terms is a complex matter. Thus the congruence of characters 17, 32 and 33 seems impressive for conifers through to angiosperms, though it could be argued that they represent redundancy owing to interlinked function. However, since character 32 is found "pre-adapted" in *Ginkgo* and cycads the case for separate assessment of the characters seems strong. The seven characters uniting seed plants may also include redundant congruences, e.g. between character 19 and 32, though again it can be suspected that some fossil seed plants may well have had pollen lacking a "pollen tube". There is no easy answer to this unit character problem, which is connected partly with a paradox in the sub-class relation and partly with the palaeontological expectation that fossils will blur the distinctions between cladistically distinct groups. Conversely, the congruence of characters 18 and 19 suggests that seeds are equivalent to pteridophyte megasporangia (Smith 1964), and fossils fill out the transition, since the megaspores of many living gymnosperms are actually smaller than their microspores. It would be redundant to add a third character (megasporangia non-integumented/integumented) to our analysis, in view of characters 18 and 19, though in a comparison with fossil out-groups an equivalent character for cupules would clearly not be redundant (Smith 1964).

However, character 41 may be simply a reflection of loss of prothallials (character 36) and, if truly redundant, represents an error in our analysis.

Figures 45–48 may provide some explicit insight into Doyle's (1978) and Takhtajan's (1969, 1976) discussions on neoteny (progenesis) and of whether circular bordered pits are generalized or not in seed plants. Wood characters have been emphasized in the literature because they appear to provide the only features that might cognitively unite gnetopsids and angiosperms, principally by the shared occurrence of vessels. When semaphoronts are compared, however, the only uniting character known to us is the Mäule reaction: a histo-chemical test for syringaldehyde. Nevertheless, by invoking the interpretation that there has been a progenetic shift, a more parsimonious result is attainable, and this permits vessels to be thought of as a synapomorphy. Secondary xylem vessels in gnetopsids are thus equated with primary xylem vessels in angiosperms. Since some angiosperms do not possess vessels in the primary xylem, however, this ontogenetic interpretation has to be carried further. The hypothesis of neoteny can then also bring other cladistically incongruent characters of gnetopsids, and of other gymnosperms, into congruence with angiosperms: characters 38, 44, 35, 37, 8, and 34 (see below). Such an interpretation, which maximizes congruence, is of interest for it represents a guess at a neater explanation involving process, and which may well be correct. Unfortunately there does not appear to be any very satisfactory independent evidence at the present time (Gould 1977), and the ability to neotenize organisms experimentally would be of profound phylogenetic interest, as is the recently developed ability to interbreed certain plant species above the "natural" genetic species level by partly overcoming their incompatibility mechanisms experimentally.

An aim of avoiding postulates about whether the primitive angiosperm was magnoliid *vs* amentiferous was to determine the possibilities of out-group comparison as a basis for suggesting such characters. Where gnetopsids appear to be the nearest sister-group (Fig. 46) the following characters might be hypothesized to occur in the primitive angiosperm: (a) opposite leaves (of various forms, e.g. as in *Austrobaileya, Trochodendron* and many monocotyledons); (b) monosulcate, polyplicate or else *Gnetum*-like porate pollen (found in Piperales); (c) secondary xylem composed of tracheids with bordered pits (*Drimys*); (d) habit anything from liane to tuberous; (e) compound strobili (e.g. *Engelhardtia, Trochodendron*). (With more or less imagination the tubular micropyle (49) of gnetopsids can be homologized with the style of *Engelhardtia* (Meeuse, *loc. cit.*; Maheshwari and Valil 1961).)

These characters do not seem to provide a clear cut basis for decision as to whether magnoliids *vs* amentiferous forms are to be considered primitive amongst angiosperms, and suggest that a more heterobathmic combination of characters than is generally proposed might be an alternative heuristic, perhaps pertaining to an as yet undiscovered extinct form or forms. Figure 47 further suggests that a *Ginkgo*, conifer or cycad-like morphotype (Mamay 1976) may be equally as feasible as a gnetopsid, magnoliid or amentiferous morphotype. Table VII shows the approximate times of appearance of the sister taxa, and indicates that such a primitive angiosperm may be predicted to have arisen at any time within the range Carboniferous to Cretaceous.

## 2. Discussion of Fossil Seed Plants

The cladistic relationships of extant seed plants provide their hypothetical relative antiquity based on morphology, without a calibrated time scale, while fossils provide calibration and also

Table VII. Approximate first fossil occurrences of the taxa shown in Figs 45–48, based partly on Harland et al. (1967). Takes into account only those forms apparently identical to or directly ancestral to living genera, e.g. *Ginkgo* but not *Sphenobaiera*, and *Asteroxylon* but not *Zosterophyllum*

| | (Angiosperms) | Gnetopsids | Conifers | Ginkgo | Cycads | Ferns | Equisetum | Lycopsids |
|---|---|---|---|---|---|---|---|---|
| Quaternary | | | | | | | | |
| Tertiary | | | | | | | | |
| Cretaceous | (X?) | | | | | | | |
| Jurassic | | | | | | | | |
| Triassic | | X? | X | X | | | | |
| Permian | | | | | X | | | |
| Carboniferous | | | | | | X | X? | |
| Devonian | | | | | | | | X |

Table VIII. Approximate first fossil occurrences of selected apomorphous character states shown in Tables V and VI

| | 1 | 4 | 5 | 7 | 10 | 14 | 15 | 16 | 18 | 19 | 24 | 32 | 34 | 35 | 37a | 39 | 40 | 48 | 49 |
|---|---|---|---|---|---|---|---|---|---|---|---|---|---|---|---|---|---|---|---|
| Quaternary | | | | | | | | | | | | | | | | | | | |
| Tertiary | | | | | | | | | | | | | | | | | | | |
| Cretaceous | | | | X | | | | | | | | | X | | | | X? | | |
| Jurassic | | | | | | | | | | | | | | | | | | | |
| Triassic | | | | | | | | | | | | | | X | X | | | X | |
| Permian | | | | | | | | | | | | | | | | | | | |
| Carboniferous | | | X? | | | X? | X? | | | | X? | X | | | | | | | X |
| Devonian | X | X | | | X | | | X | X | X | | | | | | X | | | |

details about diversity that the living world cannot provide. Table VIII extends the comparison to include approximate first fossil occurrences of characters listed in Tables V and VI; and our preliminary investigations are left very much incomplete at this point. Comparison with most fossils requires more detailed resolution of the relationships within and amongst the extant taxa of Figs 45–48, and our own research areas are restricted to cycads and a few groups of angiosperms. Intuitive assessment suggests that Caytoniales and glossopterids may be the nearest sister-groups to angiosperms (Thomas 1925; Bierhorst 1971), with most other fossils apparently having had a more distant relationship.

Within these limitations, cladistic views apparently show elements of agreement with those of Gaussen (1946), Melville (1962, 1963) and Krassilov (1973). Each of these workers has attempted to find a more or less parsimonious explanation of the large gaps between primitive angiosperms and Mesozoic gymnosperms. Gaussen compared the megasporophyll of *Caytonia* with the magnoliid conduplicate carpel (Doyle 1978), a comparison that with imagination can be extended to glossopterids following the interesting work of Gould and Delevoryas (1977) and was extended to other corystosperms by Bierhorst (1971). Melville's suggestions were more speculatively based.

Krassilov (1977) concludes that angiosperms are polyphyletic and he thus overcomes the "gap" by postulating that different angiosperm families evolved from different groups of gymnosperms. He proposed that these groups of gymnosperms be united into a paraphyletic grade group "proangiosperms" (see also Arber and Parkin (1907)).

## 3. Are Angiosperms Monophyletic (in Hindsight)?

Although some of the taxa in Figs 45–48 are united by relatively few synapomorphies, others, e.g. seed plants as a whole, are united by several. Angiosperms appear to be united uniformly by seven:

(13) Presence of axially aligned companion cells, derived in development from the same mother cells as the sieve elements (occurring as a parallelism in roots of *Equisetum*).
(24) Megaspore (embryo sac) wall lacking sporopollenin.

(30) 4–16 nuclei in megaprothallus.

(31) "Double fertilization".

(34) Pollen wall endexine not laminated as seen in TEM sections, except under germination apertures (Doyle *et al.* 1975).

(37a) Pollen-receptive stigmatic surface present, and borne on an enclosing or partly enclosing structure which is external to the integument.

(41) Pollen development has formed three nuclei by the time of fertilization, i.e. prothallials are lacking.

In view of the recent publication by Parenti (1980) it perhaps needs to be mentioned that these are hypothesized as universally shared characters, i.e. synapomorphies, that define proper sub-classes at the level of interest in hand. The two synapomorphies S2 and V6 cited by Parenti are phenetic at this level of universality. Several characters that have popularly been held in the past as synapomorphies have long since failed to survive the critical scrutiny of botanists such as Sporne. For example, the 8-nucleate embryo sac, presence of vessels, number of cotyledons, and tectate pollen wall. The word "angiosperm" itself, from the Greek for "enclosed vessel" refers to the classical defining concept of the closed carpel and ovary (more strictly, closed at anthesis). Though most angiosperms have such carpels or ovaries, in some the carpel is open at anthesis; and in a few genera such as the mignonette, *Reseda* the carpel remains open throughout its development (Sporne 1974). (In *Reseda* this can probably be interpreted on grounds of parsimony as a secondary development.) As indicated above, a number of fossil gymnospermous seed plants besides some living conifers also had more or less enclosed seeds. Where known in sufficient detail to be able to judge, they were not however closed at anthesis (Harris 1940; Harris *et al.* 1974; Krassilov 1977). The phloem of *Austrobaileya* possesses companion cells (Huber and Graf 1955) like all other investigated angiosperms, while it stands anatomically apart in having sieve cells rather than sieve elements. Thus it can no longer be suggested that sieve tubes are a synapomorphy of all 250 000 angiosperms – as a hypothesis they properly unite 249 999! Despite the obvious arbitrariness of definitional criteria (Esau 1969) it should be borne in mind that in making comparisons a dividing line has always to be put somewhere (Sneath and Sokal

1973), and other characters appear to clearly resolve *Austrobaileya* as an angiosperm in the meaning used here. Gaps within the magnoliids appear to be enormous.

This once more emphasizes that all synapomorphies are more or less conjectural guesses, and all the more so since too few angiosperms have been sampled (or in sufficient detail) to enumerate the truth. Of the characters listed above 13, 30 and 31 seem secure to our knowledge at the present time, while 41 also appears to be (based on Sporne (1974) without further reading). Character 37a is an attempt to explicitly represent Takhtajan's character (ii) and we have not critically assessed his character (i). His character (iii) appears to be of interest only in its relation in all higher plants *versus* lycopsids. Characters 24 and 34 in particular are problematical and thus of special research interest.

Character 34, mentioned also by Niklas *et al.* (1980) was proposed initially by Van Campo (1971) and sampling was later extended by Doyle *et al.* (1975) and other workers (see also Walker 1976). Recognizing that evidence is as yet incomplete, they suggest that all gymnosperms have a laminated endexine throughout the wall, whereas in angiosperms the endexine appears homogeneous except sometimes under the germination apertures. Each dark-staining lamina is in all cases recognizable because it is separated from others by laminae that appear white, termed "white lines". A major difficulty pointed out by these writers is the occurrence of uniformly laminated layers of pollen wall in several species of Annonaceae, some other ranaleans and *Gunnera peltata*. Accepted at face value this would refute the character of non-laminated endexine as a synapomorphy of angiosperms. Lugardon and Le Thomas (1974) point out however, that these angiosperm laminated layers are not necessarily comparable in position with the endexine in gymnosperms. As evidenced from the staining criterion used by palynologists to recognize homologous wall layers they are said to pertain in these angiosperms to the basal ektexine or foot layer rather than to the endexine.

Like the similar situation for the presence of vessels and bordered pits in gnetopsids *vs* angiosperms it appears that a solution to this homology problem can be imagined on cladistic grounds. Since the aim can be to recognize comparable "semaphoronts", chemical comparison may or may not be misleading. It may be equally or more useful to compare the developmental sequence of events during wall formation, as has been done by Heslop-Harrison (1971) and as practised for example by Bierhorst for wood anatomy. Heslop-Harrison's work appears to suggest that the endexine of angiosperms may pass through a laminated stage early in development. Thus the immature pollen wall of angiosperms may be equivalent to the mature wall of gymnosperms in this respect. By reference to the rule of ontogeny referred to on p. 280, this further suggests that a laminated endexine is a characteristic generalized feature of seed plants, and thus that character 34 is not in itself an angiosperm synapomorphy but rather a symplesiomorphy. Furthermore, in some angiosperms an endexine is lacking altogether (Walker 1974a).

Character 34 therefore raises comparable problems to the reduction and loss of sporopollenin from the embryo-sac wall (character 24) discussed on pp. 305. Together with the angiosperm characters 30 and 41 these four evolutionary loss/reduction "synapomorphies" are, methodologically speaking, symplesiomorphies. To this extent our methodology was imperfect (B. G. Gardiner, personal communication). Therefore only characters 13, 31 and 37a can be regarded as true methodological synapomorphies of angiosperms, while the additional characters listed by us are methodological symplesiomorphies representing preconceived input as evolutionary "synapomorphies" as dealt with in this chapter.

Character 24 is currently under investigation at the British Museum (Natural History), and pertains both to the extent of sporopollenin deposition and to the consequential micro-architecture of the megaspore wall. As outlined on p. 305, the megaspore wall in all living gymnosperms appears to be impregnated with acid-resistant sporopollenin to varying degrees of thickness and complexity of structure (Thomson 1905; J. M. Pettitt, personal communication), whilst angiosperms are thought in general to lack such an acid-resistant component in this membrane (Harris 1954). Reports in the literature frequently contradict one another and are in need of thorough re-investigation using modern EM techniques. Problem areas are particularly the gnetopsids (Maheshwari and Vasil 1961; Erdtman 1965; Martens 1971), certain conifers and most angiosperms. Particularly interesting from the evolutionary viewpoint is the correlation of this character with characters 20 and

22, most angiosperms apparently possessing a thickly cutinized nucellus which is free from the integument to the chalaza at seed maturity unlike most gymnosperms (Harris 1954); again a methodological symplesiomorphy since this resembles a megasporangium. The possibilities for imaginative speculations on homology besides critical empirical work would seem potentially huge in this neglected research area. These acid-resistant membranes are also of palaeobotanical interest, since seed cuticles are frequently well preserved as fossils and seeds themselves are not uncommon, either dispersed or attached to other organs. They are also good semaphoronts. Of special research interest are fossil seed plants that apparently lack an acid-resistant megaspore wall and in some cases possess a heavily cutinized nucellar cuticle free of the integument: notably *Caytonia*, Bennettitales (Harris 1954, 1958), *Montsechia* (work in progress) and possibly *Glossopteris* (e.g. Pant and Nautiyal 1960). In Gould and Delevoryas' excellently preserved material of *Glossopteris* a megaspore wall was observed, though it is not yet documented whether preservation was suitable to detect acid-resistance and thus by inference sporopollenin. Notably these investigated fossils appear to be intermediate in their seed cuticular characters not only between living angiosperms and gymnosperms, but also in their times of first fossil occurrence between Carboniferous and Cretaceous Barremian. In terms of evolutionary "synapomorphies" this is interesting, particularly since it clearly must compete with the opposing view that these characters are retained evolutionary "symplesiomorphies" which cannot be recognized as of special adaptive relevance to angiosperms.

Viewed as evolutionary "synapomorphies" the functional significance of internal seed membrane characters as angiosperm adaptations has yet to be more than marginally explored, as also is that of the reduced angiosperm embryo sac and of double fertilization. Conceivably the features in gymnosperms as opposed to angiosperms may correlate with a shift in location of one of the components of the sexual incompatibility mechanisms; that is, from the origin of the relevant chemicals in the nucellus and/or megaspore wall and interaction in the pollen chamber, to their origin in the pollen wall and interaction at the stigma or style (Heslop-Harrison *et al.* 1975; Pettitt 1977a,b; Heslop-Harrison 1978). Equally any megaspore-wall site of chemical origin may simply have been transferred to the thickly cutinized nucellus which has never been investigated histochemically or by EM. The tissue rejection experiments on gymnosperms by Pettitt (1979) are not only fascinating in themselves, but, as an evolutionist, one awaits eagerly their application to angiosperms.

It is extremely interesting to note that all seven "synapomorphies" are distinctly functional in aspect, and that six of them are concerned with reproduction. To derive them polyphyletically is of course possible, but it would involve a great many more parallelisms than in the cladograms offered here.

## 4. Is Darwin's Mystery Soluble?

While it is reasonable to expect angiosperm "synapomorphies" 24 and 34 to be preserved in fossils, the remaining characters are unlikely to be found preserved under normal conditions of fossilization. Flowers, however, are sometimes preserved (Basinger 1976; Tiffney 1977) and in some cases reveal a stigmatic surface and/or style, character 37a. We know no example where fossil pollen is recorded attached to such structures amongst presumed fossil angiosperms, though Harris' work on pollen in the micropyles of *Caytonia* is at the level of detail needed. (It may be difficult to be certain about results since *Caytonanthus* pollen is found adhering to all parts of the *Caytonia* "plant", including the lips of the fruits.) Characters 13, 30, 31 and 41 seem destined to remain almost undetectable, though chromosomes and pollen contents are known, very rarely, from fossils.

Since, in evolutionary cladistic terms, knowledge of all seven "synapomorphies" would be needed to elucidate the original angiosperm in terms of the meaning based on living angiosperms, the origin of angiosperms will likely remain only partly soluble in those terms. The great morphological gaps separating supposed primitive angiosperms from one another and from their nearest living sister-group suggest one of three scenarios: (a) a major punctuational event in which they suddenly crystallized out; or (b) step-wise acquisition of characters by fossils corresponding to the angiosperm/gymnosperm morphotypes defineable from Figs 45—48; or (c) evolutionary reversals of morphology in ancestors *vs* descendants on a scale not currently appreciated or recognizable, given present methods and theories. The great diversity of fossils like *Caytonia* which already occupy this gap suggests that fossils will continue to provide interesting and relevant evidence on the transition. The problem is very difficult but not necessarily mysterious.

Taking the methodologically valid synapomorphies 13, 31 and 37a, however, only 37a seems reasonably amenable to study from fossils. In that sense, *Caytonia*, *Leptostrobus* and possibly *Hirmerella* display the kind of evidence needed by the methodologist to suggest that they are very near to primitive angiosperms. In a methodological sense it could seem easy to solve the angiosperm problem and then there would seem to be no mystery at all. We have attempted, however, to remind ourselves that real solutions of evolution may be more complex than this, since the evolutionary tree involved process, besides leaving some clues as morphological patterns of derived (or any other) similarity. The interpretation of those not quite fully adequate, or not fully understood clues, in relation to what may have actually happened, remains a thorny and therefore challenging area for research. To exclude the possibility of reconciling pattern with process is however simply negative, while to attempt to reconcile them is difficult, confusing, and involves theoretical perspectives – for example on reversals – which the convinced methodologist may consider naive (and indeed often are).[2] We hope our own confusions do not obfuscate the position too much for the reader's patience.

### ACKNOWLEDGEMENTS

Several colleagues and friends have provided valuable criticisms of earlier drafts: P. D. W. Barnard, Stella Butler, B. Cornet, D. L. Dilcher, G. F. Estabrook, P. L. Forey, R. A. Fortey, B. G. Gardiner, M. Hills, C. J. Humphries, C. Patterson, J. M. Pettitt, D. E. Rosen. C. R. H. wishes to thank C. P. Palmer for the note on Occam and for his enlightening comments on philosophy of systematics, and especially Professor D. L. Hull for kind, thorough, and penetrating appraisal of the penultimate draft, which has removed at least some of the remaining nonsense. Finally our thanks to the editors for their considerable kindness expressed in numerous ways.

### NOTES

[1] The principle of parsimony is widely used in science and is often referred to as Occam's (or Ockham's) razor, after the philosopher and theologian

William (*c.* 1285–1349) of Ockham in Surrey, England (Copleston 1972). He wrote "Nunquam ponenda est pluritas sine necessitate" and "Frustra fit per plura quod potest fieri per pauciora", which translates to "Never is multiplicity to be postulated without necessity" (Kneale and Kneale 1962, p. 243) and "It is vain to do with more what can be done with fewer" (Russell 1946, p. 494). See also Sober (1975).

[2] In this chapter we have twisted the meaning of the term synapomorphy, using it in two senses: (a) the operationally cladistic *methodological synapomorphy*, based on the rules given on p. 280; (b) *evolutionary synapomorphy*. Evolutionary synapomorphy is operationally pertinent only after a cladogram has been arrived at by means of methodological synapomorphies, and to that extent our own synapomorphies for angiosperms are imperfectly explicit and imperfectly assessed. For a given taxon, evolutionary synapomorphy refers to (1) the defining methodological synapomorphies and (2) the "reversal", methodologically plesiomorphous, characters. Evolutionary synapomorphy is therefore a cluster concept, appropriate to trees rather than cladograms, and once a group is initially defined gives equal weight to all of its characters. Some simpler term than evolutionary synapomorphy may or may not be more appropriate, for the distinction is not an easy one. (Perhaps the character-centred term "synapomorphy" should be abandoned for trees, in favour of adaptation-centred terms, e.g. "cladistic adaptation" *vs* "reversal adaptation" for (1) and (2) respectively.) Whatever the terminological decision, our hypothesized equivalence is of character (in a cladogram) to adaptation (in a tree). Unlike "reversal" characters, other, continuously widespread symplesiomorphies at any one level, present less of a problem. They are, of course, synapomorphies at some higher level, where their adaptive nature would be discussed.

The cladistic method of analysis of pattern inevitably will colour perception of which characters may be adaptive, as will any other method based purely on morphological similarities. Thus we have argued strongly in this chapter towards the need for experiment and observation on process at the present day and use of independent tests, e.g. from stratigraphy and ecology, in order to provide yardsticks by which the genealogical significance of pattern may be assessed. As Hull (1965) very cogently argues, it is not necessary that such yardsticks should provide universal defining features (like synapomorphies), and therefore for them to be universally applicable. Philosophically different criteria may obtain for *interpreting* pattern in terms of reality rather than simply defining pattern. Hull's comments on phenetics in the 1960s are, in our view, very applicable to much of what is current in methodological cladistics.

REFERENCES

Adanson, M. (1763–4). "Familles des plantes". Vincent, Paris.
Andersen, N. M. (1978). Some principles and methods of cladistic analysis

with notes on the uses of cladistics in classification and biogeography. *Z. Zool. EvolForsch.* **16**, 242–255.

Arber, E. A. N. and Parkin, J. (1907). On the origin of angiosperms. *J. Linn. Soc. Bot.* **38**, 29–80.

Ash, S. R. (1976). Occurrence of the controversial plant fossil *Sanmiguelia* in the Upper Triassic of Texas. *J. Paleont.* **50**, 799–804.

Ashlock, P. D. (1974). The uses of cladistics. *A. Rev. Ecol. Syst.* **5**, 81–99.

Bailey, I. W. (1944). The development of vessels in angiosperms and its significance in morphological research. *Am. J. Bot.* **31**, 421–428.

Bailey, I. W. and Swamy, B. G. L. (1948). *Amborella trichopoda* Bail. A new morphological type of vesselless dicotyledon. *J. Arnold Arbor.* **29**, 245–254.

Bailey, I. W., Nast, C. G. and Smith, A. C. (1943). The family Himantandraceae. *J. Arnold Arbor.* **24**, 190–206.

Bailey, L. H. (1949). "Manual of cultivated plants" (2nd edn). Macmillan, New York.

Banks, H. P. (1961). Reviews of A. L. Takhtajan's "Evolutionary morphology of plants" and "Origins of angiospermous plants". *Pl. Sci. Bull.* **7** (1), 4–5.

Banks, H. P. (1968). The early history of land plants. *In* "Evolution and Environment" (E. T. Drake, ed.). pp. 73–107. Yale University Press, New Haven and London.

Baranova, M. (1972). Systematic anatomy of the leaf epidermis in the Magnoliaceae and some related families. *Taxon* **21**, 447–484.

Basinger, J. F. (1976). *Paleorosa similkameenensis*, gen. et sp. nov., permineralised flowers (Rosaceae) from the Eocene of British Columbia. *Can. J. Bot.* **54**, 2293–2305.

Bean, W. J. (1973). "Trees and Shrubs Hardy in the British Isles, 2, D–M" (8th edn). John Murray, London.

Beck, C. B. (1976). Origin and early evolution of angiosperms: a perspective. *In* "Origin and Early Evolution of Angiosperms" (C. B. Beck, ed.), pp. 1–10. Columbia University Press, New York and London.

Bessey, C. E. (1897). Phylogeny and taxonomy of the angiosperms. *Bot. Gaz.* **24**, 145–178.

Bessey, C. E. (1915). The phylogenetic taxonomy of flowering plants. *Ann. Mo. bot. Gdn* **2**, 109–164.

Bierhorst, D. W. (1958). The tracheary elements of *Equisetum* with observations on the ontogeny of the internodal xylem. *Bull. Torrey bot. Cl.* **85**, 416–433.

Bierhorst, D. W. (1960). Observations on tracheary elements. *Phytomorphology* **10**, 249–305.

Bierhorst, D. W. (1971). "Morphology of Vascular Plants". Macmillan and Collier-Macmillan, New York and London.

Bock, W. J. (1973). Philosophical foundations of classical evolutionary classification. *Syst. Zool.* **22**, 375–392.

Bock, W. J. (1979). The synthetic explanation of macroevolutionary change – a reductionist approach. *Bull. Carnegie Mus. nat. Hist.* (13), 20–69.

Bonde, N. (1975). Origin of "higher groups": viewpoints of phylogenetic systematics. *Colloques int. Cent. natn. Res. scient.* (218), 293–324.

Bonde, N. (1977). Cladistic classification as applied to vertebrates. *In* "Major Patterns in Vertebrate Evolution" (M. K. Hecht, P. C. Goody and B. M. Hecht, eds), pp. 741–804. Plenum Publishing Corporation, New York.

Bremer, K. and Wanntorp, H.–E. (1978). Phylogenetic systematics in botany. *Taxon* 27, 317–329.

Bremer, K. and Wanntorp, H.–E. (1979). Geographic populations or biological species in phylogeny reconstruction? *Syst. Zool.* 28, 220–224.

Brenner, G. J. (1976). Middle Cretaceous floral provinces and early migrations of angiosperms. *In* "Origin and Early Evolution of Angiosperms" (C. B. Beck, ed.), pp. 23–47. Columbia University Press, New York.

Buck, R. C. and Hull, D. L. (1966). The logical structure of the linnaean hierarchy. *Syst. Zool.* 15, 97–111.

Chaloner, W. G. (1970a). The rise of the first land plants. *Biol. Rev.* 45, 353–377.

Chaloner, W. G. (1970b). The evolution of miospore polarity. *Geosci. Man* 1, 47–56.

Chaloner, W. G. (1976). The evolution of adaptive features in fossil exines. *In* "The Evolutionary Significance of the Exine" (I. K. Ferguson and J. Muller, eds), pp. 1–14. Academic Press, London and New York.

Chaloner, W. G. and Boureau, E. (1967). Lycophyta. *In* "Traité de Paléo botanique, 2" (E. Boureau, ed.), pp. 435–802. Masson, Paris.

Chaloner, W. G. and Sheerin, A. (1979). Devonian macrofloras. *Spec. Pap. Palaeont.* (23), 145–161.

Chamberlain, C. J. (1935). "Gymnosperms, Structure and Evolution". University Press, Chicago.

Chandler, M. E. J. (1923). Geological history of the genus *Stratiotes*. *Q. Jl geol. Soc. Lond.* 79, 117–138.

Colless, D. H. (1967). The phylogenetic fallacy. *Syst. Zool.* 16, 289–295.

Copleston, F. C. (1972). "A History of Medieval Philosophy". Methuen, London.

Cornet, B. (1977). Angiosperm-like pollen with tectate-columellate wall structure from the Upper Triassic (and Jurassic) of the Newark Supergroup, U.S.A. *Am. Assoc. Strat. Palynol., 10th Ann. Meet., Tulsa, Abstr.*: 8–9. Reprinted 1979 in *Palynology* 3, 281–282.

Cracraft, J. (1974). Phylogenetic models and classification. *Syst. Zool.* 23, 71–90.

Cracraft, J. and Eldredge, N. (eds) (1979). "Phylogenetic Analysis and Paleontology". Columbia University Press, New York and London.

Crepet, W. L. (1979). Some aspects of the pollination biology of Middle Eocene angiosperms. *Rev. Palaeobot. Palynol.* 27, 213–238.

Cronquist, A. (1968). "The Evolution and Classification of Flowering Plants". Nelson, London.

Daber, R. (1975). Herausbildung neuer Merkmalsverbindungen bei Gefässpflanzennervaturen und Wedelstrukturen in der Devon- und Karbonzeit. *Wiss. Z. Humboldt-Univ. Berl.*, Math. -Nat. R. 24, 437–459.

Daghlian, C. P. (1978). Coryphoid palms from the Lower and Middle Eocene of southeastern North America. *Palaeontographica* (B) **166**, 44–82.

Davis, P. H. and Heywood, V. H. (1963). "Principles of Angiosperm Taxonomy". Oliver & Boyd, Edinburgh and London.

Dickison, W. C. (1975). The bases of angiosperm phylogeny: vegetative anatomy. *Ann. Mo. bot. Gdn* **62**, 590–620.

Dilcher, D. L. (1971). A revision of the Eocene flora of southeastern North America. *Palaeobotanist* **20**, 7–18. Issued 1973.

Dilcher, D. L. (1979). Early angiosperm reproduction: an introductory report. *Rev. Palaeobot. Palynol.* **27**, 291–328.

Dilcher, D. L., Crepet, W. L., Beeker, C. D. and Reynolds, H. C. (1976). Reproductive and vegetative morphology of a Cretaceous angiosperm. *Science, N.Y.* **191**, 854–856.

Doyle, J. (1945). Developmental lines in pollination mechanisms in the Coniferales. *Scient. Proc. R. Dubl. Soc.* (N.S.) **24**, 43–62.

Doyle, J. A. (1977). Patterns of evolution in early angiosperms. *In* "Patterns of Evolution, as Illustrated by the Fossil Record" (A. Hallam, ed.), pp. 501–546. *Developments in Palaeontology and Stratigraphy* 5.

Doyle, J. A. (1978). Origin of angiosperms. *A. Rev. Ecol. Syst.* **9**, 365–392.

Doyle, J. A. and Hickey, L. J. (1976). Pollen and leaves from the mid-Cretaceous Potomac Group and their bearing on early angiosperm evolution. *In* "Origin and Early Evolution of Angiosperms" (C. B. Beck, ed.), pp. 139–206. Columbia University Press, New York and London.

Doyle, J. A., Van Campo, M. and Lugardon, B. (1975). Observations on exine structure of *Eucommiidites* and Lower Cretaceous angiosperm pollen. *Pollen Spores* **17**, 429–486.

Doyle, J. A., Biens, P., Doerenkamp, A. and Jardiné, S. (1977). Angiosperm pollen from the pre-Albian Lower Cretaceous of Equatorial Africa. *Bull. Cent. Rech. Explor. Prod. Elf-Aquitaine* **1**, 451–473.

Dupuis, Cl. (1979). Permanence et actualité de la systématique: la 'systématique phylogénétique' de W. HENNIG (Historique, choix de références). *Cah. Nat.* **34**, 1–69.

Edwards, D. (1979). A late Silurian flora from the lower Old Red Sandstone of south-west Dyfed. *Palaeontology* **22**, 23–52.

Eldredge, N. and Cracraft, J. (1980). "Phylogenetic Patterns and the Evolutionary Process". Columbia University Press, New York.

Eldredge, N. and Gould, S. J. (1972). Punctuated equilibria: an alternative to phyletic gradualism. *In* "Models in Paleobiology" (T. J. M. Schopf, ed.), pp. 82–115. Freeman, Cooper, San Francisco.

Eldredge, N. and Tattersall, I. (1975). Evolutionary models, phylogenetic reconstruction, and another look at hominid phylogeny. *In* "Approaches to Primate Paleobiology" (F. S. Szalay, ed.), pp. 218–242. [*Contr. Primatol.* 5.]

Engelmann, G. F. and Wiley, E. O. (1977). The place of ancestor-descendant relationships in phylogeny reconstruction. *Syst. Zool.* **26**, 1–11.

Engler, A. and Prantl, H. (1897–1915). "Die natürlichen pflanzenfamilien". 20 vols. Englemann, Leipzig.

Erdtman, G. (1965). "Pollen and Spore Morphology/Plant Taxonomy. An Introduction to Palynology", Vol. 3. Almqvist and Wiksell, Stockholm.

Esau, K. (1969). The phloem. *Handbuch der pflanzenanatomie* 5 (2).

Estabrook, G. F. (1972). Cladistic methodology: a discussion of the theoretical basis for the induction of evolutionary history. *A. Rev. Ecol. Syst.* 3, 427–456.

Estabrook, G. F. (1977). Does common equal primitive? *Syst. Bot.* 2, 36–42.

Estabrook, G. F. (1978). Some concepts for the estimation of evolutionary relationships in systematic botany. *Syst. Bot.* 3, 146–158.

Estabrook, G. F. and Anderson, W. R. (1978). An estimate of phylogenetic relationships within the genus *Crusea* (Rubiaceae) using character compatibility analysis. *Syst. Bot.* 3, 179–196.

Estabrook, G. F. and Meacham, C. A. (1979). How to determine the compatibility of undirected character state trees. *Math. Biosci.* 46, 251–256.

Farris, J. S. (1979). The information content of the phylogenetic system. *Syst. Zool.* 28, 483–519.

Farris, J. S. and Kluge, A. G. (1979). A botanical clique. Cladistics and plant systematics. *Syst. Zool.* 28, 400–411.

Farris, J. S., Kluge, A. G. and Eckardt, M. J. (1970). A numerical approach to phylogenetic systematics. *Syst. Zool.* 19, 172–189.

Florin, R. (1951). Evolution in *Cordaites* and conifers. *Acta Horti Bergiani* 15, 285–388.

Foster, A. S. and Gifford, E. M. (1974). "Comparative morphology of vascular plants". 2nd edn. W. H. Freeman, San Francisco.

Funk, V. A. and Stuessy, T. F. (1978). Cladistics for the practising plant taxonomist. *Syst. Bot.* 3, 159–178.

Gaffney, E. S. (1979). Tetrapod monophyly: a phylogenetic analysis. *Bull. Carnegie Mus. nat. Hist.* (13), 92–105.

Gaussen, H. (1946). Les gymnospermes actuelles et fossiles. *Trav. Lab. for. Toulouse*, Tome II *Etud. Dendrol.*, sect. 1, vol. 1, fasc. 3, ch. 5, 1–26.

Gingerich, P. D. (1977). Patterns of evolution in the mammalian fossil record. *In* "Patterns of Evolution, as Illustrated by the Fossil Record" (A. Hallam, ed.), pp. 469–500. *Developments in Palaeontology and Stratigraphy* 5.

Gingerich, P. D. (1979). Palaeontology, phylogeny, and classification: an example from the mammalian fossil record. *Syst. Zool.* 28, 451–464.

Gould, R. E. and Delevoryas, T. (1977). The biology of *Glossopteris*: evidence from petrified seed-bearing and pollen-bearing organs. *Alcheringa* 1, 387–399.

Gould, S. J. (1977). "Ontogeny and Phylogeny". Harvard University Press, Cambridge, Mass.

Gould, S. J. and Eldredge, N. (1977). Punctuated equilibria: the tempo and mode of evolution reconsidered. *Paleobiology* 3, 115–151.

Griffiths, G. C. D. (1974). Some fundamental problems in biological classification. *Syst. Zool.* 22, 338–343.

Harland, W. B. *et al.* (eds) (1967). "The Fossil Record". Geological Society, London.

Harper, C. W. (1976). Phylogenetic inference in paleontology. *J. Paleontol.* 50, 180–193.

Harris, T. M. (1932). The fossil flora of Scoresby Sound East Greenland. Part 2: description of seed plants *incertae sedis* together with a discussion of certain cycadophyte cuticles. *Meddr. Grønland* 85 (3), 1–112.

Harris, T. M. (1933). A new member of the Caytoniales. *New Phytol.* 32, 97–114.

Harris, T. M. (1937). The ancestry of the angiosperms. *C. r. Congr. Av. Et. Strat. Géol. Carbon* 2, 247–249.

Harris, T. M. (1940). On *Caytonia* Thomas. *Ann. Bot.* (N. S.) 4, 713–734.

Harris, T. M. (1951). The relationships of the Caytoniales. *Phytomorphology* 1, 29–39.

Harris, T. M. (1954). Mesozoic seed cuticles. *Svensk bot. Tidskr.* 48, 281–291.

Harris, T. M. (1958). The seed of *Caytonia. Palaeobotanist* 7, 93–106.

Harris, T. M. (1960). The origin of angiosperms. *Advmt. Sci., Lond.* 17, 207–213.

Harris, T. M. (1961). Modern botanical thought: palaeobotany. *Trans. bot. Soc. Edinb.* 39, 171–180.

Harris, T. M. (1962–1963). Presidential address: the inflation of taxonomy. *Proc. Linn. Soc. Lond.* 175, 1–7.

Harris, T. M. (1979). "The Yorkshire Jurassic Flora, V: Coniferales". British Museum (Natural History), London.

Harris, T. M., Millington, W. and Miller, J. (1974). "The Yorkshire Jurassic Flora, IV: Ginkgoales and Czekanowskiales". British Museum (Natural History), London.

Hecht, M. K. (1976). Phylogenetic inference and methodology as applied to the vertebrate record. *Evolut. Biol.* 9, 335–363.

Hennig, W. (1950). "Grundzüge einer theorie der phylogenetischen systematik". Deutscher Zentralverlag, Berlin.

Hennig, W. (1965). Phylogenetic systematics. *A. Rev. Ent.* 10, 97–116.

Hennig, W. (1966). "Phylogenetic Systematics". University of Illinois Press, Urbana.

Heslop-Harrison, J. (1971). The pollen wall: structure and development. *In* "Pollen: Development and Physiology" (J. Heslop-Harrison, ed.), pp. 75–98. Butterworths, London.

Heslop-Harrison, J. (1978). Genetics and physiology of angiosperm incompatibility systems. *Proc. R. Soc. Lond.*, B202, 73–92.

Heslop-Harrison, J., Heslop-Harrison, Y. and Barber, J. (1975). The stigma surface in incompatibility responses. *Proc. R. Soc. Lond.*, B188, 287–297.

Hickey, L. J. (1977). Stratigraphy and paleobotany of the Golden Valley formation (early Tertiary) of western North Dakota. *Mem. geol. Soc. Am.* 150.

Hickey, L. J. and Doyle, J. A. (1977). Early Cretaceous fossil evidence for angiosperm evolution. *Bot. Rev.* 43, 2–104.

Hickey, L. J. and Wolfe, J. A. (1975). The bases of angiosperm phylogeny: vegetative morphology. *Ann. Mo bot. Gdn* 62, 538–589.

Hill, C. R. (1974). "Palaeobotanical and sedimentological studies on the lower Bajocian (Middle Jurassic) flora of Yorkshire". PhD thesis, Leeds University.

Huber, B. and Graf, E. (1955). Vergleichende untersuchungen über die geleitzellen der siebröhren. *Ber. dt. bot. Ges.* **68**, 303–310.

Hughes, N. F. (1976). "Palaeobiology of Angiosperm Origins". University Press, Cambridge.

Hughes, N. F., Drewry, G. E. and Laing, J. F. (1979). Barremian earliest angiosperm pollen. *Palaeontology* **22**, 513–535.

Hull, D. L. (1965). The effect of essentialism on taxonomy. *Brit. J. Phil. Sci.* **15**, 314–326; **16**, 1–18.

Hull, D. L. (1967). Certainty and circularity in evolutionary taxonomy. *Evolution* **2**, 174–189.

Hull, D. L. (1968). The operational imperative – sense and nonsense in operationism. *Syst. Zool.* **16**, 438–457.

Hull, D. L. (1970). Contemporary systematic philosophies. *A. Rev. Ecol. Syst.* **1**, 19–54.

Hull, D. L. (1979). The limits of cladism. *Syst. Zool.* **28**, 416–440.

Hull, D. L. (1980). Cladism gets sorted out. (review of Cracraft, J. and Eldredge, N. *loc. cit*). *Paleobiology* **6**, 131–136.

Humphries, C. J. (1979). A revision of the genus *Anacyclus* L. (Compositae: Anthemideae). *Bull. Br. Mus. nat. Hist., Bot.* **7**, 83–142.

Hutchinson, J. (1959). "The Families of Flowering Plants". 2nd edn. 2 vols. Clarendon Press, Oxford.

Hutchinson, J. (1964). "The genera of flowering plants". 2 vols. Clarendon Press, Oxford.

Inamdar, J. A. and Bhatt, D. C. (1972). Epidermal structure and ontogeny of stomata in vegetative and reproductive organs of *Ephedra* and *Gnetum*. *Ann. Bot.* **36**, 1041–1046.

Jardine, N. (1969). A logical basis for biological classification. *Syst. Zool.* **18**, 37–52.

Jardine, N. and Jardine, C. J. (1969). Is there a concept of homology common to several sciences? *Classif. Soc. Bull.* **2**, 12–18.

Jefferies, R. P. S. (1979). The origin of the Chordates – a methodological essay. *In* "The Origin of Major Invertebrate Groups" (M. R. House, ed.), pp. 443–477. Systematics Association Special Volume 12. Academic Press, London and New York.

Johnson, M. A. (1951). The shoot apex in gymnosperms. *Phytomorphology* **1**, 188–204.

Just, T. (1948). Gymnosperms and the origin of angiosperms. *Bot. Gaz.* **110**, 91–103.

Kidston, R. and Lang, W. H. (1921). On Old Red Sandstone plants showing structure, from the Rhynie chert bed, Aberdeenshire. Part IV. Restorations of the vascular cryptogams, and discussion of their bearing on the general morphology of the Pteridophyta and the origin of the organisation of land-plants. *Trans. R. Soc. Edinb.* **52**, 831–854.

Kneale, W. and Kneale, M. (1962). "The Development of Logic". Clarendon Press, Oxford.

Knobloch, E. (1978). On some primitive angiosperm leaves from the Upper Cretaceous of the Bohemian massiv. *Palaeontographica* (B) **166**, 83–98.

Knoll, A. H. (1980). Paleobotany turns over a new leaf. *Paleobiology* **6**, 1–2.

Krassilov, V. A. (1973). Mesozoic plants and the problem of angiosperm ancestry. *Lethaia* **6**, 163–178.

Krassilov, V. A. (1975). Dirhopalostachyaceae – a new family of proangiosperms and its bearing on the problem of angiosperm ancestry. *Palaeontographica* (B) **153**, 100–110.

Krassilov, V. A. (1977). The origin of angiosperms. *Bot. Rev.* **43**, 143–176.

Kräusel, R. and Schaarschmidt, F. (1968). *Scoresbya* Harris (Dipteridaceae?) aus dem Unteren Jura von Sassendorf. *Palaeontographica* (B) **123**, 124–131.

Lakatos, I. (1962). Infinite regress and foundations of mathematics. *Aristotelian Society* **32** (Suppl.), 155–184.

Lakatos, I. (1968). II – Criticism and the methodology of scientific research programmes. *Proceedings of the Aristotelian Society* **69**, 149–186.

Lakatos, I. (1970). Falsification and the methodology of scientific research programmes. *In* "Criticism and the growth of knowledge" (I. Lakatos and A. Musgrave, eds), pp. 91–195. University Press, Cambridge.

Lakatos, I. (1974). Popper on demarcation and induction. *In* "The Philosophy of Karl Popper, I" (P. A. Schilpp, ed.), pp. 241–273. La Salle (Open Court), Illinois.

Lang, W. H. (1937). On the plant-remains from the Downtonian of England and Wales. *Phil. Trans. R. Soc. Lond,* (B) **227**, 245–291.

Linnaeus, C. (1753). "Species plantarum".

Lombard, R. E. and Bolt, J. R. (1979). Evolution of the tetrapod ear: an analysis and reinterpretation. *Biol. J. Linn. Soc.* **11**, 19–76.

Long, A. G. (1966). Some Lower Carboniferous fructifications from Berwickshire, together with a theoretical account of the evolution of ovules, cupules and carpels. *Trans. R. Soc. Edinb.* **66**, 345–375.

Long, A. G. (1977). Lower Carboniferous pteridosperm cupules and the origin of angiosperms. *Trans. R. Soc. Edinb.* **70**, 13–35.

Lowe, J. (1961). The phylogeny of monocotyledons. *New Phytol.* **60**, 355–387.

Lugardon, B. and Le Thomas, A. (1974). Sur la structure feuilletée de la couche basale de l'ectexine chez diverses Annonacées. *C. r. hebd. Séanc. Acad. Sci. Paris,* D, **279**, 255–258.

Lyon, A. G. (1964). The probable fertile region of *Asteroxylon mackiei* K. & L. *Nature, Lond.* **203**, 1082–1083.

Magee, B. (1972). "Popper". Fontana, London.

Maheshwari, P. and Vasil, V. (1961). "*Gnetum*". Council of Scientific and Industrial Research, Botanical Monograph 1, New Delhi.

Mamay, S. H. (1976). Paleozoic origin of the cycads. *Prof. Pap. U.S. geol. Surv.* **934**.

Martens, P. (1971). Les gnétophytes. *Handbuch der pflanzenanatomie* **12** (2).

Martin, H. A. (1978). Evolution of the Australian flora and vegetation through the Tertiary: evidence from pollen. *Alcheringa* 2, 181–202.

Mayr, E. (1969). "Principles of Systematic Zoology". McGraw-Hill, New York.

Mayr, E. (1974). Cladistic analysis or cladistic classification? *Z. Zool Evol. Forsch.* 12, 94–128.

McNeill, J. (1979). Purposeful phenetics. *Syst. Zool.* 28, 465–482.

Meeuse, A. D. J. (1965). Angiosperms – past and present. *Advg. Front. Pl. Sci.* 11, 1–228.

Meeuse, A. D. J. (1972). Sixty-five years of theories of the multiaxial flower. *Acta biotheor.* 21, 167–202.

Meeuse, A. D. J. (1974). Floral evolution and emended anthocorm theory. *In* "International Bio-Sci. Monograph I" (T. M. Varghese, ed.), pp. 1–188. Hissar, New Delhi.

Meeuse, A. D. J. (1975a). Changing floral concepts: anthocorms, flowers, and anthoids. *Acta bot. Neerl.* 24, 23–36.

Meeuse, A. D. J. (1975b). Origin of the angiosperms – problem or inaptitude? *Phytomorphology* 25, 373–379.

Meeuse, A. D. J. (1977). Coincidence of characters and angiosperm phylogeny. *Phytomorphology.* 27, 314–322. Issued 1978.

Meeuse, A. D. J. (1979a). Why were the early angiosperms so successful? A morphological, ecological and phylogenetic approach (parts I and II). *Proc. K. ned. Akad. Wet.*, C, 82, 343–369.

Meeuse, A. D. J. (1979b). 5. The significance of the Gnetatae in connection with the early evolution of the angiosperms. *In* "Glimpses in Plant Research" (P. K. K. Nair, ed.), 4, pp. 62–73. Vikas Publishing House, New Delhi.

Melville, R. (1962). A new theory of the angiosperm flower. I. The gynoecium. *Kew Bull.* 16, 1–50.

Melville, R. (1963). A new theory of the angiosperm flower. II. The androecium. *Kew Bull.* 17, 1–63.

Mickevich, M. F. and Johnson, M. S. (1976). Congruence between morphological and allozyme data in evolutionary inference and character evolution. *Syst. Zool.* 25, 260–270.

Moussel, B. (1978). Double fertilization in the genus *Ephedra. Phytomorphology* 28, 336–345.

Nair, P. K. K. (1979). The palynological basis for the triphyletic theory of angiosperms. *Grana palynol.* 18, 141–144.

Nelson, G. J. (1970). Outline of a theory of comparative biology. *Syst. Zool.* 19, 373–384.

Nelson, G. J. (1972). Phylogenetic relationship and classification. *Syst. Zool.* 21, 227–231.

Nelson, G. J. (1978). Ontogeny, phylogeny, paleontology, and the biogenetic law. *Syst. Zool.* 27, 324–345.

Netolitzky, F. (1926). Anatomie der angiospermen-samen. *Handbuch der pflanzenanatomie* 2 (Pteridophyten und Anthophyten).

Niklas, K. J., Tiffney, B. H. and Knoll, A. H. (1980). Apparent changes in the diversity of fossil plants: a preliminary assessment. *Evolut. Biol.* **12**, 1–89.

Nishio, K. (1959). Investigations with the Mäule reaction on plant tissues I. *Bot. Mag. Tokyo* **72**, 384–387.

Owen, R. (1847). Report on the archetype and homologies of the vertebrate skeleton. *Rep. Br. Ass. Advmt Sci.* **1847**, 169–340.

Pant, D. D. and Nautiyal, D. D. (1960). Some seeds and sporangia of *Glossopteris* flora from Raniganj coalfield, India. *Palaeontographica*, B, **107**, 41–64.

Parenti, L. R. (1979). Review of "Essays in plant taxonomy". (H. E. Street, ed.). *Syst. Zool.* **28**, 243–246.

Parenti, L. R. (1980). A phylogenetic analysis of the land plants. *Biol. J. Linn. Soc.* **13**, 225–242.

Patterson, C. (1978). "Evolution". British Museum (Natural History), London.

Patterson, C. and Rosen, D. E. (1977). Review of ichthyodectiform and other Mesozoic teleost fishes and the theory and practice of classifying fossils. *Bull. Am. Mus. nat. Hist.* **158**, 81–172.

Pettitt, J. M. (1977a). Detection in primitive gymnosperms of proteins and glycoproteins of possible significance in reproduction. *Nature, Lond.* **266**, 530–532.

Pettitt, J. M. (1977b). The megaspore wall in gymnosperms: ultrastructure in some zooidogamous forms. *Proc. R. Soc. Lond.* (B) **195**, 497–515.

Pettitt, J. M. (1979). Precipitation reactions occur between components of the ovule tissues in primitive gymnosperms. *Ann. Bot.* **44**, 369–371.

Pettitt, J. M. and Chaloner, W. G. (1964). The ultrastructure of the Mesozoic pollen *Classopollis*. *Pollen Spores* **6**, 611–620.

Platnick, N. I. (1977). Cladograms, phylogenetic trees, and hypothesis testing. *Syst. Zool.* **26**, 438–442.

Platnick, N. I. and Nelson, G. J. (1978). A method of analysis for historical biogeography. *Syst. Zool.* **27**, 1–16.

Popper, K. R. (1959). "The Logic of Scientific Discovery". Hutchinson, London.

Popper, K. R. (1972). "Objective Knowledge: An Evolutionary Approach". Clarendon Press, Oxford.

Price, H. J., Sparrow, A. H. and Nauman, A. F. (1974). Evolutionary and developmental considerations of the variability of nuclear parameters in higher plants. I. Genome volume, interphase chromosome volume, and estimated DNA content of 236 gymnosperms. *Brookhaven Symp. Biol.* **25**, 390–421.

Riedl, R. (1978). "Order in Living Organisms – A Systems Analysis of Evolution". John Wiley, London and New York.

Rodin, R. J. (1966). Leaf structure and evolution in american species of *Gnetum*. *Phytomorphology* **16**, 56–68.

Rosen, D. E. (1978). Vicariant patterns and historical explanation in biogeography. *Syst. Zool.* **27**, 159–188.

Russell, B. (1946). "History of Western Philosophy". George Allen and Unwin, London.

Schaeffer, B., Hecht, M. K. and Eldredge, N. (1972). Phylogeny and paleontology. *Evolut. Biol.* 6, 31–46.

Schweitzer, H. -J. (1977). Die Räto-Jurassischen floren des Iran und Afghanistans, 4. Die Rätische zwitterblüte *Irania hermaphroditica* nov. spec. und ihre bedeutung für die phylogenie der angiospermen. *Palaeontographica*, B, 161, 98–145.

Seward, A. C. (1904). "The Jurassic Flora. II: Liassic and Oolitic Floras of England". British Museum (Natural History), London.

Sharrock, G. and Felsenstein, J. (1975). Finding all monothetic subsets of a taxonomic group. *Syst. Zool.* 24, 373–377.

Simpson, G. G. (1953). "The Major Features of Evolution". Columbia University Press, New York.

Simpson, G. G. (1961). "Principles of Animal Taxonomy". Columbia University Press, New York.

Singh, H. (1978). Embryology of gymnosperms. *Handbuch der pflanzenanatomie* 10 (2).

Smith, D. L. (1964). The evolution of the ovule. *Biol. Rev.* 39, 137–159.

Sneath, P. H. A. and Sokal, R. R. (1973). "Numerical Taxonomy". Freeman, San Francisco.

Sober, E. (1975). "Simplicity". Oxford: Clarendon Press.

Sparrow, A. H., Price, H. J. and Underbrink, A. G. (1972). A survey of DNA content per cell and per chromosome of prokaryotic and eukaryotic organisms: some evolutionary considerations. *In* "Evolution of Genetic Systems" (H. H. Smith, ed.), pp. 451–494. *Brookhaven Symp. Biol.* 23. Gordon & Breach, New York.

Sporne, K. R. (1948). Correlation and classification in dicotyledons. *Proc. Linn. Soc. Lond.* 160, 40–47.

Sporne, K. R. (1949). A new approach to the problem of the primitive flower. *New Phytol.* 48, 259–276.

Sporne, K. R. (1966). "The Morphology of Pteridophytes". 2nd edn. Hutchinson, London.

Sporne, K. R. (1967). "The Morphology of Gymnosperms". 2nd edn. Hutchinson, London.

Sporne, K. R. (1971). The mysterious origin of flowering plants. [*Oxford Biology Readers* no. 3.] Oxford University Press, Oxford.

Sporne, K. R. (1974). "The Morphology of Angiosperms". Hutchinson, London.

Sporne, K. R. (1977). Some problems associated with character correlations. *Plant Syst. Evol. Suppl.* 1, 33–51.

Sporne, K. R. (1980). A re-investigation of character correlations among dicotyledons. *New Phytol.* 85, 419–449.

Stebbins, G. L. (1974). "Flowering Plants: Evolution Above the Species Level". Harvard University Press, Cambridge, Mass.

Sterling, C. (1963). Structure of the male gametophyte in gymnosperms. *Biol. Rev.* 38, 167–203.

Sussex, I. M. (1966). The origin and development of heterospory in vascular plants. *In* "Trends in Plant Morphogenesis" (E. G. Cutter, ed.), pp. 140–152. Longmans, London.

Sylvester-Bradley, P. C. (Ed.) (1956). "The Species Concept in Palaeontology". Systematics Association. London.

Takhtajan, A. L. (1959). "Die evolution der angiospermen". Gustav Fischer, Jena.

Takhtajan, A. L. (1969). "Flowering Plants, Origin and Dispersal". Oliver and Boyd, Edinburgh.

Takhtajan, A. L. (1976). Neoteny and the origin of flowering plants. *In* "Origin and Early Evolution of Angiosperms" (C. B. Beck, ed.), pp. 207–219. Columbia University Press, New York and London.

Tattersall, I. and Eldredge, N. (1977). Fact, theory and fantasy in human paleontology. *Am. Scient.* 65, 204–211.

Thomas, H. H. (1925). The Caytoniales, a new group of angiospermous plants from the Jurassic rocks of Yorkshire. *Phil. Trans. R. Soc.*, B, 213, 299–363.

Thomson, R. B. (1905). The megaspore – membrane of the gymnosperms. *Univ. Toronto Stud., biol. ser.* 4.

Thorne, R. F. (1976). A phylogenetic classification of the Angiospermae. *Evolut. Biol.* 9, 35–106.

Tidwell, W. D., Simper, A. D. and Thayn, G. F. (1977). Additional information concerning the controversial Triassic plant: *Sanmiguelia. Palaeontographica*, B, 163, 143–151.

Tiffney, B. H. (1977). Dicotyledonous angiosperm flower from the Upper Cretaceous of Martha's Vineyard, Massachusetts. *Nature, Lond.* 265, 136–137.

Tschudy, R. H. and Scott, R. A. Eds. (1969). "Aspects of Palynology". John Wiley and Sons, New York and London.

Vachrameev, V. A. and Krassilov, V. A. (1979). Reproductive organs of flowering plants from the Albian of Kazakhstan. *Paleont. J.* 13, 112–118.

Van Campo, M. (1971). Précisions nouvelles sur les structures comparées des pollens de gymnospermes et d'angiospermes. *C. r. hebd. Séanc. Acad. Sci. Paris* (D) 272, 2071–2074.

Vrba, E. S. (1979). Phylogenetic analysis and classification of fossil and recent Alcelaphini Mammalia: Bovidae. *Biol. J. Linn. Soc.* 11, 207–228.

Vrba, E. S. (1980). Evolution, species and fossils: how does life evolve? *South African J. Sci.* 76, 61–84.

Walker, J. W. (1974a). Evolution of exine structure in the pollen of primitive angiosperms. *Am. J. Bot.* 61, 891–902.

Walker, J. W. (1974b). Aperture evolution in the pollen of primitive angiosperms. *Am. J. Bot.* 61, 1112–1137.

Walker, J. W. (1976). Comparative pollen morphology and phylogeny of the ranalean complex. *In* "Origin and Early Evolution of Angiosperms" (C. B. Beck, ed.), pp. 241–299. Columbia University Press, New York and London.

Walker, J. W. and Doyle, J. A. (1975). The bases of angiosperm phylogeny: palynology. *Ann. Mo. bot. Gdn* 62, 664–723.

Walker, J. W. and Skvarla, J. J. (1975). Primitively columellaless pollen: a new concept in the evolutionary morphology of angiosperms. *Science, N.Y.* **187**, 445–447.

Wettstein, R. (1901). "Handbuch der systematischen botanik". Leipzig.

Whetstone, K. N. (1978). A new genus of cryptodiran turtles (Testudinoidea, Chelydridae) from the Upper Cretaceous Hell Creek formation of Montana. *Kansas Univ. Sci. Bull.* **51**, 539–563.

White, M. J. D. (1978). "Modes of Speciation". Freeman, Reading.

Wiley, E. O. (1975). Karl R. Popper, systematics, and classification: a reply to Walter Bock and other evolutionary taxonomists. *Syst. Zool.* **24**, 233–243.

Wiley, E. O. (1978). The evolutionary species concept reconsidered. *Syst. Zool.* **27**, 17–26.

Wiley, E. O. (1979a). Cladograms and phylogenetic trees. *Syst. Zool.* **28**, 88–92.

Wiley, E. O. (1979b). An annotated Linnean hierarchy, with comments on natural taxa and competing systems. *Syst. Zool.* **28**, 308–337.

Wodehouse, R. P. (1935). "Pollen Grains". McGraw-Hill, New York.

Wodehouse, R. P. (1936). Evolution of pollen grains. *Bot. Rev.* **2**, 67–84.

# 11 | Systematics in Biology: A Fundamental Comparison of Some Major Schools of Thought

ALAN J. CHARIG

*Department of Palaeontology, British Museum (Natural History), London, England*

**Abstract:** The most important points in the article are the following. "Cladistics" has polarized into (a) classical Hennigian systematics, with an hierarchical classification based entirely on the reconstructed phylogeny, and (b) the newly developed "transformed cladistics" which, ignoring evolution and all other time-related phenomena, simply arranges organisms into nested pairs on their existing characters as parsimoniously as possible; the "natural order" thus obtained may, if so desired, be interpreted as reflecting the phylogeny. "Transformed cladistics", herein re-named *natural order systematics,* is strongly criticized on several counts and seems to serve little purpose.

Like Hennigian systematics, conventional (herein called *Simpsonian*) systematics produces an hierarchical classification based on the reconstructed phylogeny, but the classification depends also on other considerations — notably "grade" characters (*symplesiomorphies* of Hennig). The Hennigian procedure for *phylogeny reconstruction* resembles the Simpsonian but is preferable because it is formal and more disciplined. On the other hand, the preferred *classification* is the Simpsonian, mainly because it allows paraphyletic taxa — clearly definable as complement sets — and is therefore more practical.

Incidentally, there is no reason to change the generally accepted definitions of the various terms applied to groupings. A monophyletic group includes the most recent common ancestor of all its members, a polyphyletic assemblage does not. Further, a monophyletic group or taxon is holophyletic if it includes all the descendants of that common ancestor, paraphyletic if it does not.

Systematics Association Special Volume No. 21, "Problems of Phylogenetic Reconstruction", edited by K. A. Joysey and A. E. Friday, 1982, pp. 363—440, Academic Press, London and New York.

GENERAL INTRODUCTION

The 1970s saw a rapid growth in interest among biologists, especially zoologists, in the development of cladism, cladistics or phylogenetic systematics. Towards the end of the decade the interest seemed to snowball; this resulted in a strong polarization of specialist systematists into cladists and anti-cladists, who were soon indulging in heated controversy *ad nauseam* and producing an endless spate of esoteric publications. The vast majority of biologists, however — those who made little or no use of systematics in their daily work — displayed a lamentable ignorance of the subject, sometimes misunderstanding much of it completely but more often professing a massive and contemptuous indifference thereto. Thus do they continue to react until this very day.

One group of zoologists who have been particularly affected by the polemics is the palaeontologists, whose main *raison d'être* is the elucidation of the history of life on Earth and who are deeply involved in matters of phylogeny and classification; their involvement is enormously increased and complicated by the introduction of an additional dimension into the problem, that of time. It seemed to me, as an ordinary working palaeontologist, that many of the issues were becoming unnecessarily complex, so extraordinarily complex in some instances that they were quite beyond my comprehension. Not only had classification become an end in itself instead of a means to an end but even the method of classification was becoming an end in itself; one could not see the wood for the phylogenetic trees. Further, some of the protagonists of the opposing schools of thought were clouding their views with emotion, almost as though the controversy concerned religion instead of science. I therefore decided, as a layman rather than as a specialist in systematics or philosophy (which I am not) that I would attempt a clarification of some of the major issues in question, more for my own satisfaction and use than for anybody else's. Some of my points have been made before, perhaps many times, though generally less simply; some may be so obvious as to be hardly worth making. I nevertheless venture to set them down on paper in the hope that they may help some other poor souls find a way out of the confusion and that my lack of sophistication will therefore be excused. I am aware too that I shall be accused of having over-simplified some of the issues concerned, but I considered such simplification unavoidable if my text was to be reasonably comprehensible and not too long.

It is worth giving brief consideration to the meaning of the word "clade", from which are derived other words such as "cladism" and "cladistics". "Clade" was used in Middle English (from 1100 or 1150 until about 1450) and in early Modern English (until 1604) to mean a disaster or a plague (Latin *clades,* a disaster); the opponents of cladistics might well applaud that meaning. Today, however, the word has entirely different meanings, owing its origin to the Greek κλαδος, a branch or young shoot. Willi Hennig, who might justifiably be called the "Father of Cladistics", used "clade" in two different senses: in the sense of Cuénot (1940), in which it is approximately equivalent to the word "phylum", and in the terminology of Julian Huxley (1957), in which it means a "delimitable monophyletic unit". It is the latter meaning which is generally accepted in present-day systematics. The word "cladistic", however, carries a very definite implication of branching and should not be used in any other sense.

The numerical and other phenetic approaches to taxonomy are outside the scope of the present symposium on "Phylogenetic Reconstruction" and are therefore outside the scope of this article. My original intention was to make a simple and straightforward comparison between "cladistic" and "evolutionary" systematics, but it soon became apparent that the situation was far too complicated for that:

1. Parts of the controversy over cladistics are purely semantic; they do not really exist. Different people use the same word to mean entirely different things, thus rendering much of the argument utterly futile.

2. Even more confusing, it is often not realized that the so-called cladists vary enormously in their beliefs and attitudes; within the last few years this heterodoxy has culminated in the emergence of an altogether new approach – an approach which Platnick has dealt with in an article (1980) entitled "Philosophy and the transformation of cladistics" and which might therefore, with reason, be referred to (for the time being) as "transformed cladistics" (Patterson 1980, p. 239). Among its supporters are certain authors in the present volume (Forey and Patterson). Proponents of this

new systematics regard it as a natural development and continuation of the cladistic school; they look upon themselves as the intellectual successors of Hennig, still call themselves cladists and consider that they represent the mainstream of modern cladistics. Yet some of the fundamental concepts of "transformed cladistics" are very different from those of the classical Hennigian school, being diametrically opposed to the teachings of Hennig in certain respects; "transformed cladistics" is neither Hennigian nor phylogenetic — nor even cladistic in the proper sense of that word (simply "branching"). Few of the "transformed cladists" seem to realize this, for practically none of them distinguishes in discussion or argument between their "transformed cladistics" on the one hand and Hennigian cladistics on the other; they appear not to recognize the existence of two separate schools of thought. Indeed, it even seems that their younger members are sometimes ignorant of the basic tenets of classical Hennigism, as expounded by Hennig himself in 1950 and (more completely) in 1966. To add still further to the confusion, some "cladists" clearly adopt intermediate positions, curiously inconsistent, placed at various points between the classical and "transformed" schools, and different degrees of "transformation" may be found among the members of the latter.

3. The naming of the three schools of thought in question is variable, illogical and downright confusing.

4. The terminology employed in the application of these systems is also used inconsistently.

In the light of this it seems to me absolutely essential to distinguish clearly between the different approaches to systematics that are to be discussed and to choose standardized, unambiguous names by which to call them — descriptive names as far as possible. We should likewise standardize the specialized terminology used in their application.

Any explanation here of ordinary Hennigian cladistics or of "evolutionary" systematics would be entirely superfluous; the literature is already full of works on those subjects. But before I go further I must offer some explanation of "transformed cladistics", which is too new and too little published upon to be known and understood except by its own adherents; indeed, as I have just implied, many systematists attending the present symposium seem to be unaware of its very existence. The only works I know of which give any sort of account of its principles are the above-mentioned paper by Platnick (1980) and a short, very general exposition of "Cladistics" by Patterson (1980).

"TRANSFORMED CLADISTICS" (NATURAL ORDER SYSTEMATICS)

## 1. Introduction

As my earlier comments (p. 366) imply, no two supporters of "transformed cladistics" agree on everything, and the ideas of each one of them are changing continually with the passage of time; a series of conversations with various "transformed cladists", attendance at their lectures and perusal of their articles are liable to leave the would-be initiate utterly confused. I make no criticism of this, for it is healthy and desirable that the school should eschew the rigid, dogmatic approach and allow its members the intellectual freedom to develop their philosophy further. But it does make it very difficult to discuss their ideas or to argue against them. Any successful assault against one of their positions is usually countered with the observation that the speaker dissents strongly from the outmoded views of his fellow-supporters who still defend that position or that the position has already been evacuated completely. In fact, to do battle with such people is like fighting an octopus or a hydra; or perhaps a jellyfish provides a more appropriate comparison, for there is nothing really solid and immovable with which to come to grips.

However, those of my colleagues in the British Museum (Natural History) who advocate "transformed cladistics" are reasonably united in their approach and they have kindly explained it to me. If I have understood them correctly, Sections 3 and 4 below provide a simple exposition of their type of systematics.

## 2. Nomenclature

Many of the fundamental doctrines of "transformed cladistics" are opposed to those of Hennig, so it cannot be called Hennigian; for example, it is manifestly stated by its own proponents that its ideas are certainly not based on phylogeny, which means that it cannot be called phylogenetic either. Because there appeared to be no other name for this new type of systematics I was tempted to refer to it as "neocladistics" — until I suddenly became aware that it was not even cladistic in the proper sense of that word, a fact which its own supporters seem to have overlooked. Another idea which crossed my mind was to name it for one of its principal proponents and coin the word "Nelsonian". However, I think it much better to use a term which specifies what I consider to be its most important characteristic. I therefore propose to call it, for reasons which will become obvious below, *natural order systematics*.

## 3. Fundamental Philosophy

The one fundamental attribute which, above all others, distinguishes natural order systematics from the classical Hennigian variety is that the adherents of the former take a Popperian view of evolution. Popper (1976) wrote that "it is therefore important to show that Darwinism is not a scientific theory but metaphysical". Following this, the proponents of natural order systematics maintain that evolution is an unproven hypothesis (some of them believe that it is also unprovable), i.e. it is not a phenomenon which can be recognized directly through the senses; all the evidence for it is circumstantial and is theoretically capable of alternative interpretations, however far-fetched and unlikely the latter may be. (Compare a process of criminal law, in which the defendant has not been seen or heard by a truthful witness to commit the offence but could nevertheless be convicted by a jury on circumstantial evidence, including a mass of corroborative detail; the possibility that he is innocent can never be dismissed entirely.) It is therefore considered that evolution should play no essential part in systematics; the same applies to any other factors which include the dimension of time, such as palaeontological sequences.

Natural order systematists appear to ignore Popper's further comment to his remark quoted above, which continues "But its value for science as a metaphysical research programme is very great". It should also be noted that very recently, in 1980, Popper claimed that an article by Halstead published in the same year had misrepresented his views; it was a mistake, so he said, when "some people think that I have denied scientific character to . . . palaeontology, or the history of the evolution of life on Earth", and he also stated that "the description of unique events can very often be tested by deriving from them testable predictions or retrodictions." I do not doubt his word; yet I am obliged to affirm that Halstead's quotations from Popper's book and from Patterson (1978) certainly misled me, despite the fact that they are much too extensive to be regarded as in any way out of context. Perhaps I am too naïve, too unversed in the niceties of the philosophy of science to understand them properly.

## 4. Procedure

The attitude of the natural order systematists towards evolutionary theory results in their using the systematic procedure detailed below.

(a) *Arrangement.* The following operation is called "cladistic analysis" by "cladists" of all persuasions; for the reasons stated below (pp. 374–376) I prefer to call it "character distribution analysis".

Within the group of living organisms to be considered certain characters are chosen and the distributions of the various states of each character are noted. Working on the principle that one proceeds from the most general to the more particular, the characters with their respective states are arranged accordingly into a hierarchy comprising nested sets of two, after which the organisms are likewise arranged into sets in precisely the same fashion, the sets being defined by their common possession of the particular character-states. However, the use of different characters for this purpose results in arrangements which may be either the same (*congruent*) or different; which characters and which consequent arrangement should be preferred are determined on one criterion only, namely *maximal congruence* or *parsimony* (i.e. one chooses the arrangement yielded by the greatest number of characters and requiring the rejection of the fewest). The simplest case imaginable involves three species (A, B and C); we wish to determine which two should be classified more closely with each other than with the third. The procedure is to compare the number of characters in which A and B share a common character-state, different from that possessed by C, with the corresponding number of characters in which only A and C are the same and then likewise for B and C; the highest number of arrangements with the same pairing is deemed to give the maximal congruence and therefore the "correct" solution, with the minimal (most parsimonious) rejection of alternative arrangements. Thus the approach is essentially probabilioristic. No account whatever is taken of evolutionary phenomena like parallelism, convergence or back-mutations or of such related subjects as stratigraphical (= chronological) and geographical distribution; the time dimension, so it is claimed, is not to be considered at all.

This procedure differs from the procedure followed in Hennigian systematics in two respects. First (this has no practical consequences), the Hennigian arranges the organisms *directly* into nested pairs, rather than through the intermediary of the characters and their character-states. (Thus the nodes in Hennig's cladograms represent ancestral species, hypothesized if not actual, while the nodes in the cladograms of the natural order systematists − who have no real necessity to use cladograms at all − represent shared character-states; for some Hennigians the nodes can also represent speciation events.) Secondly, the Hennigian does *not* ignore such phenomena as parallelism and convergence; on the contrary, he rejects arrangements which do not accord with transformation series established through palaeontological sequences and which therefore appear to demonstrate such phenomena (for a good example see Hennig 1966, pp. 144−145).

(*b*) *Classification.* The next step is to produce an hierarchical classification of the organisms, based on the arrangement chosen; this is very simple in so far as it consists only of naming the nested sets of organisms (which thus become the taxa) and ranking them according to their positions in the hierarchy. The lowest-ranking taxa are the species; if the arrangement of the natural order systematist is expressed as a cladogram (the customary practice, though in my view undesirable), then all species are regarded as terminal taxa and placed at the ends of the "twigs", never at the nodes as fossil species may be in a Hennigian cladogram. Neither the arrangement nor the classification carries any implication whatever of any sort of phylogenetic relationship. All that the disciples of natural order systematics are doing in their character distribution analysis is searching for a *pattern*, to which some biological significance may or may not be attached. Indeed, the only significance which they claim for their chosen arrangement and their taxa is that they reflect the "natural order", they represent the "correct" or "most natural" grouping of the organisms concerned. The "natural" classification of a group, of course, is simply the one which is supported by the greatest quantity of evidence, preferably consilient evidence (evidence of widely differing sorts from several diverse areas of knowledge). It is the most natural in the sense that it accords with the distribution of the greatest number of characters; in other words, it is the classification obtained by maximal congruence, *ipso facto*. Ruse (1979, p. 533) wrote

> A natural classification is one which shows how divisions and groupings made on the basis of characteristics of one kind are exactly the divisions and groupings which would be made on the basis of characteristics of another kind.

To emphasize his point he also quoted Whewell (1840 [1], p. 521):

> And the Maxim by which all Systems professing to be natural must be tested is this:- that the *arrangement obtained from one set of characters coincides with the arrangement obtained from another set*" [Whewell's italics].

Finally, to cite the very recent writings of two leading natural order systematics (Nelson and Platnick 1980, p. 342): "to the extent that the resulting classifications agree, they are judged to be natural . . . ." Thus it would seem that this supposedly "cladistic" approach to classification is really no more than a measure of "overall" or aggregate similarity. (Note how I avoid using the word "overall" as in "overall similarity"; I was forcibly fed in my youth on a diet of Fowler and Sir Ernest Gowers.) This is yet another point on which natural order systematists are diametrically opposed to the Hennigians, for the latter are generally inclined to pour scorn on to "overall similarity" as a basis for classification; by contrast, the natural order systematists seem to have much in common with pre-Darwinian naturalists like Ray and Linnaeus on the one hand and with modern pheneticists on the other.

From this it follows that the taxa of natural order systematics have no absolute value; they are merely the sets which happen to show maximal congruence and they do not necessarily possess *any* biological significance, phylogenetic or otherwise. Nevertheless the arrangement and the taxa will often coincide with the cladogram and the taxa of a Hennigian classification, and, in consequence, the "natural order" may be yet another expression of the supposed phylogeny (provided, of course, that the fact of evolution has been accepted). Whether the "natural order" of natural order systematics coincides with the pattern of evolution cannot be verified absolutely, for the latter is known to us only through circumstantial evidence and is incapable of proof; even more important, the chief type of that circumstantial evidence is drawn from a phylogenetic approach (Hennigian *or* "evolutionary") to the characters of the organisms in question, interpreted subjectively, i.e. possibly the same characters as those on which the arrangement of natural order systematics is based. Any differences between the "natural order" and the supposed phylogeny can be due only to the differences between the two approaches (maximal congruence unadulterated and maximal congruence modified, respectively) and to differences in subjective interpretations (see below, pp. 373–374). On the other hand, the apparent coincidence can be falsified on other types of circumstantial evidence, e.g. palaeontological evidence from stratigraphical sequences.

Apart from all this, the taxa of natural order systematics could also be interpreted as demonstrating the plan of Creation. In the practical field they can be used for such purposes as cataloguing and information retrieval, though without any proper theoretical basis. No one has yet succeeded in working out any other uses for them or how to disentangle the various types of factor which produced the "natural order".

It is, however, claimed that these taxa will generally coincide with – or at least approximate to – the popular groupings of organisms employed by the layman, e.g. "insects", "birds", "mammals". Whether or not the taxa have any real existence outside the mind of the observer is a philosophical question which we need not answer here; it is irrelevant to the present discussion and, in any case, it may well be unanswerable.

Having expounded the principal features of natural order systematics as I understand them, I shall now comment upon this new school of philosophy and practice.

## 5. Criticism

(a) *Lack of complete objectivity.* One of the major advantages often claimed for natural order systematics lies in its objectivity. First, there is no purposeful selection or weeding out of the characters to be employed in the distribution analysis; secondly, the recognition of the "natural" arrangement is based entirely upon the single criterion of maximal congruence (parsimony), no attention whatever being paid to such matters as a lack of conformity with the stratigraphical or other types of chronological data, the probability of parallelism or convergence, the evaluation of characters or anything else that involves subjective judgments. All that is true; nevertheless, in both respects, the claim of total objectivity is false. Neither the employment of characters for the analysis nor the determination of which arrangement shows maximal congruence is a wholly objective procedure.

On the former topic, it is evident that the characters to be employed in the distribution analysis should be of as many widely different sorts as possible (in order to provide consilient evidence); but it is impossible to avoid exercising some degree of selection ("weighting") when making the analysis. The "evolutionary" systematist *chooses* his characters for analysis, perhaps subconsciously or intuitively, but he chooses those which are generally believed to be good indicators of phylogeny — usually morphological or perhaps biochemical — and which, so he thinks, will therefore give him a good classification; he avoids those which, so experience has taught him, are often irrelevant — such as the colours of the plumage in birds. Both the Hennigian systematist and the natural order systematist are bound to do the same to some extent, it is an inevitable consequence of their own learning. Yet, if they are to be absolutely impartial and objective in their analysis, they should try to include characters of *every* sort; if we use birds again as our example, they should include not only the coloration of their plumage but also the month of the year when the eggs are laid and whether or not the flesh is good to eat. In any case, a certain degree of selection is imposed on every systematist by the availability or otherwise of the characters.

Secondly, making the "count" of characters which yield congruent arrangements and thus determining which arrangement shows maximal congruence require arbitrary judgments at several levels and are therefore subjective procedures. What constitutes a character? (This is not the same question as "Which characters shall we use?") What constitutes an identity of two character-states as opposed to non-identity, when are they to be counted the same and when are they different? Most difficult of all — when two or more different characters are genetically and/or functionally linked into a single suite, should they be counted together as one or separately as several for purposes of determining maximal congruence? Upon that decision, when two alternative arrangements are competing closely for numerical superiority, may depend the final result. Further, in some cases of linked characters an important source of potential error may lie in our ignorance of the very existence of the linkage.

It seems to me that natural order systematics achieves a slightly greater degree of objectivity than does Hennigian systematics, but only in so far as the arrangement showing maximal congruence is not tested against evidence from other sources — evidence which is itself of a partly subjective nature — and accepted or rejected accordingly. Natural order systematics is still subjective to a significant extent and secures no real advantage by its deviation from Hennigian procedures.

*(b) Inconsistency with regard to evolution and time.* As stated above, the philosophy of natural order systematics does not recognize evolution as a proven fact and permits neither evolutionary theory nor the time dimension to play any essential part in its procedures. Yet every natural order systematist whose works I have read or to whom I have spoken obviously believes, as strongly as I do, that evolution *has* occurred; they believe this, despite the lack of proof of evolution and its possible unprovability. While I cannot object to this slightly paradoxical conjunction, I do find myself bewildered by the curious inconsistency which those same people display in their resulting approach to systematics.

On the one hand, they insist that the character distribution analysis should ignore such evolutionary phenomena as parallelism and convergence, and their procedures expressly forbid the use of time-related evidence — in particular, palaeontological sequences and transformation series — to help elucidate the true pattern of natural order. On the other hand, it is surprising to note that they still call

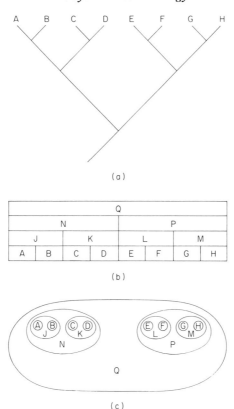

Fig. 1. Three different ways of representing an hierarchical arrangement of eight
species A–H; J–Q are the names of higher taxa. (a) Dendrogram; has
evolutionary (phylogenetic) implications which, in natural order systematics, do not necessarily ensue. (b) Tabulated hierarchy. (c) Venn diag-
grams of nested binary sets.

their analysis a "cladistic analysis", employing also such terms as
"homology" and "synapomorphy" which must surely be meaningless
if not used in an evolutionary context. Further, they frequently
express the results of that "cladistic analysis" in graphic form as a
dendrogram (Fig. 1a); this implies some sort of branching sequence, a
*progression,* with occasional dichotomies. But natural order systematics, as explained by its proponents, does not include any such
concept of a branching sequence as a necessary part of itself. It merely
arranges characters and the groups of organisms which possess those
characters into nested sets of two, without any implication of a phylogenetic relationship; and it proposes an hierarchical classification,

based on that arrangement, in which the nested sets are called taxa. Thus the use of a dendrogram to represent either the arrangement of characters or the classification of organisms can badly mislead the reader, for any tree-like representation inevitably carries some implication of branching (cf. "family tree") and therefore suggests an evolutionary sequence which, in the view of the natural order systematists, is both unproven and unprovable and does not merit consideration. While it is true that most natural order systematists use cladograms without any intentional implications of phylogenetic relationships, this practice still misleads the uninitiated – it certainly misled me (see p. 394) – and causes confusion. The very words "clade", "cladistic" and "cladogram" – according to their proper etymological derivation (see p. 365) – carry the suggestion of a branching sequence; "dendrogram" (derived from the Greek word for tree, δενδρον) does the same, though probably to a lesser degree. Alternative, non-cladistic, non-dendritic ways of representing a "natural order" arrangement are suggested in Fig. 1b and c.

It is also strange to discover that the natural order systematists, when considering what biological significance may attach to the results of their "cladistic analysis", clearly rate the evolutionary significance above all others. Most of them, so it seems to me, still regard the dendrogram (or cladogram) obtained from their analysis as the equivalent of a tentative or provisional phylogenetic tree, reflecting the probable course of unprovable evolution. Not only do they use cladograms to express phylogeny, however, but also for other purposes based on time-relations, e.g. in studies of vicariance, the phenomenon of sister-groups replacing each other in space (Hennig 1966, p. 169). Thus it was written (Nelson and Platnick 1980, p. 342): "To the extent that biological and geological area cladograms agree, they can also be regarded as real reflections of history." At first sight all this appears rather illogical if evolutionary phenomena such as parallelism and convergence have been deliberately excluded from consideration when selecting the "correct" arrangement.

Thus, although it will not accept evolution as a fact, natural order systematics still uses evolutionary terminology and a type of graphic representation which has evolutionary implications; and its results are employed in various ways which assume that evolution has indeed occurred. To return to our crime simile, it is rather like a detective finding a corpse with a badly battered skull yet refusing to issue a positive statement that a murder has been committed because the evidence is only circumstantial and therefore falsifiable. (He is not a fool, however, and privately has no doubts that he has a case of homicide on his hands.) He examines the evidence, which appears to point towards various persons (X, Y or Z) as the murderer; he names the suspect X towards whom the greatest *quantity* of evidence is directed (how do you quantify evidence?), irrespective of the *quality* of that evidence, and says that he has made the "correct" decision (correct in what respect if there has been no murder?); he nevertheless refuses to charge the suspect. There is further evidence suggesting that, in fact, only Z could have committed the murder because both X and Y are too small and weak to have smashed in the victim's skull so forcefully; but the detective refuses to use that evidence because it presupposes that a murder has indeed been committed, a proposition which cannot be verified absolutely. The facts about this case are then fed into the police computer as data for a research project on homicide in general and on its incidence in different regions, despite the fact that, officially, the murder is deemed not to have taken place!

I find this approach confused and inconsistent, at least in so far as the terminology and graphic representation are concerned, and I therefore regard it as unacceptable in a supposedly scientific method. If there is a genuine desire that all reference to evolution should be expunged from natural order systematics, then terms such as "homology" and "synapomorphy" must not be employed (see Hennig's definition of these, 1966, pp. 93 and 89 respectively, both embodying the concept of transformation as "real historical processes of evolution"). The analysis can then be based on shared characters only, without any consideration of the possible *reason* for their common possession by different organisms or groups of organisms (e.g. the common ancestry which the use of the above terms implies). Further, no use should be made of dendrograms or of words such as "cladistic" or "cladogram", with their implications of progression and branching. On the other hand, the application of the *results* of the analysis to evolutionary studies should not be objected to, for that application is merely a testing of a hypothesis, an investigation of one possible reason for the pattern of nature. If such application were *not* permitted, if all studies related to evolution and the time dimension were banned, then what point could there possibly be to natural order systematics?

Platnick (1980, p. 545) summed it up very neatly: "systematists must analyze characters in an attempt to find order in nature, and that once order has been found, we may, *if we wish* [my italics], assume that it's [*sic*] the result of evolution". This represents another striking contrast to the attitude of Hennig; for whereas Hennig's approach to systematics was unequivocally centred on evolution and phylogeny, Platnick treats those subjects as "optional extras".

Alternatively, it seems to me that the initial refusal of the natural order systematists to recognize the fact of evolution is not much more than a statement of faith (or lack of it!) and makes little difference to the way in which the "natural order", once discovered, is employed. The practical differences lie in the method of its discovery, in the rejection of parallelism and related phenomena and in the prohibition of evidence from the fossil record. But even if one cannot accept evolution as a fact, would it not still be possible to stick to classical Hennigism with its central rôle for phylogeny, emphasizing (after Popper) that evolution (Darwinism) may be regarded as "a metaphysical research programme"?

(*c*) *Non-coincidence of taxa with popular groupings.* The claim that the taxa of natural order systematics (like those of Hennig) will generally coincide with the popular groupings of organisms employed by the layman is very far from true; in many instances they do not even approximate to the vernacular categories. Many of the popular groupings of animals referred to in everyday conversation are what Hennigian terminology would call "paraphyletic" (e.g. "fishes") or even "polyphyletic" (e.g. "dinosaurs"), and are therefore inadmissible to a "cladist". There is even less conformity in the plant kingdom; the ordinary person without much botanical knowledge is likely to group plants into such "unnatural" assemblages as "trees", "herbaceous plants" or "root crops". In any case, whether there is coincidence or not has no relevance to anything and is of no importance whatever.

(*d*) *Circular reasoning.* Having made his character distribution analysis and having applied only the criterion of maximal congruence when choosing the "correct" (better, "preferred") arrangement from among the various possibilities, the natural order systematist ends up with a classification which, hardly surprisingly, is characterized only by maximal congruence. True, it is claimed that such a classification is typified by the greatest aggregate ("overall") similarity between forms classified together and also yields the greatest information content, but those statements are just other ways of saying the same thing.

Each grouping of organisms is defined by a number of homologies. While Hennig (1966, p. 93) thought that homologies were different transformation stages of the same original character, and Bonde (1977, p. 779) considered homology as an abstract concept of relationship to be the same thing as synapomorphy, Patterson (personal communication) defines homology simply as one of a number of characters used in defining a natural grouping – which brings us round full circle. The principal test of a homology is said to be the congruence of its distribution with that of other characters (Patterson 1980, p. 236).

The groupings of natural order systematics are somehow "related" to each other if they share certain common characters – so we all believe. But, if we do not accept the fact of evolution (as meaning descent with modification), it is not easy to understand how "related" can mean anything more than "sharing common characters"!

(e) *What does the natural order tell us?* This leads us into my last and most important criticism of natural order systematics. Let us consider the sort of information that we might expect to obtain from the results of a character distribution analysis.

The species of living organisms are characterized by specific features of many different kinds – not only of morphology but also of physiology, biochemistry, genotype, behaviour, geographical distribution, chronological (= stratigraphical) distribution and so on. Some of the categories merge into each other and not all characters are easy to classify; but that is not important.

The distribution of one single character-state through a range of different species, if it be considered in isolation from any other data, means absolutely nothing. The respective distributions of the various states of any two or more characters may well be entirely random and independent of each other; on the other hand, they may coincide more frequently or less frequently, to a statistically significant extent, than would be expected if that were the case, i.e. they may form a *pattern of coincidence*. Complete coincidence of the distribution of the states of two or more characters is called *congruence*. Except when we are dealing with very simple cases, the mathematical probability of there being a purely fortuitous congruence of such distributions is extremely small. Thus, in the simplest case imaginable, of two characters each with two character-states (say the absence and the presence of a particular condition) distributed at random among three species, the probability of the distribution of the two different states of the second character among the three species being the same as it is for the first is 1 in 4. But as soon as the number of characters involved, or the number of species involved (or both) begins to increase the probability diminishes rapidly almost to zero. To give a concrete example, in the case of three characters each with three character-states distributed among eight species, the chance of an exact coincidence is slightly less than one in a million – to be precise, 1 in 1 195 742.25.

It follows from this that whenever two or more distribution analyses based on different single characters give the same arrangement, no matter whether there are two such characters or a couple of hundred, the congruence of the resulting patterns will probably indicate some common factor, some causal relationship between the characters concerned. Only in those very simple cases which involve no more than two or three characters is there any significant likelihood that the congruence is purely accidental and does not suggest any causal connection. Causality can be of different types, each producing an independent pattern; the patterns when superimposed will tend to obscure and interfere with one another, they may even themselves coincide and reinforce each other to a certain extent, yet in some instances they will intersect beautifully and may be clearly distinguishable by direct observation of the resulting incongruence. Thus an elongated snout, a long sticky tongue, large salivary glands, reduced teeth (or none at all), powerful limbs, strong claws and the myrmecophagous habit all happen to occur together in certain species of mammal which, in some of their other characters, differ fundamentally from each other and are therefore assigned to five different major taxa.

In many cases of congruent character distribution it is difficult to avoid the conclusion that there is a direct functional connection between the characters. One example of this is afforded by certain Arctic mammals which have a circumpolar distribution and a thick white woolly coat, another instance is the various ant-eating mammals possessing the several characters just mentioned; the correlation between the characters is obvious. The functional connection is established by observation in nature, experience and common sense; its mechanism, whether it be evolutionary (natural selection) or Divine purpose or anything else, is totally irrelevant to the present argument. Characters involved in this type of relationship are often called adaptive characters, a term which I dislike because (a) it suggests that characters indicating phylogenetic relationships cannot be adaptive and (b) it has a somewhat teleological flavour. In other cases of congruent character distribution there appears to be no direct functional connection between the characters, as between (in ornithischian dinosaurs) the possession of a predentary bone and of an opisthopubic pelvis; each might be functionally connected to some other factor, as yet unknown, but they do not seem to have any connection with each other. An evolutionist would attribute this coincidence to community of ancestry, a creationist to the order in the mind of the Creator, and there could be other explanations which we cannot even imagine; but there has to be an explanation of *some* sort.

But what is the explanation of the "natural order"? We already know what the natural order comprises (see p. 371): having been obtained from the character distribution analysis by maximal congruence and *only* maximal congruence, it is the arrangement which is supported by the greatest number of coincident character distributions and which shows the greatest aggregate ("overall") similarity. We have also seen (p. 378) that some natural order systematists (e.g. Platnick 1980) accept that the natural order may be regarded as the result of evolution − "if we wish". Most, perhaps all of them, regard it in that light; but they are not *obliged* to do so. In fact, however, while the natural order *could* indicate the result of evolution, it could also indicate something entirely different, or nothing at all, or perhaps it could represent a synthesis of the various sorts of similarity between organisms. Let us analyse the situation further.

Within a class of organisms there could be several different groups of characters, the characters in each group functionally connected among themselves; this would fragment the potential coincidence of character distribution within that same class of organisms and make it less conspicuous. By contrast, there could not have been more than one actual phylogeny (as opposed to mistaken attempts to derive it), so that any shared characters acquired through community of ancestry should all operate to produce the same pattern of coincidence. Thus it does indeed seem that phylogenetic (genealogical) relationships — if relationships with an evolutionary basis are acceptable in principle — would afford the most likely explanation of the so-called "natural order".

However, it is not as simple as that. It is in the search for the phylogeny, so we have implied, that the criterion of maximal congruence (parsimony) really comes into its own; but even there it cannot be regarded as paramount. There seems to be no reason why an arrangement chosen simply on the grounds of maximal congruence should of necessity, when expressed as a cladogram, give an exact indication of the evolutionary history. If two species A and B share more common character-states with each other than does either with a third species C, then it is true that A and B are the two species out of the three which are the most likely to share a common ancestry from which the third is excluded. But it remains very possible, and is doubtless the case in many actual instances, that most of the similarities shared by A and B are linked functional adaptations without "sister-group" significance (like the long list of common adaptations to "fully improved" stance and gait shared by all "dinosaurs" (see Charig 1972, p. 153)). In such instances A and C, or B and C, may be the fruits of the more recent dichotomy (see Fig. 2).

To put it another way, there seems to be no *prima facie* reason why the totality of those characters which give the maximal congruence arrangement is more likely to afford an indication of common ancestry than it is to suggest functional similarity, or indeed vice versa. A corollary of this is that an arrangement based on the supposed phylogeny will sometimes cut sharply across the maximal congruence arrangement, and the same is true — probably more frequently — of arrangements based on similarity of functional adaptations. Again, relationships of either sort may often be indicated by degrees of congruence which are far from maximal. One cannot assume that the natural order, showing the greatest aggregate similarity, indicates nothing more or less than the course of evol-

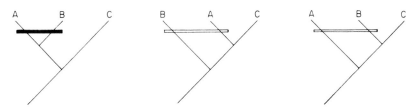

Fig. 2. Cladograms representing the three possible ways in which species A, B and C could be genealogically related to each other, on the assumption that all branching is necessarily dichotomous. The horizontal bars denote character-states shared by A and B but absent in C; the black bar is a synapomorphy, the white bar is a parallelism (or, if it were assumed to be present also in the common ancestor of all three species, a symplesiomorphy).

utionary history. This is particularly true when we are dealing with fossils, in which, so often, the evidence is woefully incomplete; the similarity due to fortuitous homoeomorphy or to functionally connected suites of characters (when measured in terms of a simple numerical count of those shared characters which support a given arrangement) may well outweigh and effectively mask the similarity due to common ancestry.

We are therefore forced to the conclusion that the "natural order", different parts of which must inevitably be based on the distribution analysis of different groups of characters, is nothing more than a hotchpotch of groupings reflecting the phylogeny (which natural order systematists do not necessarily accept anyway), groupings reflecting similarity of functional adaptations, groupings reflecting non-functional homoeomorphy, groupings reflecting any other type of causal agency that one might hope to invent and perhaps groupings due to nothing more than chance similarity. That is the essence of my criticism.

The unfortunate consequence of this is that no useful information can be extracted from the natural order unless the individual component groupings can be separated from each other by the use of other evidence, generally time-related, and their nature established by reference thereto. For example, if it were suspected by an evolutionist that a certain grouping indicated community of ancestry, he might examine transformation series from stratigraphical sequences to help confirm (or falsify) his suspicions; and the belief that another grouping was due to similarity of functional adaptations could be supported (as suggested above) by observation, experience and common sense. This has to be done, as it is in the phylogenetic systematics of Hennig; the natural order *per se* is otherwise quite useless.

On the other hand, if one refuses to acknowledge the fact of evolution, then one really *does* have a problem; with the best will in the world, I simply cannot – in those circumstances – conceive of any use at all for "natural order". Nor can I understand why the natural order systematists, whose avowed intent is to discover relationships of every sort, should seek those relationships only in the pattern of maximal congruence and ignore all other congruent arrangements.

## 6. *Conclusions*

My criticisms below (pp. 414–415) concerning the application of the principle of maximal congruence (parsimony) to systematics in general are relevant here too; they are even more relevant here than in other types of systematics, for the procedure of natural order systematics that yields the "natural order" relies *entirely* upon that principle.

Those comments apart, what other, more important conclusions may be drawn regarding the doctrines of natural order systematics? First, in a not very successful attempt to achieve higher objectivity natural order systematics has entirely abandoned Hennig's intention of correlating classification with phylogeny and has thereby sacrificed its own usefulness towards the establishment of a general reference system in biology. Secondly, its whole approach – especially towards evolution and the time factor – displays a remarkable degree of inconsistency. And thirdly, the information it yields is nothing more than a measure of aggregate similarity, the sum total of an indeterminate mixture of phylogenetic, adaptive and fortuitous similarities together with any others that could conceivably exist; not only is it impossible to disentagle those various categories but it is expressly forbidden to do so! The foundations of this unfortunate system seem to lie in over-generalization, over-simplification and circular reasoning.

In any case, natural order systematics is unlikely to make any impact upon the vast majority of biologists because they find it impossible to comprehend either its confused and misleading terminology and graphic representation or its constantly shifting ideas. As I see it, the system can serve no useful purpose whatever; I believe that it is nothing more than an ephemeral manifestation of the iconoclastic fringe, which, whether or not this article of mine has any effect upon its readers, will soon disappear from the biological scene. It is certainly not worth getting excited about (Patterson 1980).

OTHER TYPES OF SYSTEMATICS

## 1. Introduction

Having decided that numerical and other phenetic approaches to taxonomy are outside the scope of this chapter, and having eliminated natural order systematics as an undesirable complication, we are left with the classical "phylogenetic" and the traditional "evolutionary" varieties. It was in order to compare these two that I began writing the present chapter.

## 2. Different Types of Systematics

(*a*) *Nomenclature.* The nomenclature currently applied to the various types of systematics, systematists, classifications etc. dealt with in this volume is in a terribly confused state, some of the terms employed being completely misleading. Thus the name "phylogenetic" is applied to the so-called "cladistic" varieties (or at least to some of them) and "evolutionary" to the conventional, a contrasting usage which suggests to me that phylogenetic systematics is in no way evolutionary and that evolutionary systematics is not phylogenetic. It is obvious, however, that both those propositions are totally false (the second one is discussed below) and I should ideally prefer to see neither of those adjectives used to qualify any of the nouns in question. Other adjectives which have been employed to describe conventional systematics are "traditional", "gradistic", "synthetic", "eclectic" and "syncretistic", but I find none of them especially appropriate.

The introduction of "transformed cladistics" has not helped matters, for most workers (if aware of its existence) still regard it as a branch of Hennigian systematics, phylogenetic systematics or cladistics; yet it is the exact antithesis of phylogenetic systematics, from which it follows that it cannot be described as Hennigian, and it is certainly not cladistic. Simpson (1978) distinguished between *canonical cladistics* (i.e. those of Hennig, Hennigian cladistics) and *non-canonical* (i.e. those of Nelson), and I have already proposed in this article (see above, p. 368) that Platnick's term "transformed cladistics" should be replaced by the term *natural order systematics*.

Application of the term "cladistics" to what has been called phylogenetic systematics is in any case also entirely inappropriate; "evolutionary" systematics is cladistic too, for both varieties are based upon a branching system.

So what should we use? For the older systems, alas, I cannot think of any suitable appellation of a qualitative or attributive nature and can only suggest that we use terms derived from the names of the principal proponents thereof. Thus, while "cladistic" systematics of the classical variety would continue to be called Hennigian, conventional systematics could be called *Simpsonian*. Aware that the views of Simpson and Mayr do not always mirror each other exactly and not wishing to choose between them on any other grounds, I have decided on "Simpsonian" on nothing more than the grounds of euphony; I, at least, should not find the word "Mayrian" so easy to pronounce. Professor George Gaylord Simpson has kindly agreed to my using his name in this fashion.

I therefore propose that all the other terms just mentioned be dropped (except as indicated below) and that the non-phenetic varieties of systematics be referred to as *Simpsonian, Hennigian* and *natural order systematics* respectively. Such usage would be clear, unambiguous and descriptively correct. However, where a general term embracing both Hennigian and natural order systematics is urgently required, I shall still employ the word "cladistics", written in inverted commas; its use is so prevalent that it would be pointless to suggest an alternative, even though (as already mentioned) Simpsonian systematics is also cladistic in the proper sense of that word and natural order systemstics is not. Likewise the term 'cladists'', again in inverted commas, might be used to include both Hennegians and natural order systematists.

(*b*) *Classification of types of systematics.* Let us consider the three types of systematics and classification under discussion here:

Simpsonian, Hennigian and natural order. In all three a character distribution analysis leads to various alternative arrangements into sets of the character-states of the organisms in question and/or of the organisms which possess those character-states. Let us ask ourselves the familiar question: which two of the three are more closely related to each other than is either to the third?

I hope to show that the primary division of these schools of thought should be – unexpectedly – into what I propose to call *clado-evolutionary systematics* on the one hand, comprising both the Simpsonian and the Hennigian varieties, and *natural order systematics* on the other. In the two types of clado-evolutionary systematics the classification is based upon a branching diagram representing the phylogeny; natural order systematics is utterly different in that it has nothing whatever to do with either phylogeny or evolution.

Platnick (1980, pp. 544–545) tried very hard to draw distinction between "cladists" and evolutionists in their attitude towards the rôle of evolution in taxonomy; he clearly included Hennigian systematists in the former category, as well as members of the "transformed cladistics" school to which he himself belongs. He declared that the methods of the "cladists" were first to discover a "natural hierarchic system", a "natural order", without reference to evolutionary theory, and afterwards, *if they wished,* to explain the existence of that natural order by means of evolutionary theory, to assume that the order in nature was the result of evolution. He also quoted Hennig in an attempt to imply that Hennig's own attitude had been the same: "the task of the phylogenetic system is not to present the result of evolution, but only to present the phylogenetic relationships of species and species groups" (1966, p. 194); *once established,* however, those relationships could be explained "by assuming common descent" (1966, p. 15). In other words, natural order (= classification) comes first, evolution second. This approach was implicitly contrasted with the fundamentally different approach of the evolutionists, in which evolution comes first and classification second; according to Platnick, Simpson suggested "that the taxonomist must first construct a phylogenetic tree, and then somehow arbitrarily chop it up to form a classification" (quite right too!). Throughout the whole process of constructing phylogenetic trees the Simpsonian systematists are fully aware that evolution has taken place. The implication is that this knowledge prejudices the evolutionists in the construction of their phylogenetic trees, while the "cladists", of course, are quite impartial and scientific in the derivation of their "natural order".

I find Platnick's views on these matters difficult to accept. First, since Hennig was explicitly referring to *phylogenetic* relationships, he was obviously using an evolutionary approach (how can there be phylogeny without evolution?) Platnick's first quotation from Hennig (1966, p. 194) ends with the phrase (omitted by Platnick) "on the basis of the temporal sequence of origin of sister groups", which unequivocally indicates that Hennig's concept of phylogenetic relationships at least embodied a time-factor — something which the natural order systematists deliberately exclude. Perhaps what Hennig really meant to imply was that evolution could not be *proved* to have taken place, even though he had used it as the corner-stone of his entire philosophy; I can see nothing wrong with that attitude, which is also the attitude of the Simpsonian systematists. It could be that Hennig's difficult style, or his translators' choice of language, confused the issue to some extent. But whatever the reason, I cannot otherwise make sense of Hennig's statements. Secondly, I find it hard to believe that any "cladist", even a 100% red-blooded natural order systematist like Platnick, is able, while working out his "natural order", entirely to divest himself of the knowledge (or the strong suspicion) that evolution has actually occurred. The difference is nothing more than word-play; in practice there can be no difference at all.

Some people will be surprised at my noting a resemblance between the Simpsonian and the Hennigian approaches to systematics. But the basis of Hennig's "cladistic analysis" is that nested set of synapomorphies enable us to ascertain the probable phylogeny; and that does not differ at all from what those of us who like to think we are good palaeontologists were doing long before we had ever heard of Hennig or cladistics. We did it, of course, in a much looser, vaguer, less disciplined manner, without the specialized terminology and occasionally making erroneous use of a symplesiomorphy; that is why I warmly welcome the advent of Hennigian discipline and formalized procedure.

It is especially important to note that any clado-evolutionary system comprises two separate and altogether different stages, the attempted ascertainment of the phylogeny and the proposal of a formal classification based thereon. Many people have been (and still are) confused over systematics in general for no other reason than their failure to grasp this simple distinction, which was explained very clearly by Mayr (1974, p. 97) and is essential to any understanding of Simpsonian systematics as well as of the subject that we usually call "cladistics".

The first procedure of clado-evolutionary systematics, phylogeny reconstruction, is an operation which is intended to elucidate the relevant part of the branching pattern of organic evolution. Many different arrangements are obtainable from the character distribution analysis, that which is supported by the greatest number of characters (i.e. by the criterion of maximal congruence or parsimony) being generally preferred; but other valid factors are also taken into consideration, possibly before and certainly after the analysis. Thus arrangements are rejected if they do not conform to, say, transformation series from stratigraphical sequences. Character-states of which the distribution supports the preferred arrangement are regarded as synapomorphies (shared derived characters). The chosen arrangement is then converted into some sort of branching diagram or dendrogram ("phylogenetic tree") which purports to depict the history of the living organisms, a series of events which actually took place; the form of the "tree" is therefore objective fact, even though our knowledge of much of it is incomplete or uncertain. It may be regarded as reflecting both our knowledge and our ignorance of the phylogeny.

The second procedure is the classification of the organisms concerned, effected by the dividing up of the phylogenetic tree into taxa and the subsequent ranking of the latter into a hierarchy; both those tasks are carried out according to a preferred set of rules or guidelines. This is where the Simpsonian method and the Hennigian method differ from each other fundamentally. Since there is a choice of different ways in which those tasks may be performed and a consequent choice of results, the classification is to a certain extent arbitrary and subjective – far more so for the Simpsonian than for the Hennigian, who, being obliged to follow the branching pattern of his cladogram much more closely, has much less freedom of choice.

Those two procedures, common to the two clado-evolutionary varieties of systematics, are the important features which distinguish them from natural order systematics. Natural order systematists, by

# Table 1. Comparison of different types of systematics

| | CLADO-EVOLUTIONARY SYSTEMATICS | | NATURAL ORDER SYSTEMATICS |
|---|---|---|---|
| | SIMPSONIAN SYSTEMATICS | HENNIGIAN SYSTEMATICS | |
| Other names used | CONVENTIONAL, traditional, "EVOLUTIONARY", orthodox; gradistic, synthetic, eclectic, syncretistic | "PHYLOGENETIC", cladistic (canonical) | "TRANSFORMED CLADISTIC", "cladistic" (non-canonical) |
| Arrangement produced by character analysis is chosen on basis of: | MAXIMAL CONGRUENCE (= PARSIMONY) modified to varying degrees by SUBJECTIVE CONSIDERATIONS (attempted recognition of parallelisms and convergences through transformation series – e.g. palaeontological sequences – and through likelihood of characters being purely adaptive; also back-mutations) | | MAXIMAL CONGRUENCE ALONE (but note that this too involves many subjective judgments) i.e. essentially probabilioristic |
| Arrangement consists of: | some sort of BRANCHING DIAGRAM (dendrogram) representing PHYLOGENETIC RELATIONSHIPS, though much of it remains uncertain or unknown | | simply HIERARCHICAL PATTERN OF NESTED PAIRS, not representing the phylogeny, not necessarily connected with it in any way but reflecting the "NATURAL ORDER" (?); therefore misleading to translate it into a branching diagram i.e. NON-CLADISTIC |
| | dendrogram branches DICHOTOMOUSLY or POLYCHOTOMOUSLY | i.e. CLADISTIC — dendrogram ALWAYS branches DICHOTOMOUSLY (apparent polychotomies represent unresolved series of dichotomies) | |
| | therefore sets with any number of members | therefore binary sets ("sister-groups") only | binary sets only |
| Classification | based on PHYLOGENETIC DENDROGRAM ("family tree") but influenced also by OTHER CONSIDERATIONS, e.g. evolutionary grade; higher taxa are arbitrarily chosen segments of tree | derived directly from PHYLOGENETIC DENDROGRAM ALONE; higher taxa comprise all monophyletic groups shown on that dendrogram (but classification may be simplified for reasons of practicality) | THE HIERARCHICAL PATTERN ITSELF (i.e. the classification is really obtained from the character analysis in one stage rather than two) |
| | a taxon does not necessarily include all its own descendants (HOLOPHYLETIC or PARAPHYLETIC) | each taxon includes ALL its own descendants (HOLOPHYLETIC) | |

| Information obtainable from classification | classification gives SOME INFORMATION ON PHYLOGENY and DOES NOT CONTROVERT IT | ENTIRE PHYLOGENY automatically retrievable from classification (unless latter has been simplified, in which case only part of phylogeny is retrievable) | associations into sets may have various biological significances (genealogical, Creational, adaptive, biogeographical) or none at all (purely fortuitous); the different types of significance are not easy to separate, therefore main indication is of AGGREGATE SIMILARITY. However, significance may be partly revealed by congruence of pattern with patterns in other groups and with geographical cladograms; i.e. system MIGHT have phylogenetic implications, might not! But then other, non-maximally congruent arrangements may also have some significance |
| | gives also OTHER INFORMATION, e.g. on aggregate ("overall") similarity and evolutionary grade | gives NO OTHER INFORMATION, no indication either of aggregate similarity or of evolutionary grade | |
| | *Note.* Every taxon should originate *somewhere* within the taxon of next superior rank to which it is immediately subordinate, but beyond that the information supplied by the classification is vague; the taxon in question may originate simultaneously with the superior taxon (in which case it is the stem-taxon of the latter) or at any subsequent point therein. For example, the class Aves originates somewhere within the Vertebrata, but there is no means of telling from the classification that it originates in the sister-class Reptilia, in the subclass Archosauria and so on perhaps to an even narrower area of origin | | |

contrast, strongly deny that the hierarchical arrangement produced by their character distribution analysis is necessarily of any phylogenetic significance; indeed, evolution plays no essential part in their approach to systematics, the only admissible type of evidence being the unweighted analysis of character distribution and the only criterion of preference being parsimony. The hierarchical arrangement of the organisms is itself the classification, although it may be simplified if necessary; the sets in the arrangement have only to be named as taxa and given appropriate ranks.

Simpsonian and Hennigian systematics, though both evolution-orientated, differ from each other in two fundamental respects:

1. They differ in their approach to character distribution analysis, not only in various details but also in that the Hennigian procedure requires a more disciplined approach (which, as I have said, I greatly prefer).

2. More important, they differ radically in the way in which they divide up the branching pattern or phylogenetic "tree" into named taxa and rank those taxa into a hierarchy (see below, pp. 417–435). In this case I find the conventional, Simpsonian method infinitely preferable, but clearly my preference is not shared by everybody.

Incidentally, it is proper to describe the making of a classification in the Hennigian manner as "cladistic", since it is done *entirely* according to the branching of the phylogenetic dendrogram.

## 3. Causes of Confusion

(a) *Variability of approach in "cladists".* I have already observed that "the so-called cladists vary enormously in their . . . attitudes" (p. 365), that "this heterodoxy has culminated in the emergence of . . . 'transformed cladistics'" (*ibid.*), that "some 'cladists' clearly adopt intermediate positions . . . between the classical and 'transformed' schools, and different degrees of 'transformation' may be found among the members of the latter" (p. 366), and that the existence of so many shades of opinion within the "transformed cladistics" school makes it "very difficult to discuss their ideas or to argue against them" (p. 367). The same is true, of course, of discussion relating to classical Hennigan ideas; it seems even truer in those distant days when I was still blissfully unaware of the gestation and subsequent birth of natural order systematics and was arguing (so I thought) about classical Hennigism with natural order systematists to whom "cladistics" meant something entirely different. Often, just when I thought I was winning, I found myself defeated by such disclaimers as "Well, I personally don't accept that particular article of faith" or "Yes, I agree with you that Hennig was talking nonsense on that score" or "That is no longer part of our current thinking". Nevertheless I must admit that over the last few years I have enjoyed – I really mean that word – innumerable discussions and arguments with cladistically inclined colleagues; and I would much rather find this rich diversity of opinion within the movement than have to face the unthinking acceptance of a unified dogma.

To complicate matters still further, even their own nomenclature has not been standardized; when the same word means different things to different people, misunderstandings can arise which exacerbate the apparent contradictions between the systems. For example, Gardiner *et al.* in their celebrated paper "The salmon, the lungfish and the cow: a reply" (1979) called the three diagrams at the foot of p. 175 "cladograms" (see Fig. 3); they call any branching diagram a cladogram, considering that the word does not carry any phylogenetic implications (Patterson 1980, Nelson unpublished). Other "cladists", however – I instance Bonde and Cracraft, both as recently as 1977 – have used "cladogram" to mean something much closer to "phylogenetic tree", stating explicitly that a cladogram may symbolize phylogenetic relationship (Bonde 1977, p. 773) or clearly implying that it does so (Cracraft 1977, p. 17). (The main differences between any sort of branching diagram drawn up by a Hennigian to symbolize phylogenetic relationships and a Simpsonian phylogenetic tree are that the former is entirely dichotomous in its branching, does not necessarily include ancestral forms at the nodes, and, according to Cracraft, does not distinguish between common ancestry and ancestor–descendant relationships.)

Reverting to Gardiner *et al.* (1979), the statement that "evolutionary systematists prefer cladogram *c*" might therefore suggest to a disciple of Bonde or Cracraft that evolutionary (i.e. Simpsonian) systematists believe lungfish and salmon to share a more recent common ancestor with each other than they do with cows, until he

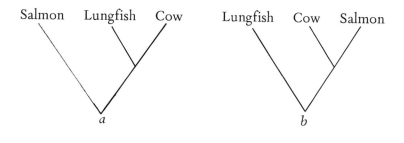

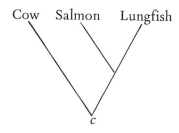

Fig. 3. Three cladograms. After Gardiner *et al.* (1979, p. 175).

turns the page to p. 176 and discovers that what evolutionary
systematists really believe is "that lungfish and salmon are more
similar [to each other] than [are] lungfish and cows". This latter
statement of their belief I know to be true, from close personal
acquaintance with both the evolutionary systematists mentioned
(Parrington and Halstead); while they agree with the cladists that
cladogram *a* gives the most likely indication of lines of *descent*, they
would still prefer to *classify* the lungfish and the salmon together
because their shared primitive characters produce a greater aggregate
similarity. I myself was a victim of the aforementioned confusion,
for when I first read the paper in question I was unaware that the
word "cladogram" was used by some *without* phylogenetic implications.

(*b*) *Re-definition of words.* We have already seen that use of the same
word in different ways can lead to confusion in the reader. Hennig
(1965, 1966) not only coined several new words (including *paraphyletic*) but also re-defined existing words, in particular *phylogeny,*

*relationship* and *monophyletic* (see Mayr 1974, pp. 100–105). Two interconnected consequences of these ambiguities are that many of the more important elements in the controversy over cladistics have no real existence and that some of the arguments are utterly futile.

Hennig's re-definition of these words is all the more remarkable when considered in the light of his own comments in the same book (1966, p. 27):

> Much can be said in favor of changing the name of a concept when the interpretation of the concept changes. Because of the "power of language over thought" the name long continues to be associated with obsolete ideas concerning the content of the concept that bears the name.

The meaning of *relationship* is considered later in this article (pp. 422–423). *Monophyly* and related concepts are discussed in some detail in the section immediately below (pp. 395–410).

### (c) Monophyly, polyphyly, paraphyly and holophyly

(i) *Introduction*. The meanings of these words (especially *monophyly*) are surrounded by a great deal of confusion. The first two were introduced by Haeckel more than a century ago (see Haeckel 1874), the third by Hennig in 1965 and the last by Ashlock in 1971. In recent years many authors have tried to re-define the words or to clarify their meanings, while others have simply discussed some of them or all of them; among these have been "cladists" of all types (including Hennig 1965, 1966 etc.; Nelson 1971, 1973; Farris 1974; Bonde 1977; and Platnick 1977, 1980), "anti-cladists" (including Simpson 1961, 1971 etc.; Mayr 1974 etc.; and Ashlock 1971, 1972, 1979) and others whom I should prefer not to categorize (Charig 1976). Although the wordings of the definitions, even the thinking behind them, are very different in nearly every case, the resulting meanings – in practical terms – are usually much the same in a vague, general sort of way; most workers understand them. (Thus, for example, almost everyone agrees that Reptilia is a paraphyletic group and that Mammalia is not.) On the other hand, there are some notable differences. The most important difference is that Hennig and consequently all members of the "cladistic" school restrict the use of the word *monophyletic* to describe those groups of monophyletic origin which include *all* the descendants, without exception, of a single common ancestor; this contrasts with the usage of earlier authors and of later, "anti-cladist" systematists who, considering only the origin of the group, make no such qualification. To avoid confusion Ashlock (1971) proposed the term *holophyletic* to replace *monophyletic* in its Hennigian (restricted) sense. Nevertheless the confusion has steadily worsened and has now reached a stage where the original purpose of introducing these words, to simplify explanation and thus form a useful part of the language, has got lost in a sea of complexity; the need for such terms has been largely forgotten and, as has happened all too frequently elsewhere, the esoteric terminology seems to have become an end in itself.

*(ii) General discussion.* Platnick (1977, p. 196) performed a valuable service in pointing out that Hennig had displayed an inconsistency in his definitions of these terms; for whereas he had defined monophyly by reference to genealogical relationships (Hennig 1966, p. 73), paraphyly and polyphyly were defined by reference to character types, symplesiomorphous and convergent respectively (Hennig 1966, p. 146, figs 44 and 45). In any case, Platnick thought it wrong that these terms should be defined by reference to characters, mainly because of the "eternal fallibility" of our judgments of the status of various character-states. Even more important, he noted that, because groups may be defined on the presence of two or more character-states, then, if one of those character-states was acquired by symplesiomorphy and another by parallelism

> the group originally delineated is by Hennig's definitions both paraphyletic and polyphyletic, despite the fact that the three species are connected by a unique set of genealogical relationships (Platnick 1977, p. 197).

Platnick gave no example of such a group. I have tried hard to think of one; the best I can do is to envisage a group of primitive mammals approximating to the Marsupialia, in which the symplesiomorphous character-state would be the possession of a yolk-sac (choriovitelline) placenta, as opposed to a chorioallantoic placenta, and the parallelism would be the female's possession of a pouch (marsupium), probably derived independently in several lineages (Sharman 1976).

I am wholly in sympathy with Platnick's comments — except in so far as I do not regard all these terms requiring definition as being parallel and co-ordinate and therefore do not object to their being defined in different ways. However, I cannot approve of the attempts by Nelson (1971), Farris (1974) and by Platnick himself (*op. cit.*) to provide satisfactory definitions of paraphyly and polyphyly, with the intention that they should replace Hennig's unsatisfactory definitions; they seem to have involved themselves in all sorts of unnecesary complications. The basic concepts, in fact, are very simple. Polyphyly is in a sense the *opposite* of monophyly: the members of a polyphyletic grouping do not all arise out of a single stem-species but originate from two or more stem-species. (The two or more stem-species will themselves have a common ancestor, of course, but that common ancestor and all the forms in between are not regarded as part of the grouping.) A paraphyletic group is a group of monophyletic origin which does not include all its own descendants. These concepts are in general use and most people are perfectly happy with them; even "cladists", of all shades of opinion, use the words quite cheerfully in those "normal" senses. Incidentally, it follows from these simplified definitions that a group cannot be both paraphyletic and polyphyletic at once.

All three workers mentioned in the preceding paragraph accept Hennig's restriction on the meaning of the term "monophyly", agreeing that a monophyletic group *must* include all its own descendants. With regard to paraphyly and polyphyly, Nelson's "Redefinitions" (1971) are almost entirely new – as the title implies. His definition of a paraphyletic group (translated into my terminology – see below, p. 406) is a clade from which *only one* younger, smaller clade nested within it has been excluded; by contrast, he defined a polyphyletic group as a clade from which *two or more* younger, smaller clades nested within it have been excluded. This means that his two categories "paraphyletic groups" and "polyphyletic groups" complement each other, they are together equivalent to the whole of what everyone else would call "paraphyletic groups". (I do not know what name Nelson would give to what *I* call "polyphyletic assemblages".) Platnick (1977, p. 195) criticized these definitions, saying that they "are internally sound but have undesirable consequences"; I do not understand his meaning. Nelson did not explain the thinking behind his proposed terminology, I cannot guess it, and I cannot imagine what purpose his distinction could possibly serve – other than the avoidance of problems with the existing definitions of "paraphyletic" and "polyphyletic"! It is best forgotten.

Putting aside this idiosyncratic approach of Nelson's, I am forced to the conclusion that the problems over the definitions of these terms are essentially due to Hennig's practice of establishing artificial groups (i.e. paraphyletic and polyphyletic) upon the distribution of single character-states. Most "cladists" have continued that practice ever since. The particular cladogram on which such a group is based will have been obtained by making a "cladistic" analysis of as many characters as possible and choosing the arrangement which gave maximal congruence, the cladogram which is supported by more characters (in the distribution of their respective character-states) than is any alternative; it was this preference that relegated the character-state defining the group in question to the status of a symplesiomorphy or a convergence and, at the same time, rendered that particular group artificial (i.e. not a natural group).

More precisely, the primary cause of the confusion is undoubtedly Hennig's definitions and diagrams (1966) correlating synapomorphy with monophyly, symplesiomorphy with paraphyly and convergence with polyphyly; they seem to have misled everyone. They confused me because the correlations were so neat, so aesthetically satisfying that I never thought to question them; others probably felt the same. But they are quite wrong, even if we ignore the fact (first pointed out by Nelson (1971)) that in Hennig's fig. 45 there is no essential difference between the diagrams illustrating polyphyly and paraphyly respectively. It is true that one can delimit a monophyletic group on the possession of one synapomorphous character-state; it is also true that one can delimit a polyphyletic group on the possession of a shared convergence, but a convergence may always be resolved into two or more independently derived character-states which are subjectively deemed to be similar. As for a paraphyletic group, it is evident that the shared possession of one symplesiomorphous character-state alone is not enough to delimit it.

In fact, it is abundantly clear that what Hennig really meant by the term "paraphyletic group" — and this is the very useful meaning generally accepted by nearly everyone ever since — is conveyed by the simple definition "a group of monophyletic origin (i.e. which arises from and includes a single stem-species) which does not include all its own descendants". Such a group has at least two boundaries, one "below" and one or more "above", all marked by the transition from a plesiomorphous character-state to the corresponding apomorphy. The fact that a systematist delimiting the group has to *choose* which character-changes are to be conjoined together in this fashion is what renders the paraphyletic group "artificial", i.e. not a natural group. An excellent example of this is afforded by the group Amphibia; its upper limit is marked by the transition from the plesiomorphous anamniote condition to the apomorphous amniote condition. The anamniote condition alone, however, is not sufficient to delimit the Amphibia; if it were, then an amoeba, a butterfly and a parsnip would all be amphibians, for none of them possesses an amnion. We need also the tetrapod synapomorphy of possessing a pentadactyl limb, i.e. the lower limit of the group Amphibia is marked by the transition from the plesiomorphous fin to the apomorphous limb. An amphibian, remember, is simply a tetrapod which is not an amniote! Thus the state of being an amphibian requires the fulfilment of *two* conditions.

However, as Hennig pointed out (1966, p. 89), plesiomorphy and apomorphy are relative concepts, referring to a *transformation series*; every plesiomorphous character-state is also apomorphous at a different, somewhat lower level — that is, lower "topographically" on the cladogram, higher taxonomically, earlier chronologically. Thus, in theory at any rate, a paraphyletic group could be delimited satisfactorily on one character-state alone if there were actual instances where the plesiomorphous character-state used for fixing its upper boundary was also apomorphous at a lower level conveniently placed for situating the lower boundary of the group; i.e. where there was a series of at least three, increasingly apomorphous character-states a paraphyletic group could be delimited on its possession of one of the middle ones. But does this ever happen? We say that a paraphyletic group is delimited on two or more character-*changes*, but can it ever be said that two character-changes really affect the same character? It is all a matter of opinion. As an example of this we could again cite the method of reproduction in mammals, considering a group of primitive mammals more or less co-extensive with the marsupials and characterized by viviparity of the sort found in, say, kangaroos. This is apomorphous in contrast with the plesiomorphous condition of oviparity but is plesiomorphous in contrast with the apomorphous condition of typical "placental" viviparity. On the other hand, the connection is rather tenuous. One might argue, quite reasonably, that oviparity versus viviparity on the one

hand and the absence of a chorioallantoic placenta versus its presence
on the other are manifestations of two entirely separate characters.

Incidentally, any clado-evolutionary system of classification
(applied to extinct organisms, real or hypothesized, as well as to
those living today) requires the drawing of boundaries across transformation series. This is as true of the Hennigian system, where
boundaries are drawn only "below" taxa, as it is of the Simpsonian,
where they are also drawn "above". It is obvious that entirely
separate characters — those which are connected neither genetically
nor functionally — are unlikely to change their states simultaneously
from plesiomorphous to apomorphous; a "diagnosis" of a taxon
citing several such characters *in relation to the same boundary* is
therefore liable to cause problems in the classification of fossil forms
lying near that boundary.

The erroneous statement, "A paraphyletic group can be delimited
on a symplesiomorphy" is what has led many "cladists" to regard
paraphyletic groups with contempt and to look upon them as "non-
groups"; they are delimited, so they say, on a negative character, the
absence of a feature. Thus, for example, the Amphibia are called a
non-group simply because they belong to the category of organisms
which lack an amnion — like amoebae, butterflies and parsnips!
Such people ignore the need to possess a tetrapod limb (or secondarily to have lost it).

Let us return to the distinction between paraphyletic and polyphyletic groupings. I have shown that in order to delimit a grouping
of either type satisfactorily we need information on the distribution
of at least two separate character-states; it is therefore clear that the
distribution of one single character-state alone is insufficient to
determine the category of an artificial grouping, i.e. whether we
should call it paraphyletic or polyphyletic in the generally accepted
sense of those words, even if we had adequate knowledge of that
character-state. (In fact, its presence in the hypothetical common
ancestor can only be guessed at.) The determining factors must
surely be the members' genealogical relationships (both to each
other and to forms outside the group), the totality of the distributions of the character-states that we decide to use in establishing
the group, and, in consequence, which members we *choose* to
include within the group. The only criteria on which the group
should be judged are the simple statements "A paraphyletic group is

a group of monophyletic origin which does not include all its own descendants" and "A polyphyletic assemblage is not of monophyletic origin but arises from two or more stem-species". The more nested groups are excluded from a larger grouping to form a paraphyletic grouping, the more symplesiomorphous character-state distributions (not congruent with each other) will be needed to help delimit that grouping; thus, in order to delimit Reptilia, one character-change is needed to exclude Aves from Amniota and one to exclude Mammalia.

The mistake made by Farris and Platnick in drawing their distinction between paraphyletic and polyphyletic groups was to follow Hennig and base it on the distribution of one single character-state, which, as we have seen, is not enough; and, in any case, their result is quite different from what Hennig seems to have intended. As Platnick himself implied, the answer to the question "Paraphyletic or polyphyletic?" might well be different if a different character were used. The procedure of these authors, in essence, is to (a) delimit a group on its members' possession of a particular character-state; (b) try to assess whether the distribution of that character-state indicates symplesiomorphy or parallelism/convergence (the basic question here is whether the unknown common ancestor also possessed it or did not possess it, respectively); and then (c) call the group paraphyletic or polyphyletic accordingly. Farris complicated the matter still further by devising a method for carrying out the most parsimonious assessment of symplesiomorphy versus parallelism/convergence, his statement of which (according to Platnick 1977, p. 198) "is couched in terms that non-numerically orientated cladists may have difficulty in following". Platnick therefore evolved a simplified version of Farris' method which, though still intimidating at first sight, is nevertheless within my own limited powers of comprehension. What this method actually does is as follows. Imagine several species connected by an accepted cladogram, some (but not all) of which possess a certain character-state distributed in an incongruent manner. Consider the most recent common ancestor of all the species possessing that character-state: is it more likely that that unknown ancestor also possessed that character-state and passed it on without a break to all those species ("paraphyly" of Farris) or is it more likely that it did not fulfil those conditions ("polyphyly")? In any case, Farris' method does not work in every instance, as Ashlock pointed out (1979, p. 444).

Thus, when people like Farris and Platnick try to distinguish paraphyly from polyphyly, they are really asking the wrong question altogether – if they want an answer in terms of what those words usually signify. Paraphyly and polyphyly, in their generally accepted meanings, cannot be opposed in this manner; to ask whether a group is paraphyletic or polyphyletic is an absurd question, rather like asking "Is this a book of poetry, or is it a technical work?" Really we should be asking two separate questions: "Is it technical or literary?" and, if it is literary, "Is it poetry or prose?" In the same way Farris and Platnick should be asking about the group concerned: "Is it of monophyletic or polyphyletic origin?" and, if monophyletic, "Is it paraphyletic or holophyletic in its composition?"

In any case, even if a "cladist" knows whether a given artificial grouping is paraphyletic or polyphyletic, either in the generally accepted sense of those words or in the different sense of Farris and Platnick or in the even stranger sense of Nelson, what does it signify to him? *Does it matter?* After all, "cladists" reject both paraphyletic *and* polyphyletic groups, so why should they want to distinguish between them? Their attempts to do so are totally unnecessary as well as just plain wrong.

Platnick almost hit upon the major problem caused by Hennig's incorrect approach when he commented (see above) upon the possibility of an artifical group, judged upon the distribution of single character-states, being both paraphyletic and polyphyletic at the same time; it was this that led me to discover the anomaly, and I am only surprised that Platnick did not realize it himself. But, on the contrary, he continued to use Hennig's illogical, inconsistent approach and to write about it as though he had not found anything in it to which he might object, as though he had never made the criticism. This accounts for such strange ideas of his as the suggestion that "one could define a [paraphyletic] group consisting only of the platypus and man" (1980, p. 544); I would have thought that such a remarkable grouping could only be polyphyletic, unless – in addition to excluding every other mammal group that diverged from the two lineages extending back from platypus and man to their most recent (presumably Triassic) common ancestor – one also *included* that most recent common ancestor and every species that actually lay on those lineages between the common ancestor and the two modern forms. A typical paraphyletic group would be all mammals *except* platypus and man; it could be recognized by the usual mammalian synapomorphies and two particular symplesiomorphies (say, no duck-like bill, no higher intelligence).

The most recent discussion of these definitions is by Ashlock (1979). My attention was not drawn to his paper — which I consider to be easily the best of any on this controversial topic — until after I had written the rest of this section of mine; I therefore derive great comfort and reassurance from the fact that I seem to have come quite independently to many of the same conclusions as he, indeed there is little cause for dissent between us. Ashlock tackled the problem from a very different angle and expressed himself in a very different way, his article being characterized by its admirable clarity. I do not feel, however, that what I have written has been entirely superfluous, in so far as I believe that I have added one useful point to the argument: I have put my finger on what I consider to be the prime cause of much of the confusion which has prevailed hitherto — the basing by the "cladists" of their definitions of paraphyly and polyphyly upon the distribution of a single character-state. Meanwhile Ashlock made the important point that, apart from all the other objections to it, Hennig's modified terminology is suitable only for "cladists" and does not meet the requirements of other sorts of systematists — like him and me.

In my attempt to clarify these issues further I need to make certain important points which, though they seem self-evident to me, have not been stressed by any of the "cladists" mentioned above; some of those authors, perhaps, have even failed to realize them themselves. These are as follows:

1. The words under discussion, ending in "-phyly" (noun), "-phyletic" (adjective) or "-phyletically" (adverb) show by their etymology (Greek φυλον, tribe or race) and by the contexts in which everyone had used them until recently that they are concerned entirely with ancestry and phylogeny. They are therefore meaningless when applied to systems which do not have an evolutionary basis, e.g. the natural order systematics at present so fashionable in certain circles. It is especially interesting to note that natural order systematists are very active — indeed predominant — in the present discussion of the meaning of these words,

even after their "transformation" to the belief that evolution plays no essential rôle in systematics.

2. The words listed (despite their common root in "-phyly") are not all parallel, co-ordinate terms; they do not all name or describe alternative conditions. To explain that statement it is necessary once more to emphasize the essential two-stage nature of any clado-evolutionary system, be it Simpsonian or Hennigian or anything else.

(*iii*) *Comments on monophyly.* The word *monophyletic* relates only to the first stage of clado-evolutionary systematics, phylogeny reconstruction, for it describes the origin of a natural group or clade; indeed, any natural group (natural in a genealogical sense) is *ipso facto* of monophyletic origin, consisting of a single ancestral inter-breeding population (real or hypothesized) and all its descendants. From this it follows unequivocally that any species, fossil or extant, may be regarded as the "stem-species" of a natural group. Taxa above the species level (see below) have no part to play in phylogeny reconstruction except in so far as (a) clades are used in phylogeny reconstruction, (b) most clades correspond to named taxa and (c) they may therefore be called conveniently by those same names. The ranking of taxa has no connection whatever with phylogeny recon-struction; in particular, ranking is wholly irrelevant to the concept of monophyly, and the definition of monophyly should contain no mention of taxonomic ranks. This conflicts with the views of Simpson (1961, p. 124).

(*iv*) *Comments on paraphyly and holophyly.* On the other hand, the word *paraphyletic* relates also to the second stage of clado-evol-utionary systematics, namely classification; for its more frequent application is to *taxa*, sets of kindred organisms formally defined in an arbitrary manner by the classifier and formally named by him according to the International Codes of Zoological or Botanical Nomenclature.

Most present-day clado-evolutionary systematists, Simpsonian or Hennigian, would agree that a taxon ought to be of monophyletic origin and should therefore constitute one single continuous segment of the evolutionary "tree". The Hennigian systematist also believes that a taxon *must* comprise the whole of the clade in question, i.e. it must be what Ashlock (1971) called *holophyletic* as well as possessing monophyletic origin. But the Simpsonian systematist opposes that strongly held belief, maintaining that for various reasons (see below, p. 430) it is sometimes desirable – and therefore permissible – that one or more of the younger, smaller daughter-taxa (=clades) nested within the parent taxon (=clade) should be ranked as highly as the latter or even more highly; in such cases the hierarchical structure of the classification necessitates that those younger descendant taxa be excluded from the parent taxon, which, though still of monophyletic origin, is consequently described as *paraphyletic*. This demonstrates the Simpsonian belief that taxa do not necessarily correspond to natural groups (which exist in reality, outside the human mind); on the contrary, say the Simpsonians, taxa may sometimes be entirely artificial concepts (existing only in the minds of systematists).

The word *paraphyletic* – but not *holophyletic* – is also applied to groups. Paraphyletic groups are incomplete clades which have not yet been formally designated as taxa.

(*v*) *Comments on polyphyly.* The word *polyphyletic* is used to describe an assemblage which is not of monophyletic origin, consisting of two or more separate clades; such as assemblage is not natural, it too is artificial, and it has no claim to be regarded as a taxon. Not only is that " non-taxon" polyphyletic, but its origin – in so far as an artificial assemblage can be said to have an origin – is also polyphyletic. When it is discovered, or strongly suspected, that a taxon hitherto believed to be of monophyletic origin is actually polyphyletic, then that "taxon" should be ruled invalid and rejected from the formal classification; this should be done as soon as it is realized that there is no reasonably good evidence of immediate sister-group relationships between its component parts. (Nevertheless an equivalent term in the vernacular may still be useful outside the classification; thus, while we should not refer to "Dinosauria", we may yet find it convenient to talk or write about dinosaurs in an informal manner.)

(*vi*) *Definitions.* For the sake of clarity I shall now propose a formal definition of all the terms under discussion; I hope that my readers will find it simple, unambiguous and in conformity with generally accepted usage.

1. A group which originates from a single stem-species and which includes that stem-species within itself is said to be of *monophyletic* origin (referred to, for brevity, as a "monophyletic group"). Every species, fossil or extant, may be regarded as a potential stem-species, a species on which such a monophyletic group might be based.

2. A group of monophyletic origin which includes within itself all the descendants of the single stem-species is called a *natural group* or *clade*. Such a group has only one boundary, marked by the character-change which defines the origin of its stem-species and thereby determines the extent of the group itself. The group is considered to be "natural", for it exists independent of the observer.

3. A group of monophyletic origin consisting of a clade from which one or more younger, smaller daughter-clades nested within it have been excluded is called a *paraphyletic* group. Such a group has at least two boundaries, one "below" — marked by the origin of the stem-species of the group itself — and one or more "above" — marked by the origin(s) of the stem-species of the daughter-clade(s) excluded from it. The group is therefore "artificial", for it exists only in the mind of the systematist who made the selection — in arbitrary fashion — of the clade or clades to be excluded (and likewise in the minds of those who accept his grouping). Nevertheless all ancestral groups and all stem-groups are paraphyletic by definition.

4. Consider a group of monophyletic origin, a complete clade. Suppose then that the stem-species be excluded, perhaps together with other previously included species, so that the group ceases to be of monophyletic origin and is therefore no longer one single group; it automatically becomes an assemblage of *polyphyletic* origin (referred to, for brevity, as a "polyphyletic assemblage"). A polyphyletic assemblage consists of two or more groups of monophyletic origin (clades and/or paraphyletic groups) and must have the same number of stem-species. Such an assemblage is arbitrary and artificial.

Table II. Comparison of different types of groups and taxa

GROUPS employed in phylogeny reconstruction:

| Of MONOPHYLETIC origin, i.e. originating from a single stem-species (▼) known or hypothesized | | Of POLYPHYLETIC origin, i.e. originating from two or more stem-species |
|---|---|---|
| i.e. comprising one clade, complete or incomplete | | i.e. comprising two or more separate clades, complete or incomplete |
| The whole clade or natural group | Excluding one or more smaller clades nested within it | |

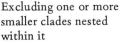

| CLADE | PARAPHYLETIC GROUP | POLYPHYLETIC ASSEMBLAGE |
|---|---|---|

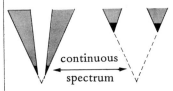

continuous ←—— spectrum ——→

PROXIMATE possibly showing parallel evolution    REMOTE probably showing convergent evolution

| Eligible for designation as VALID TAXA for use in classification | | INVALID as TAXA; not to be used in classification |
|---|---|---|

TAXA employed in classification:

| HOLOPHYLETIC TAXON | PARAPHYLETIC TAXON, excluded clades must be given equivalent or higher rank | |
|---|---|---|
| *Favoured by all systematists* | *Not permitted by the Hennigian school, much used by the Simpsonian* | *Not approved by any school of systematics, to be rejected if already in use* |
| NATURAL GROUPS and TAXA | ARTIFICIAL GROUPS and TAXA | |

5. Any group of monophyletic origin is eligible for designation as a *taxon,* a formal classificatory unit, by a Simpsonian systematist. An assemblage of polyphyletic origin is not eligible for designation as a taxon. Therefore all taxa are monophyletic, at least within the framework of clado-evolutionary systematics; to speak of a "monophyletic taxon" is a tautology.

6. A taxon consisting of a complete clade is called a *holophyletic* taxon.

7. A taxon consisting of a paraphyletic group is called a *paraphyletic taxon*; the younger, smaller daughter-clades nested within the parent clade but specifically excluded from it must be given equivalent or higher rank. Some systematists find paraphyletic taxa useful in classifications; they are nevertheless unacceptable to Hennigian systematists.

8. Artificial groups and assemblages, despite their artificiality, may be defined in terms of clades; and clades (terminal species included) are all natural entities.

9. All these terms are absolute terms, "either-or" terms; there are no intermediates. Thus groups are invariably of monophyletic origin, assemblages of polyphyletic origin. Groups may be either natural (complete clades) or artificial (paraphyletic groups); assemblages, by definition, are all artificial. Taxa may be either holophyletic or paraphyletic.

Perhaps I myself shall be accused of the same offence that I have laid at Hennig's door, namely that of re-defining existing words, in respect of my definition of *paraphyletic.* That would be a justifiable accusation, up to a point; but I have really done no more than give formal definition to a meaning which is not only the meaning accepted by virtually everyone at present but was also, in effect, the meaning that Hennig himself probably intended. Unfortunately Hennig expressed himself rather badly by defining it in terms of the distribution of a single character-state.

*(vii) Degrees of polyphyly.* A distinction may be drawn between two types of polyphyly:

1. *Proximate* polyphyly, in which the most recent ancestor

common to all the component groups is not much older than those groups themselves; i.e. not much else, apart from that ancestor, has been excluded from the larger clade of which it would constitute the stem-species. A polyphyletic assemblage of this type could be extended backwards a little in time to include the common ancestor (and thereby acquire a monophyletic origin), become a clade (or paraphyletic group) and consequently become eligible for formal designation as a valid taxon. Thus the question of whether we are to have a holophyletic (or paraphyletic) taxon or — alternatively — a polyphyletic assemblage consisting of two or more separate groups depends entirely upon our decision as to how far back we place its origin(s). And that, for the palae-ontologist, depends in turn upon his arbitrary decision as to which particular character-change is to be regarded as marking the origin of the taxon.

Similar characters which have originated independently in two lineages shortly after their point of divergence, i.e. in an assemblage showing proximate polyphyly, are said to have evolved *in parallel*.

2. *Remote* polyphyly, in which the most recent ancestor common to all the component groups is very much older than the groups themselves; i.e. not only has that ancestor been excluded from the larger clade of which it would constitute the stem-species but many subsequent species too have been excluded likewise. The components thus appear to have originated from stocks so widely separated from each other that it would be unthinkable to unite them by extending the assemblage back to the common ancestor.

Similar characters which have originated independently in two lineages so widely separated are said to have evolved *convergently*.

The distinction proposed here between proximate and remote polyphyly is not of course an absolute distinction; the separation between the components of a polyphyletic assemblage is a matter of degree, so that the terms suggested represent the extremes at either end of a continuous spectrum. The same applies to the distinction between parallelism and convergence.

It should be mentioned that other authors have tried to draw a distinction between two sorts of polyphyly. For example, only very recently Crompton and Jenkins (1979, pp. 68–69) distinguished two different meanings of the word. One referred to an instance of something very like proximate polyphyly, concerned with the arbitrary positioning of a taxonomic boundary between two classes (reptiles and mammals). The other usage obviously signified remote polyphyly: different groups of one class (mammals) were derived independently from separate stocks of another (reptiles, in particular therapsids), those stocks giving rise also to other groups of therapsids.

(d) *Emotive use of words and circular reasoning.* These methods are sometimes employed by "cladists" of either school when trying to press their case. They should *not* use them, of course, not even unwittingly; as scientists they should argue objectively and with logic. One typical ploy is to describe characters used by Simpsonian systematists as "primitive", "adaptive", "grade characters" and so on; those terms, while not pejorative in themselves, are spoken or written in various ways that imply disdain on the part of the speaker and thereby produce a pejorative effect. "Cladists" have often said to me, àpropos my use of a character in classification, "Ah, but that character is *primitive*"; I used to let them put me down by this, but now I reply "So what?" Unfortunately most of us will accept the disparaging overtones unthinkingly, even though they cannot be justified. "Cladists" tend also to argue in circular fashion, using as points in their arguments propositions that would be valid only if their case were already proven. The two methods outlined above are often employed in conjunction.

Thus Gardiner *et al.* (1979, p. 176) disparaged the classifications of their taxa were paraphyletic) and — again in part — "were recogised . . . by pre-Darwinian systematists"; yet those authors did not of their taxa were paraphyletic) and — again in part — were "recognized . . . by pre-Darwinian systematists"; yet those authors did not specify what was *wrong* with overall similarity, grades and pre-Darwinian systematics, taking it for granted that everyone else had as low an opinion of them as they had. Many of their readers, on the other hand — especially those whom they were trying to convince — would, if they thought about it, find no immediately obvious reason to regard any of those concepts with such disfavour. The authors' position might be summed up by the statement "Evolutionary systematics is bad because it uses grades etc.", with the implication that grades etc. are bad in themselves. But that implication, being itself an unproven assumption, provides a perfect example of the well-known fallacy of "begging the question" — what the logicians call *petitio principii*. It could even be interpreted as suggesting a dislike of the concepts in question simply *because* of their connection with the despised evolutionary systematics; they would be saying, in effect, that "Evolutionary systematics is bad because it uses grades etc., which must be bad because they are used in evolutionary systematics"!

Incidentally, it should be noted that the rise of natural order systematics has now brought both "overall similarity" and pre-Darwinian attitudes back into favour with many of the so-called cladists, including Gardiner and at least some of the *al*. Their beliefs have changed fundamentally in a matter of months.

## 4. Cladistic Analysis (Character Distribution Analysis)

(*a*) *My essential approval, with reasons.* My respective attitudes towards "cladistic analysis" and cladistic classification are very different one from the other; I am almost wholly *for* the analysis and almost wholly *against* the classification. Two of my cladist friends have told me that I am a "cryptocladist", that I am "halfway there already". Yet my initial reaction to the procedures of Hennigian analysis was one of *déja vu*, nothing new in this, this is what we were all doing thirty years ago without the specialized jargon. How else could one draw up a family tree except on the grounds of common ancestry, and what better evidence of common ancestry than shared characters? The only innovation of "cladistic analysis" seemed to be its insistence on attempting to impress the uninitiated by calling those shared characters "synapomorphies" (silly word, so I thought) and using other apparently pseudotechnical terms — mostly unnecessary. But soon I began to realize that the more formal procedures of cladistic analysis were incomparably superior to the vague, slapdash, woolly methods used by us non-cladists. I read manuscripts for friends, I read articles sent to me for refereeing, I read published papers, and I quickly became conscious of the fact that many of my non-cladistic colleagues — hard-working intelligent colleagues with high reputations — were collecting magnificent specimens in the field, preparing them beautifully, describing them very fully and with the utmost clarity, and then writing accounts of their phylogenetic relationships that were nothing more than meaningless waffle. Taxa were not properly defined, relationships were proposed on the basis of primitive characters or negative characters (including reductions and losses) or characters which could easily have evolved independently in several different lines. One would-be author even suggested a close relationship between two fossil reptiles because both had unshortened tails! My own experience as a Ph.D. student illustrates both sides of the coin in question. When writing my dissertation (Charig 1956) I suggested a close relationship between two fossil reptiles because they both possessed a unique type and arrangement of dermal scutes — a perfectly proper procedure, leading to a classification which no one has ever queried. But I also proposed that two of my new thecodontians were each ancestral to a major group of dinosaurs on the grounds, in each case, of a general similarity — similarities which, today, I cannot easily substantiate with hard and fast synapomorphies, even though I still believe that I may have been intuitively correct. In summary, then, I am strongly in favour of the disciplined approach which forms a necessary part of a proper character distribution analysis.

(b) *Misunderstandings due to confusion of Hennigian systematics with natural order systematics.* This source of misunderstanding has already mentioned (pp. 365–366, 392–394); so many "non-cladists", not realizing how very widely natural order systematics has diverged from the classical Hennigian variety or even that the former philosophy exists, have found themselves completely baffled by the apparent inconsistency of the cladistic approach. Thus they have been puzzled by certain aspects of cladistic analysis, in particular the attitude of many "cladists" towards parallelism. Once again we find a good example in the famous paper on "The salmon, the lungfish and the cow: a reply" (Gardiner *et al.* January 1979), which predates the earliest published statement of the principles of "transformed cladistics" (Platnick January 1980) by twelve months. Gardiner *et al.* made the remarkable statement (p. 176) that parallelism "is outside the scope of cladism", which seems not to conform with the received wisdom as laid down by Hennig (1966) and consequently outraged not only me but also most of my like-thinking "non-cladistic" friends. Steeped in classical Hennigism, we could not be expected to see how a refusal to consider the effects of parallel evolution could be justified; surely, so we reasoned, the fact that we often do not know *which* characters are parallel or convergent, *which* of the alternative possible cladograms reflects the phylogeny, does not entitle us to ignore the problem; and the very existence of alternative cladograms proves that the problem exists. (It is precisely in cases like this that we have a better chance of reaching a likely solution through an approach involving the dimension of time, like the "stratophenetic" approach of Gingerich and Schoeninger (1977).) Only later did I realize — many others have yet to see the light — that Gardiner and his fellow-authors were writing, not as Hennigians, but as natural order systematists; only then did their writings make some sort of sense, even though I could not accept their philosophy.

(*c*) *Significance of congruence to an evolutionist.* The possession by a species of *any* feature is explicable only as being due either to some sort of evolutionary process — "descent with modification" — or to direct Creation; the latter interpretation is even more metaphysical than the former. If it be assumed that evolution has indeed occurred, then any two species must possess a most recent ancestor, common to the two of them, at some point in time. If those two species have one or more features in common, then there are but three possible explanations for their both having acquired each of them:

1. The common ancestor already possessed that feature and they have inherited it therefrom. Its joint presence in both species is therefore by "community of ancestry" and it may be of use in phylogeny reconstruction. (The *reason* for the common ancestor's possession of the feature and the *mechanism* of inheritance are of no importance in this connection.)

2. The two species have each acquired the feature independently since the lineage split, evolving it by adaptation to a similar function or to a similar mode of life in a similar environment. Its joint presence in both species cannot serve in phylogeny reconstruction but indicates only the similarity of adaptation, i.e. it is interpreted (in the language of evolutionary biology) as an adaptive parallelism or convergence. The *mechanism* of adaptation, whether it be evolutionary (natural selection) or Divine purpose or anything else, is again totally irrelevant.

3. The two species have each acquired the feature independently since the lineage split, evolving it in a purely fortuitous fashion for no apparent reason. Its joint presence in both species indicates nothing.

If it be assumed that evolution has *not* occurred, or (as is claimed by the natural order systematists) that it is outside the scope of systematics, then the pattern of distribution of any feature — if not entirely random — can be given a meaningful interpretation only in terms of adaptation and purposeful Creation.

(*d*) *Panchen's objection.* It is also worth noting the claim made recently by Panchen (1982) that the principle of parsimony is not a valid criterion of a character distribution analysis ("Cladistic analysis"). It is certainly *difficult* to apply it to analysis of this sort, whether it be in Simpsonian systematics, Hennigian systematics or natural order systematics. The list of characters available for any given analysis is bound to be very incomplete; indeed, the characters actually employed in any given instance can be no more than a tiny sample of *all* the existing characters of the organisms in question. Panchen contends that, were we able to use all those characters, we might find that the distribution of the shared character-states in the sample (for instance, $x\%$ of characters with character-states common to A and B, $y\%$ to A and C, and $z\%$ to B and C) differed greatly from what the distribution would be if all the characters were used; in other words, the proportions found in the sample need not be representative of the whole. It follows from this, in Panchen's opinion, that there is no justification for using the distribution of the character-states in a mere sample for the purpose of determining maximal congruence. (Of course, it is highly unlikely — nay, virtually impossible — that a complete list of characters could ever be drawn up; and further, even if it could, there would be no way of recognizing it as such.)

Panchen is undoubtedly correct, and his objection must be sustained if we are to regard biological systematics as the natural order systematists would have us regard it — like logic or mathematics. But in biology and biological systematics absolute proof is generally lacking and often impossible — this applies particularly in historical biology, including palaeontology — so that most biologists and systematists are content to base their conclusions on *reasonable probability,* being ready to change them if new evidence comes to light; their procedures are probabilioristic. Fairly comprehensive lists of shared character-states must surely have some reasonable probability of indicating maximal congruence, no matter how many more might remain unobserved, unless there is good reason to suspect bias – usually towards adaptive characters. Whether the arrangement which gives maximal congruence is of any particular value, phylogenetic or otherwise, is – as we have seen – an altogether different question.

*(e) Dichotomous branching.* The Hennigian insistence that phylogenetic branching is invariably dichotomous is clearly nonsensical if interpreted literally. Imagine, for example, a species of lizard inhabiting a range of mountains that is suddenly transformed into a group of islands by a rapid marine incursion; the quasi-simultaneous reproductive isolation of the several lizard populations, one to each island, would lead to quasi-simultaneous speciation. In any case, the impossibility of a lineage splitting into more than two daughter-stocks at once could not be upheld unless speciation were an instantaneous process – a supposition that is equally nonsensical.

Most Hennigians, however, are prepared to admit that dichotomous branching is nothing more than a "methodological principle" (e.g. Bonde 1977, p. 775) and it is true that its application makes no difference in character distribution analysis. Where it *is* objectionable is in the production of cladograms for didactic exhibitions and publications, for those cladograms will obviously mislead the unsophisticated student.

*(f) Recognition of ancestors.* Hennigians refuse to accept that ancestral species can be recognized; they must have existed, they may be hypothesized, but they cannot be recognized. They believe that no known species – extant or fossil – can be recognized as the ancestor of another taxon, and that all fossil species should be represented on the cladogram as terminal twigs. Thus, for example, one should not state that "*Archaeopteryx lithographica* is the ancestor of the birds", even if it could be demonstrated that that species bore a much closer anatomical similarity to later birds

than is actually possible at present. It should only be claimed that *A. lithographica* seems to be not far removed from the line of their direct ancestry.

Strictly speaking, this approach is correct; there *is* no way of proving unequivocally that one species is the ancestor of another. But biology is not an exact science, a science in which the only statements of value are those which can be proved beyond any doubt whatever. If that were the case then very few statements in biology would be acceptable, certainly no phylogenies; for we have no infallible way of recognizing valid synapomorphies as such, and *any* attempt at phylogenetic reconstruction can be wrong. Indeed, if one adopts the rigid point of view that biology ought to be an exact science, then the whole procedure of character distribution analysis becomes worthless.

In fact — and I have only just made this point (p. 415) — the whole of biological science is based very largely upon reasonable probabilities; absolute certainties play only a small part in biology. If we did not consider reasonable probabilities we should make no progress at all. This is particularly true of palaeontology, where wholly objective data are especially rare and include only the first of four categories of information statements:

1. Positive information which is entirely objective, e.g. "*Aetosaurus ferratus* occurs in the Upper Trias of Germany". There can be no doubt about this because the name *A. ferratus* was given to type-material from the Stubensandstein, the species *ferratus* is the type-species of the genus *Aetosaurus*, and the Stubensandstein forms part of the Upper division of the type-section of the Trias.

2. Positive information which contains one or more subjective judgments, e.g. "*Brachiosaurus brancai* occurs in the Kimmeridgian of Tanzania". Two subjective judgments have been made here, namely that the species *B. brancai* is congeneric with the type-species of the genus *Brachiosaurus*, *B. altithorax* from the Morrison Formation of western Colorado, and that the Tendaguru Beds of Tanzania are of Kimmeridgian age.

3. Negative information, e.g. "*Heterodontosaurus tucki* has been found only in the Lower Jurassic of southern Africa". Much information in palaeontology, especially in palaeobiogeography, concerns the limitation of the ranges of taxa, in time and/or space.

Apart from the difficulties of defining the taxa themselves (since the whole evolutionary tree is a continuum in time and space), limits of their ranges can never be known but only *suspected* because of their absence beyond those limits — and absences can, *per se*, only be disproved and never verified. In other words, the statement that *H. tucki* has been found only in the Lower Jurassic of southern Africa should not really be taken to imply that it did not exist at any other time or in any other place — it means only precisely what it says, namely that it has *not yet* been found in rocks of any other age or in any other part of the world. (Once again the correlation of the rocks, in this case with the Lower Jurassic, is purely subjective; most workers still regard them as Upper Triassic.)

4. Speculation, unsupported by any real evidence of any sort and based only on observations of what are believed to be analogous cases, usually of organisms living at the present day, e.g. "*Archaeopteryx* could not have flown well because it did not have a large keel to its sternum". Apart from doubts as to the applicability of the analogy, the analogous observation may itself be wrong — as in the example given.

Since I accept the "reasonable probability" approach to biological science, I believe that it is perfectly proper to postulate *hypothetical* ancestors lying within major taxa; the lower the hierarchical ranking of the parent taxon, i.e. the more precise the taxonomic position of the hypothetical ancestor, then the smaller the probability that that placing is correct. Thus the common ancestor of all mammals must surely be a synapsid, almost as surely a therapsid, a little less certainly a cynodont, and less certainly still a member of the family Chiniquodontidae — although no serious palaeontologist would doubt any of those statements (*pace* Gardiner 1982). On the other hand, I should not be so foolish as to choose any particular genus of chiniquodontid as the putative mammal ancestor, even though I might be tempted to specify *Probainognathus* as the closest to it.

## 5. Comparison of Simpsonian and Hennigian Classifications

(a) *Aims of classification.* The central point of this article is the controversy between the supporters of Simpsonian classification on the one hand and those of Hennigian classification on the other. But that controversy, so it seems to me, can *never* be properly resolved; it can never be resolved because the adherents of the two conflicting sides set out from two entirely different sets of initial assumptions and have entirely different aims — usually without even realizing that the differences exist. Patterson (1978, p. 124) wrote that "classifications have two purposes — to express evolutionary relationships, and to act as *aides-mémoire* or simple summaries of knowledge". Both schools of thought share those two purposes, but they differ enormously in where they place the emphasis, so that some systematists may stress the importance of one of them to the virtual exclusion of the other. What *are* the essential differences? In this instance it is better to reverse our usual order and start with the Hennigian school.

(*i*) *The Hennigian classification.* The Hennigian systematist believes that the main purpose — if not the only purpose — of a classification of living organisms is to reflect their phylogenetic relationships, their "family tree", this purpose invariably taking priority over every other. Thus Hennig wrote in 1966 (p. 64) "For us the phylogenetic system is a medium for presenting the phylogenetic relationships of species (primarily the living species) as accurately and clearly as possible." In fact, the Hennigian's ideal classification is nothing more or less than a verbalization of the entire phylogenetic cladogram; again I quote from Hennig, this time from the English summary to his 1974 paper (p. 294): "both types of presentation [i.e. cladogram, hierarchical classification] being exactly equivalent and convertible". Thus the Hennigian classification concentrates entirely on the evolutionary history of the organisms, past and (in the case of some fossils) future — on the *apparent* evolutionary history as evinced by "cladistic analysis"; but it may well ignore their present characters, separating forms which are phenetically (perhaps even genetically) identical and lumping together forms which are very different. According to Hennig it is not the characters that matter, on the relationships that they indicate.

(*ii*) *The Simpsonian classification.* The Simpsonian systematist, by contrast, believes that existing characters are important *per se,* they *do* matter; the classification, *while entirely consistent with the phylogeny,* should reflect those characters too. I stress "entirely consistent with the phylogeny", for "cladists" frequently claim that a Simpsonian classification is unscientific — they contemptuously call it a "cook-book classification", based on common characters that have nothing to do with relationship, one might just as well segregate the organisms into edible and non-edible! This is quite absurd. A Simpsonian classification is based on an arbitrary division of the phylogenetic tree into separate segments, each taxon being equivalent to one (and only one) continuous segment of the tree and therefore to a phylogenetic group — i.e. to *one* monophyletic group (*sensu* Hennig) from which however smaller monophyletic groups may have been lopped off to make it paraphyletic. Incidentally, I am quite unable to fathom the Hennigians' obsessive dislike of such paraphyletic groups, but I shall return to that later (pp. 429–432).

(*iii*) *Summary of comparison, who favours which, and why.* To summarize the comparison, a Hennigian classification is based *entirely* upon the past history of the organisms concerned, a Simpsonian classification gives also some indication of their present condition. I shall illustrate *my* attitude towards this comparison by telling a little story as an analogy, a story based on an excruciatingly weak pun. A man comes home from work one day and sits down to his evening meal; his wife places before him a bowl of some revolting-looking mush. "What's this, woman?" he roars at her. "It's bean soup", she replies. "*Bean* soup?" he yells. "It's *bean* soup? I don't want to know what it's been, I want to know what it is now!" That, exactly, is my attitude towards biological classification.

To be fair, we must admit that our own system of surnames, the names by which we ourselves are generally known in our human society, is more "Hennigian" than "Simpsonian". Each of the names "Crookshank", "Thatcher", "Scott", "Williamson" tells us something about one distant paternal ancestor of the person bearing it (or her husband) — about his physical peculiarities, occupation, geographical origin or *his* immediate genealogy — but it tells us nothing whatever about the person himself. The examples given are in the English language, but other European tongues are no different in this respect (Legrand, Kirpichnikov, Frankfurther, Pugh = ap Huw). In other words, the "information content" of the "general reference system" used in human affairs is absolutely zero except in that the possession of a surname in common does indicate the possibility of a family relationship. On the other hand, a close family relationship may well exist even between persons with different names.

It is clear that, irrespective of what we are classifying, there can be no such things as correct or incorrect systems of classification. ("Correct" and "incorrect" may be applied to the *drawing up* or the *actual use* of a classification, but not to the criteria on which it is based.) Ideally we should all use the same classification; but classifications are always subjective, and which particular variation a given worker will choose on any given occasion is likely to depend to some extent upon the purpose for which he needs it − for example, on whether he is interested in what the animal (or plant) has been or what it is now. In practice his decision will probably depend more on which group of organisms he is studying. Those who work on organisms from a single time-horizon − in effect that means neontologists as opposed to palaeontologists − are liable to be more "cladistically" inclined; so are those who work on groups that show a great deal of diversity at a low taxonomic level, like insects or fishes, especially if those groups have also a large modern representation. Again, the systematist's choice of classification may well be affected by where he works and with whom (certain institutions, notably the American Museum of Natural History, seem to have been influenced far more strongly than others by the Hennigian philosophy) and it may even depend upon how old he is. Senior citizens like me tend to choose Simpsonian classifications, younger men often prefer Hennig; for just as we old conservatives have an instinctive reaction against change of any sort, we feel threatened by it, so − equally deplorably − is there a tendency in every progressive iconoclast to jump on to the trendsetters' bandwaggon.

## (b) *Alleged advantages of a Hennigian classification*

(i) *Complete objectivity and immutability.* First, it is true that a complete and correct cladistic classification would be absolutely objective and therefore standardized and unchangeable. In practice, however, we could not construct such a classification until we knew the complete phylogeny, without error, which in turn would require the infallible recognition of parallel and convergent characters; and, in fact, the little that we do know of the phylogeny is known only with varying degrees of uncertainty, for we are often unable to recognize which characters are true synapomorphies and which are merely parallel or convergent. Thus our approach to the phylogeny, whether we be "cladists" or not, is inevitably subjective, and in consequence the same must apply to any classification based on that phylogeny. I agree with Hennig's contention (1966, p. 28) that the "investigation and presentation [of phylogenetic relationships between organisms] as completely as possible became an unavoidable scientific task"; though it is an "endless task" with an unattainable goal, it should nevertheless be pursued with energy. But I still cannot see why there should be any logical obligation to base the *classification* entirely upon the phylogeny, perfect or imperfect; indeed, Hennig himself qualified his contention (concerning the need to elucidate the phylogeny) with the statement (*loc. cit.*) "whether one elevated the phylogenetic system to the position of a general reference system for biology or not".

(*ii*) *Retrievability of cladogram.* Secondly, it is also true that the ideal Hennigian classification would – by its very nature – permit instant, unambiguous and total retrieval of the entire cladogram (complete and perfect to the last dichotomy) and hence of the phylogeny too. I cannot understand, however, why anyone should wish to verbalize a cladogram into a classification simply in order to be able to put it back into graphic form as the same cladogram that he had started with! Such a person, indulging in so pointless an exercise, would inevitably remind me of Eeyore the Donkey in *Winnie-the-Pooh*; every child remembers how Eeyore picked up his burst balloon with his teeth and placed it carefully in his empty honeypot, took it out again, put it back and then went on "taking the balloon out, and putting it back again, as happy as could be" (Milne 1926). Further, a complete cladistic classification, reflecting the phylogeny in full, is (as I have just pointed out) an unrealizable dream. Even if it *were* attainable, even for a limited grouping (say Mammalia), it would be utterly impractical for most purposes because of its vast size, unwieldiness and various other factors, notably the Hennigian prohibition from the hierarchy of paraphyletic groups (including such useful groupings as Sarcopterygii, Amphibia, Reptilia and Thecodontia – see below, p. 430); convinced "cladists" have admitted as much in conversation. Indeed, the only effective use of such a classification would be to record the phylogeny, which can be done equally well — or even better — by the cladogram itself.

*(iii) Presentation of phylogenetic relationships.* Thirdly, such a classification would consist of taxa in nested sets (each higher-category taxon beginning with a hypothetical stem-species from within an older, larger taxon and continuing until extinction) and, since it would permit the total retrieval of the cladogram, it would indeed present the phylogenetic relationships of species very well. But is not this excellence of presentation, at least in part, the same thing as the total retrievability of the cladogram? A classification of this type would also permit comparisons of the closeness of relation-ships of species; but it would do that *only if we defined closeness of relationship in the Hennigian manner,* measuring it by the comparative recency (the absolute chronology is unnecessary) of the latest common ancestor of each pair of species concerned. The undeniable fact that this is the only way in which the phylogenetic relationships between each pair of species (in a group of three or more) can be compared *objectively* does not mean that it is necessarily the only way or even the best way of comparing phylogenetic relationships of any sort; for a major weakness of the Hennigian method is that it recognizes only the sequential order of the dichotomies in the phylogenetic tree and ignores the chronological distances along the branches of the tree, not only between latest common ancestors (at the dichotomies or nodes) and species (at the ends of the twigs) but also between one dichotomy and another. Thus a pair of species which are nearly or completely contemporaneous and are phenetically much alike (A, B in Fig. 4a) may share a less recent common ancestry than another pair much more widely separated chronologically and much less similar (B, C), yet *ipso facto* A and B will be regarded by the cladists as less closely related than B and C, the similarities of A and B being rejected as "plesiomorphous" (i.e. so old-established and primitive as to be of no significance in this connection). A might be a Devonian actinopterygian, B a Devonian lungfish and C a present-day cow. Indeed, it seems to me that only when all the taxa being compared are on the same time-plane (A′, B′, C in Fig. 4b — e.g. modern salmon, modern lungfish and modern cow respectively) can there be much significance in the "recency of common ancestry" concept, in so far as only then must that concept provide an infallible

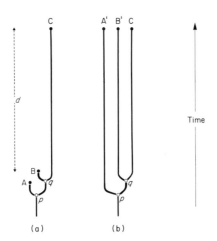

Fig. 4. Three species and their genealogical relationships, with the vertical axis representing time. (a) Three asynchronous species. (b) Three contemporary species. For further explanation see text (pp. 422–423, 426–427).

comparison of the chronological separation between any two taxa with that between any other two. Whatever else might be said about them, it is an indisputable fact that B' and C have had less time in which to evolve divergently from their common ancestor, morphologically and in other ways, than have A' and B'. It is therefore obvious that the idea of comparing relationships in this fashion was originally conceived for Recent taxa only, not for fossil species and certainly not for higher taxa ranging through long periods of time.

(iv) *Lack of confusion of two criteria in developing classification.* This is the one advantage of a Hennigian classification that cannot be disputed. A Hennigian classification is based entirely on the phylogeny and it reflects the phylogeny, more precisely the order of branching and the comparative recency of common ancestry; it can reflect nothing else. Consider Fig. 4a; if a Hennigian systematist classifies B and C together into a taxon which excludes A, then this can mean only that B and C share a common ancestor ($q$) which is not ancestral to A. It cannot mean that B and C have a greater aggregate similarity to each other than either has to A, nor could it indicate that they share similar adaptations of independent derivation; there can be no confusion. A Simpsonian classification, by contrast, does depend on information of two types, some relating to the phylogeny of the organisms concerned and some to their present characteristics; it cannot often be possible to unravel them completely, which means that each will obscure the other to some extent.

(v) *The classification as the general reference system.* Hennig (1966, p. 239) ended his book as follows:

> The phylogenetic system ... has the same universal significance for all phylogenetic research, as has, for example, the topographic map as the foundation for all other possible or desirable cartographic presentations in the fields of geography and other geosciences. The latter statement, then, is the essence of our contention that the phylogenetic system may be regarded, for inherent reasons, as the general reference system of biology.

It seems to me that Hennig's geographical analogy serves most admirably – not to prove his point, however, but to refute it. The context makes it abundantly clear that the two quoted references to "the phylogenetic system" refer to the phylogenetic system of classification rather than to the phylogeny itself or to the cladogram, in which case I disagree strongly with Hennig's comparing its use in biology to that of the topographical map in the geosciences.

I concur with the view that the basis of reference of all geoscientific studies is the topographical map, a graphical representation of geographical relationships in space, but that does not of itself constitute a *system* of reference. On the other hand, a biological classification (like Hennig's "phylogenetic system") *is* a system of reference, though not a general one; the *basis* of reference of all studies in the life sciences is the phylogenetic cladogram, a graphical representation of biological relationships in time. Although we *can* use the topographical map when describing the position of some geographical feature or Man-made structure on the surface of the Earth, pinpointing that position by the use of numerical cartographic co-ordinates, we more frequently employ for that purpose an arbitrary division of the Earth's surface into regional units – terrestrial or marine – which we then arrange into an hierarchical classification. (There are several different ways of doing this: politically, ecclesiastically, postally and so on.) The territorial divisions are based on the topographical map, in details the two may even coincide as when a river forms a national boundary, but they are certainly not the same thing; other relevant factors are taken into consideration when the boundaries are being decided upon. Imagine trying to address a letter in cartographical terms, say with a numerical map reference; it might get to its destination eventually, but it would prove most impractical for the postal sorters!

In the same way we *could* use the phylogenetic cladogram when describing the systematic position of biological organisms, fossil or extant, by devising some type of numerical reference system to pinpoint their positions on the cladogram; but for practical purposes it is far more convenient to divide up the "tree of life" into arbitrary units, namely taxa, arranged into an hierarchical classification to suit our convenience.

To put it in a nutshell:

1. In neither geography nor biology is there any such thing as one general reference system, there is only a *general foundation of reference* upon which various hierarchical systems of classification can be based.

2. That foundation is the topographical map in geography and the phylogenetic cladogram in biology.

3. Each hierarchical system of classification, although constructed primarily upon the basis of the appropriate foundation of reference, is also modified — in the light of other relevant factors — to meet various practical requirements.

(*c*) *Alleged disadvantage of a Simpsonian classification*: *inability to quantify "relationship"*. Hennigians claim that Simpsonian classification should not be used because, in that method, relationships cannot be quantified. This strikes me as very odd, for, as I understand it, it is *their* system which should be criticized on this count. It is *their* system, as I have shown above, which provides only a comparison of recency of common ancestry and ignores the distances along the branches of the phylogenetic tree, distances which in fact *can* be given some sort of quantification in either type of cladoevolutionary systematics. An approximate indication may be given (inversely) in terms of the time-interval between the bifurcation of the common ancestral stock into two separate lineages, one leading to each of the two taxa concerned, and the first appearance of each taxon in the fossil record; though rates of evolution may differ enormously, the probability is that a longer chronological separation will lead to a greater divergence in genotype and phenotype than will a shorter one. Thus, in Fig. 4a (p. 423), the species A and B are likely to have retained more morphological features in common than are shared by B and C. This means that the magnitude of the evolutionary divergence between B and C will depend to some extent on their chronological separation, on the lengths of the two respective periods during which — partly concurrently — B and C were able to diverge from their hypothetical common ancestor $q$ and from each other; the chronological separation between those two species should therefore be measured not directly ($d$) but along the branches of the phylogenetic tree, either as the totalled time-intervals $qB + qC$ or as the mean interval $\dfrac{qB + qC}{2}$ . (Such time-intervals could be calculated by reference to an absolute time-scale, e.g. in Harland, Smith and Wilcock 1964.) For the "sister-groups" of the Hennigians, this time-interval — totalled or mean — will be zero. Theoretically it would also be possible to quantify "phylogenetic distance" by comparing numbers of intermediate dichotomies or nodes, of successive intermediate species (same thing to a Hennigian!) or of intermediate generations, although it is difficult to see how this could be done in practice.

But, in any case, do relationships *need* to be quantified? If there is a cladogram to give us the genealogical relationships (as evinced by the order of branching) and a subjectively based hierarchical classification to provide us with a convenient system of reference, why should we want to make accurate measurements of the degree of relationship, or a precise comparison of one relationship with another?

### (d) Further disadvantages of a Hennigian classification

(i) *Polarity in relationship.* Another strange and disturbing aspect of the Hennigian concept of relationship is its polarity, defying the axiom that reciprocal relationships are not only opposite but also equal. Thus, in the simple diagram shown on p. 423 (Fig. 4a), the unknown stem-species $q$ is considered to be more closely related to its remote descendant C than it is to its immediate ancestor $p$, for $q$ and C are both members of a holophyletic taxon based on $q$ from which $p$ is excluded. Yet, whereas $p$ is not closely related to $q$, $q$ is very closely related to $p$ because both are members of a slightly larger holophyletic taxon based on $p$ as the stem-species. If we wished to avoid the use of hypothetical ancestors we could say that B is more closely related (in the Hennigian sense) to C than it is to A; this makes it seem highly paradoxical that all three are nevertheless regarded as being closely related to each other because all are members of the holophyletic taxon based on the stem-species $p$, within which B appears to be closer to A than to C in every respect — except, of course, in recency of common ancestry!

(*ii*) *Delimitation of taxa with respect to time.* The delimitation of taxa with respect to time is another area in which the simple Simpsonian systematist like myself finds Hennigian classification difficult to comprehend. The official line, as laid down by Hennig himself, shows a strange inconsistency; there is a differing approach towards the delimitation of species, which extend from node to node of the phylogenetic "tree", and the delimitation of higher taxa (family-group and above) which can end only in total extinction. Halstead (1978, p. 760) was clearly unaware that the Hennigian creed distinguished between species and higher taxa in this manner; he evidently thought that species too were required to include *all* their own descendants, so that we ourselves would belong to the same species as the unknown rhipidistian fish from which the tetrapods must have evolved. Between species and "higher taxa" lies a grey area, covering taxa of the genus-group, concerning which I have not been able to obtain any information either from the works of Hennig himself or from repeated questioning of any of his disciples; why this should be I do not know.

One objection to the Hennigian concept of a palaeospecies, extending from node to node of the phylogenetic tree, is to the fact that it may lump together very different individuals into the same species and, at the same time, separate closely similar ones. That, however, is a general and unavoidable failing of any system of classification in which boundaries have to be drawn in arbitrary fashion across a continuously varying spectrum; it applies to Simpsonian classifications as much as to Hennigian, and in relation to higher taxa as well as to species. (The same criticism may be made, of course, of many non-biological classifications.) There is a strong temptation to draw boundaries where the widest gaps occur in the fossil record; but those gaps are haphazard in their occurrence and it is far more sensible to place the boundaries where — as far as we can determine it, or at least estimate it — the rate of evolution is at its fastest.

A second objection to Hennigian procedures, a more valid one, is to the fact that specific boundaries may be drawn *only* at nodes, no matter how much a species may have evolved in between. Thus, in the words of Panchen (1979):

> Unlike all other taxonomists, the cladists characterise phylogeny solely as the result of speciation or cladogenses (the splitting of one species into two or more in evolution). They ignore the results of evolutionary change within evolving species (phyletic evolution or anagenesis).

It is evident that the application of Hennigian classification to a lineage which had undergone prolonged anagenesis without clado-genesis would produce an unacceptable degree of variability within a single species.

The third objection is the corollary of the second, namely to the fact that specific boundaries *must* be drawn at every node. Thus, at each dichotomy, the parent species must be terminated and separate species status conferred on both the resultant lineages, even though one of those lineages might be absolutely indistinguishable from the parent stock. A "cladist" has given me the argument that the origin of a daughter-species requires the removal of a number of individuals from the parent population and thereby leaves a residual population with an altered gene pool, justifying the bestowal of a new specific name on that residual population too. But there are two strong objections to this. First, there are no grounds for supposing that, at the moment when speciation began, the group of individuals which were about to speciate must certainly have possessed a gene pool which differed appreciably from that possessed by the remaining individuals; the two groups may have had *identical* gene pools, both absolutely representative samples of the gene pool of the parent species as a whole. Secondly, a strange anomaly arises if it be supposed that the entire daughter population might perish instead of speciating. Imagine a flock of birds belonging to a continental species, blown offshore by a storm; they might land on a sparsely populated oceanic island and evolve there into a new species, but it is also possible that they might all drown in the sea. In one case a Hennigian would find it necessary to give a new name to the residual population on the mainland, in the other case he would not, even though the alternative fates of the vagrant flock could not make one iota of difference to the genetic and phenetic constitution of the mainland population in the two cases.

I do not see how the gene pool of the parent population can be altered by such speciation events, or at least not to the extent that it could warrant a change in the name of that parent species. If it could, then surely a similar change in name would be required as a result of *any* event which removed a significant number of individuals from a population; one could instance the fourteenth-century outbreak of bubonic plague (the Black Death) which spread from China to Europe and, in 1348–49, wiped out from one-third to one-half of the population of England alone. Despite the fact that those who succumbed to that epidemic might well have differed from the survivors in their aggregate genetic constitution, the disaster has not been used to justify a change in name of the human species *before* the Black Death from *Homo sapiens* to *Homo* something else!

(*iii*) *Prohibition of paraphyletic taxa.* In the majority of groups of organisms the most conspicuous differences between the Hennigian and the Simpsonian classifications are due to the Hennigian prohibition of paraphyletic taxa which, so it seems to me, is the fatal flaw in the Hennigian approach. It is surely axiomatic in any system of classification that the objects or concepts to be grouped together as one class shall exhibit some recognizable degree of homogeneity, be it of structure or use, relationships or colour, behaviour or politics, odour or origin or whatever property it is that forms the basis of the classification; the larger the group (in the case of biological classification the higher the rank of the taxon) the greater will be the variability within the group but the more fundamental will be the characters that unite its members. If we can declare, with a reasonable degree of certainty, that an extremely large, varied and essentially different clade (say the mammals) originated from somewhere within a particular family (say the Chiniquodontidae) among the reptiles, then it seems convenient, from several points of view, not to include that vast new clade within the parent family as a mere

subfamily thereof but to elevate it to much higher rank – in this
particular case, to class status as the Mammalia. Only thus can we
accord proper recognition to the enormous difference between the
Mammalia and their reptilian forbears; only thus can we recognize
the great importance of the mammals; and only thus can we leave
ourselves enough "elbow room" in the hierarchy to classify the
mammals further, into ever-smaller nested taxa within the class. But
when we elevate the Mammalia to class status we are obliged to
exclude them from the family Chiniquodontidae and from all higher
taxa which include the Chiniquodontidae, up to and including the
taxon of equivalent rank – the class Reptilia. Thus all those taxa will
now represent incomplete clades, i.e. we have rendered them paraphyletic. That, however, in no way diminishes their usefulness. All
ancestral taxa, all stem-taxa are *ipso facto* paraphyletic.

I do understand the contention of the Hennigian systematists, that
paraphyletic groups have no part to play in phylogeny reconstruction
(using the latter term to mean the construction of a cladogram of
phylogenetic significance); I agree with them that phylogeny reconstruction should be based on *clades,* natural monophyletic groups.
But I find it difficult to comprehend some of their other attitudes
towards the same subject, as when they say that paraphyletic groups
have no unique or independent history (see, for example, Bonde
1977, pp. 763, 768); would one say that the Norman Kings of
England "have no independent history" simply because they gave
rise to the Plantagenets and we choose to distinguish the latter as a
separate dynasty? Another favourite statement of the "cladists"
(Hennigian or natural order systematists, it matters not) is that
paraphyletic taxa cannot be defined, that they are "non-groups",
uncharacterizable or at best characterizable only by lack of characters. They abhor such useful terms as Sarcopterygii, Amphibia,
Reptilia, Therapsida or Thecodontia, pouring scorn and derision on
anyone who employs them. Thus Patterson (1980, p. 239) believes
that

> general statements such as 'reptiles were ancestral to birds and mammals'
> . . . are meant to convey information, yet . . . the only information con-
> veyed is that the groups held to be descendant (. . . birds, mammals, for
> instance) are characterizable, and the groups held to be ancestral are not.

This is nonsense. It must surely be true that any paraphyletic

Fig. 5. For explanation, see text.

taxon can be defined in terms of a clade, from which one or more younger, smaller clades, nested within the bigger one, have been excluded (see above, p. 408); if all those clades can be defined, then the paraphyletic taxon can be defined too. To put it another way: if two sets can be defined, a smaller set characterized by possession of feature B nested wholly within a larger set characterized by possession of feature A, then surely we are allowed to define a third set (the outer area in Fig. 5) as possessing A but not B? In mathematical terminology such a set is called a *complement set*; it should not be denied validity on the extremely dubious grounds that it is "characterizable only by lack of characters". Thus "Reptilia" is simply a convenient way of saying "Amniota minus Aves and Mammalia", "Thecodontia" is a handy term for "Archosauria minus half a dozen or so monophyletic daughter-orders nested within it".

In any case, I cannot accept Patterson's contention that no useful information is conveyed by statements referring to paraphyletic groups. Presumably he would argue that the statement "Amniota originated from the Amphibia" is a meaningless statement, simply because the Amphibia are a paraphyletic taxon; nevertheless he would agree that the Tetrapoda are a natural monophyletic group, a clade, within which are nested the Amniota as a subordinate clade. This means – if an evolutionary interpretation be permitted – that the Amniota originated within the Tetrapoda. But they could not have originated from themselves; they must have originated from *other* tetrapods. And the only other tetrapods, by definition, are the Amphibia!

| | |
|---|---|
| Amniota originated from the Amphibia | NOT ACCEPTABLE |
| But Amniota are a subordinate clade of the Tetrapoda | ACCEPTABLE |
| i.e. Amniota originated *within* the Tetrapoda | ACCEPTABLE |
| i.e. Amniota must have originated from *other* Tetrapoda | ACCEPTABLE |

i.e. Amniota must have originated from (Tetrapoda minus
Amniota)                                                    ACCEPTABLE
i.e. Amniota must have originated from the Amphibia!

The paraphyletic taxon is especially useful in that it enables us to
refer to ancestral groups and stem-groups. We are now back to the
old argument concerning vertical versus horizontal classification, in
which we have no space to get involved at present. Some Hennigian
systematists (perhaps some natural order systematists too) do
recognize the practical value of paraphyletic taxa and have told me
that they have no objections to their being used in the formal classification, at least for the time being, but they would prefer that their
names be written in inverted commas!

(*iv*) *Absolute ranking of taxa.* Another characteristic of the
Hennigian system is its remarkable attitude towards the assignment
of absolute rankings in the hierarchy to the higher taxonomic
groupings, in particular Hennig's proposal that rank should depend
on absolute age. A detailed study of this subject, however, is not
required here, mainly because the subject appears to be meaningless
in any biological sense. It could even be said that it is without any
sort of theoretical meaning whatever and can be only of practical
significance.

I therefore hold that only practical considerations should be
allowed to influence the assignment of rankings. The rank of a taxon
should enable it to fit conveniently into its particular hierarchy,
sufficiently elevated in position to permit its adequate subdivision
where that is likely to be necessary; and, at the same time, it should
also serve to give some indication of the distinctiveness, importance
and perhaps size of the grouping that it represents. Those two
desiderata may either coincide or conflict; where they conflict it is
necessary to effect an arbitrary reconciliation.

(*e*) *Comparison of usefulness for practical purposes*

(*i*) *Classifications.* The Simpsonian systematist finds, as a matter
of experience, that his traditional classifications are better than
cladistic classifications for virtually every *practical* purpose for which
classifications are required, e.g. the expression of aggregate ("overall")

similarity, the provision of names by which to refer to groups possessing such similarity, the construction of reference systems for the retrieval of material from storage in museums and of information from libraries, the facilitation of memorizing and the making of predictions. Only in the expression of evolutionary relationships are conventional classifications sometimes less effective, and even there they have the merit of indicating the extent of our ignorance.

(*ii*) *Cladograms versus spindle diagrams.* Cladograms certainly have their uses, but there are many reasons why the conventional spindle diagrams or radiation charts ("Romerograms") are far better for most purposes, not only for conveying information but also for indicating our lack of it. The cladogram has many disadvantages:

1. It gives no indication of the known time ranges of the taxa represented (and therefore fails to indicate other chronological factors, e.g. the contemporaneity or otherwise of two specified taxa, the possibility of one taxon originating from another).

2. It gives no indication of the relative abundance (as known from the fossil record) of a specified taxon at a specified time.

3. It suggests that the phylogenetic tree ramified by a series of dichotomies in an order for which the evidence may not be adequate and which may therefore be wrong.

4. It is based on an analysis of shared derived characters (synapomorphies) which, in the procedures of some workers, makes no allowance for parallel or convergent evolution and may therefore be totally misleading.

5. It is generally featureless and unattractive and therefore quite unsuitable for easy remembering.

Incidentally, I strongly support Halstead's preference for family trees rather than cladograms in public exhibitions of natural history material (1978, p. 760); they are so much easier for the general viewer to understand.

(*f*) *Conclusions.* In essence, Simpsonian systematists are pragmatists; they regard a classification as a tool, a means to an end. They agree with the Hennigians that the classification should be consistent with the phylogeny, and they deplore the erection of taxa (polyphyletic groups or assemblages) which cannot be represented by a single continuous segment of the phylogenetic tree; nevertheless they consider that the practicality, convenience and usefulness of the classification in various biological contexts should be the main criteria in its selection and are more important than its ability to reflect the entire phylogeny. Hennigians, by contrast, are idealists; they regard a classification as an end in itself, a manifestation of the genealogical relationships, and they consider that objectivity and parsimony should be the chief criteria in its construction. Unfortunately their ideal is largely unattainable in practice. Patterson made an excellent summary of the situation when he wrote (1978, pp. 126–127):

> Whether or not biologists should follow Darwin's advice, and make their classifications genealogical, is still controversial. The decision rests on a choice between a traditional classification whose main virtue is that it is stable and easily memorized; and a more complicated classification, changing with new discoveries, whose main virtue is that it is a scientific theory of relationships.

Neither approach is "correct", I repeat that there is no such thing as a "correct" classification, and it would surely be better if each school of thought were to recognize that the classification preferred by the other was devised for an entirely different primary purpose.

I have recently re-read Mayr's analysis and severe criticisms of cladistic classification (1974) and Hennig's reply thereto (1974, 1975). Mayr's article, despite a few weaknesses, is in my opinion quite admirable; Hennig's is largely irrelevant and generally inadequate. Hennig ended the English summary to his 1974 paper with the words

> On the basis of MAYR's arguments the primacy of the phylogenetic system as the general reference system in biology can by no means be disputed.

Why he chose to make such a dogmatic and totally unfounded assertion is beyond my understanding, for I find little to justify it in any of his arguments. However, I have no objection to the use of Hennigian classifications by those who find them helpful; as far as I am concerned they may do as they wish. I ask only that those of us who prefer traditional classifications be permitted to do as *we* wish — without being regarded as reactionary conservatives opposed to progress, as non-scientists whose outdated views are hardly worth the trouble of refutation. After all, some of us are very worried by the problems which are created by the necessarily differing approaches of palaeontology and neontology, but we are trying to solve them by the development of new refinements in our methods, retaining always the attitude that a classification is a practical system of reference rather than the expression of a philosophy.

### GENERAL CONCLUSIONS

I have also re-read parts of Hennig's *magnum opus* (1966) and was particularly struck by the nature of his criticisms (pp. 10 *et seq.*) of what he called "typological systems"; he contrasted the latter — to their detriment — with his own and other "phylogenetic systems". His main target, however, was not the Simpsonian systematics that he so often attacked elsewhere (and indeed regarded as "typological", for it was not phylogenetic in his sense). Rather were his criticisms directed against a number of authors who had supported various truly "typological" approaches. The views of Klingstedt (1937) seem to be worthy of especial note; Klingstedt distinguished three steps in taxonomic work, which may be listed (in simplified form) as follows:

1. *Description of species* and arbitrary arrangement thereof.

2. *Ordering of species into a typological system of classification*, based on all their characters.

3. *Phylogenetic interpretation* of that classification.

An important characteristic of these "typological" approaches was that phylogeny entered the subject only as an afterthought, as a speculative interpretation of the typological classification already arrived at. This view had been put forward much earlier by Holdhaus (cited in Karny 1925), while Horn (1929) believed that the task of the systematist is to order, without saying anything as to how the ordering came about. This brought home to me very strongly that natural order systematics affords an excellent example of such an approach and that it is the absolute antithesis of Hennig's ideas. It struck me even more forcibly that, were Hennig writing today, it would be natural order systematics and phenetics that he would be railing against, not conventional systematics of the Simpsonian type;

how absurd that natural order systematics and Hennigian systematics should be bracketed together as "cladistics"! I certainly agree with Hennig that the statement that "it is possible to carry on science (including systematics) without making assumptions" (p. 11) is itself a false assumption.

All this serves to reinforce my opinion that the fundamental division of systems of systematics should be into:

(a) *Typological systems,* of which the essential stages are (1) species description and (2) typological classification; these may be followed, if desired, by (3) phylogenetic interpretation. Such systems include phenetics and natural order systematics.

(b) *Clado-evolutionary systems,* of which the essential stages are (1) phylogeny reconstruction and (2) classification consistent therewith. These include Simpsonian and Hennigian systematics.

I further agree with Hennig that the best systems of systematics are those based on phylogeny, as in (b) above, what I call clado-evolutionary systems. However, I would reject the *purely* phylogenetic (i.e. "Hennigian") system that Hennig himself always insisted on and prefer something rather different. I recommend the combination of a mainly Hennigian approach to character distribution analysis with a Simpsonian classification, as outlined immediately below.

A "CODE OF PRACTICE" FOR SYSTEMATISTS

My own attitude towards phylogeny reconstruction and classification, the "Code of Practice" I propose to adopt henceforth in my systematic studies, is as follows:

1. I recognize that we have no absolute proof of the theory of evolution, by direct evidence of the senses; all the available evidence is merely circumstantial. However, there is no scientifically acceptable evidence against evolution and no other theory fits the known facts so well. I therefore accept it as a working hypothesis of immeasurable heuristic value.

2. I believe that the branching pattern of the phylogeny should be reconstructed as accurately as possible, the reconstruction being

based primarily upon a character distribution analysis conducted in a strictly disciplined Hennigian manner. The evidence obtained from the analysis may be modified and augmented in the light of evidence obtained from any other valid source, e.g. in palaeontology from what is known as the "stratophenetic" approach.

3. The reconstructed phylogeny may be conveniently expressed, as completely as possible, by means of a branching diagram (dendrogram). *The dendrogram itself* provides a "topographical" reference system in biology.

4. The dendrogram, representing continuity in time and space, may be divided into taxa in an arbitrary fashion in order to fulfil, as well as possible, the general requirements of a biological classification. Thus the classification should impart, as far as is consistent with division in that manner, the most important characteristics of each taxon at the time of its existence as well as the broad outline of its evolutionary history. Each taxon should correspond to a single continuous segment of the dendrogram; this means that it must be of monophyletic origin, not polyphyletic. Further, each taxon may include either the whole of the clade (holophyletic taxon) or only part thereof (paraphyletic taxon); in the latter case smaller clades nested within the clade are excluded from it, having been given equivalent or higher rank.

ACKNOWLEDGEMENTS

I am grateful for the helpful information, advice and criticism given me in the compilation of this article by Dr C. G. Adams, Dr H. W. Ball, Prof. C. B. Cox, Dr A. R. I. Cruickshank, Dr P. L. Forey, Dr R. A. Fortey, Dr L. B. Halstead, Dr K. A. Joysey, Dr T. S. Kemp, Dr L. D. Martin, Dr Angela C. Milner, Dr A. L. Panchen, the late Dr F. R. Parrington, Dr C. Patterson, H.-D. Sues, C. A. Walker and K. N. Whetstone. They do not all share my opinions.

I also thank my wife for typing several drafts of the manuscript, Miss Sandra Chapman for drawing the diagrams, and Dr Angela Milner for helping in innumerable ways.

A two-page summary of certain points in this present article has already been published in *Biologist* (Charig 1981), in reply to a short article on "Cladistics" in the previous number of the same journal (Patterson 1980).

## REFERENCES

Ashlock, P. D. (1971). Monophyly and associated terms. *Syst. Zool.* 20, 63–69.

Ashlock, P. D. (1972). Monophyly again. *Syst. Zool.* 21, 430–438.

Ashlock, P. D. (1979). An evolutionary systematist's view of classification. *Syst. Zool.* 28, 441–450.

Bonde, N. (1977). Cladistic classification as applied to vertebrates. *In* "Major Patterns in Vertebrate Evolution" (M. K. Hecht, P. C. Goody and B. M. Hecht, eds), pp. 741–804. Plenum Press, New York and London.

Charig, A. J. (1956). "New Triassic archosaurs from Tanganyika, including *Mandasuchus* and *Teleocrater*". Ph.D. dissertation, University of Cambridge.

Charig, A. J. (1972). The evolution of the archosaur pelvis and hind-limb: an explanation in functional terms. *In* "Studies in Vertebrate Evolution (Essays Presented to Dr F. R. Parrington, F. R. S.)" (K. A. Joysey and T. S. Kemp, eds), pp. 121–155. Oliver & Boyd, Edinburgh.

Charig, A. J. (1976). "Dinosaur monophyly and a new class of vertebrates": a critical review. *In* "Morphology and Biology of Reptiles" (A. d'A. Bellairs and C. B. Cox, eds), pp. 65–104. Academic Press, London and New York.

Charig, A. J. (1981). Cladistics: a different point of view. *Biologist* 28, 19–20.

Cracraft, J. (1977). Phylogeny debated. *GeoTimes* 22 (11), 16–17.

Crompton, A. W. and Jenkins, F. A., Jr. (1979). Origin of mammals. *In* "Mesozoic Mammals: The First Two-thirds of Mammalian History" (J. A. Lillegraven, Z. Kielan-Jaworowska and W. A. Clemens, eds), pp. 59–73. University of California Press, Berkeley.

Cuénot, L. (1940). Remarques sur un essai d'arbre généalogique du règne animal. *C. r. hebd. Séanc. Acad. Sci. Paris,* D, 210, 23–27.

Farris, J. S. (1974). Formal definitions of paraphyly and polyphyly. *Syst. Zool.* 23, 548–554.

Gardiner, B. G. (1982). Tetrapod classification. *Zool. J. Linn. Soc.* (in press).

Gardiner, B. G., Janvier, P., Patterson, C., Forey, P. L., Greenwood, P. H., Miles, R. S. and Jefferies, R. P. S. (1979). The salmon, the lungfish and the cow: a reply. *Nature, Lond.* 277, 175–176.

Gingerich, P. D. and Schoeninger, M. (1977). The fossil record and primate phylogeny. *J. hum. Evol.* 6, 483–505.

Haeckel, E. (1874). The gastraea-theory, the phylogenetic classification of the animal kingdom, and the homology of the germ-lamellae. *Q. Jl microsc. Sci.,* N. S. 14, 142–165, 223–247 [translated by E. P. Wright].

Halstead, L. B. (1978). The cladistic revolution – can it make the grade? *Nature, Lond.* 276, 759–760.

Halstead, L. B. (1980). Popper: good philosophy, bad science? *New Scient.* 87, 215–217.

Halstead, L. B., White, E. I. and MacIntyre, G. T. (1979). [Reply to Gardiner *et al.*] *Nature, Lond.* 277, 176.

Harland, W. B., Smith, A. G. and Wilcock, B. (eds) (1964). "The Phanerozoic time-scale: a symposium dedicated to Professor Arthur Holmes". *Q. Jl geol.*

*Soc. Lond.* **120** (supplement). Geological Society of London, London. [see pp. 260–262]

Hennig, W. (1950). "Grundzüge einer Theorie der phylogenetischen Systematik". Deutsche Zentralverlag, Berlin.

Hennig, W. (1965). Phylogenetic systematics. *A. Rev. Ent.* **10**, 97–116.

Hennig, W. (1966). "Phylogenetic Systematics". University of Illinois Press, Urbana.

Hennig, W. (1974). Kritische Bemerkungen zur Frage "Cladistic analysis or cladistic classification?" *Z. zool. Syst. EvolForsch.* **12**, 279–294.

Hennig, W. (1975). "Cladistic analysis or cladistic classification?": a reply to Ernst Mayr. *Syst. Zool.* **24**, 244–256.

Horn, W. (1929). On the splitting influence of the increase of entomological knowledge and on the enigma of species. *4th Int. Congr. Ent., Ithaca* **2**, 500–507.

Huxley, J. S. (1957). The three types of evolutionary process. *Nature, Lond.* **180**, 454–455.

Huxley, J. S. (1958). Evolutionary processes and taxonomy with special reference to grades. *Acta Univ. Upsaliensis* 1958 (6), 21–39.

Karny, H. H. (1925). Die Methoden der phylogenetischen (stammesgeschichtlichen) Forschung. *In* "Handbuch der biologischen Arbeitsmethoden" (H. Abderhalden, ed.), vol. 9 (3), pp. 211–500. Urban & Schwarzenberg, Berlin.

Klingstedt, H. (1937). A taxonomic survey of the genus *Cyrnus* Steph. including the description of a new species, with some remarks on the principles of taxonomy. *Acta Soc. Fauna Flora fenn.* **60**, 573–598.

Mayr, E. (1974). Cladistic analysis or cladistic classification? *Z. zool. Syst. EvolForsch.* **12**, 94–128.

Milne, A. A. (1926). "Winnie-the-Pooh". Methuen, London.

Nelson, G. J. (1971). Paraphyly and polyphyly: redefinitions. *Syst. Zool.* **20**, 471–472.

Nelson, G. J. (1973). "Monophyly again?" — A reply to P. D. Ashlock. *Syst. Zool.* **22**, 310–312.

Nelson, G. J. and Platnick, N. I. (1980). A vicariance approach to historical biogeography. *BioScience* **30**, 339–343.

Panchen, A. L. (1979). [*In* "The Cladistic Debate Continued".] *Nature, Lond.* **280**, 541.

Panchen, A. L. (1982). The use of parsimony in testing phylogenetic hypotheses. *Zool. J. Linn. Soc.* (in press).

Patterson, C. (1978). "Evolution". British Museum (Natural History), London.

Patterson, C. (1980). Cladistics. *Biologist* **27**, 234–240.

Platnick, N. I. (1977). Paraphyletic and polyphyletic groups. *Syst. Zool.* **26**, 195–200.

Platnick, N. I. (1980). Philosophy and the transformation of cladistics. *Syst. Zool.* **28**, 537–546.

Popper, K. R. (1976). "Unended Quest. An Intellectual Autobiography". Fontana/Collins, Glasgow.

Popper, K. R. (1980) [*In* "Letters: Evolution".] *New Scient.* **87**, 611.

Ruse, M. (1979). Falsifiability, consilience, and systematics. *Syst. Zool.* **28**, 530–536.

Sharman, G. B. (1976). Evolution of viviparity in mammals. *In* "Reproduction in Mammals. Book 6: The Evolution of Reproduction" (C. R. Austin and R. V. Short, eds), pp. 32–70. Cambridge University Press, Cambridge.

Simpson, G. G. (1961). "Principles of Animal Taxonomy". Columbia University Press, New York and London.

Simpson, G. G. (1971). Concluding remarks: Mesozoic mammals revisited. *In* "Early Mammals" (D. M. Kermack and K. A. Kermack, eds) [*Zool. J. Linn. Soc.* **50**, suppl. 1], pp. 181–198. Academic Press, London and New York.

Simpson, G. G. (1978). Variation and details of macroevolution. *Paleobiol.* **4**, 217–221.

Whewell, W. (1840). "The Philosophy of the Inductive Sciences, Founded Upon Their History". 2 vols., 1st edition. Parker, London.

## NOTE ADDED IN PROOF

For my readers' interest I append details of three works which receive no mention in my article, despite the fact that they are highly relevant thereto and generally very important. Two of them were published too late for me to use them. The third (by Holmes) was published in 1980, after the Symposium but while I was still finishing my chapter; I did not learn of its existence, however, until the author kindly sent me a copy in 1981.

Arnold, E. N. (1981). Estimating phylogenies at low taxonomic levels. *Z. Zool. Syst. Evol-Forsch.* **19**, 1–35.

Holmes, E. B. (1980), Reconsideration of some systematic concepts and terms. *Evolut. Theory* **5**, 35–87.

Wiley, E. O. (1981). "Phylogenetics: the theory and practice of phylogenetic systematics" Wiley, New York, Chichester.

# Index of Key Words